FRAMED TIME

FRAMED

GARRETT STEWART

UNIVERSITY OF CHICAGO PRESS · CHICAGO AND LONDON

TIME

TOWARD A POSTFILMIC CINEMA

CINEMA AND MODERNITY • A SERIES EDITED BY TOM GUNNING

Garrett Stewart is the James O. Freeman Professor of Letters in the English Department at the University of Iowa, and the recipient of Guggenheim and NEH fellowships. His most recent books include *The Look of Reading: Book, Painting, Text* and *Between Film and Screen: Modernism's Photo Synthesis*, both published by the University of Chicago Press.

The University of Chicago Press, Chicago 60637
The University of Chicago Press, Ltd., London
© 2007 by the University of Chicago
All rights reserved. Published 2007
Printed in the United States of America

16 15 14 13 12 11 10 09 08 07   1 2 3 4 5

ISBN-13: 978-0-226-77415-2 (cloth)
ISBN-13: 978-0-226-77416-9 (paper)
ISBN-10: 0-226-77415-5 (cloth)
ISBN-10: 0-226-77416-3 (paper)

Library of Congress Cataloging-in-Publication Data
Stewart, Garrett.
    Framed time : Toward a postfilmic cinema / Garrett Stewart.
      p. cm. — (Cinema and modernity)
    Includes bibliographical references and index.
    ISBN-13: 978-0-226-77415-2 (hardcover : alk.paper)
    ISBN-13: 978-0-226-77416-9 (pbk. : alk. paper)
    ISBN-10: 0-226-77415-5 (hardcover : alk.paper)
    ISBN-10: 0-226-77416-3 (pbk. : alk.paper)
    1. Motion pictures. 2. Space and time in motion pictures. I. Title.
    PN1994.S8174 2007
    791.43'684—dc22
                                                                        2006031119

♾ The paper used in this publication meets the minimum requirements of the American National Standard for Information Sciences—Permanence of Paper for Printed Library Materials, ANSI Z39.48-1992.

ONCE MORE FOR NATAŠA

CONTENTS

FOREWORDS { *Acknowledgments* IX
Introduction: On Optical Allusion I

1 LEXEME TO PIXEL: AN EXPERIMENT IN NARRATOGRAPHY 20

The Golden Bowl
The House of Mirth
Citizen Kane

2 TRICK BEGINNINGS AND THE EUROPEAN UNCANNY 54

Memento   Insomnia   Run Lola Run
Three Colors: Blue   Three Colors: Red
The Double Life of Véronique   The Red Squirrel
Lovers of the Arctic Circle   Time Regained
Simon the Magician   Heaven   Swimming Pool

3 OUT OF BODY IN HOLLYWOOD   86

The Matrix   Dark City
The Manchurian Candidate   Abre Los Ojos
Vanilla Sky   A.I. Artificial Intelligence
The Sixth Sense   The Others
Jacob's Ladder   Adaptation   Identity
One Hour Photo

4 TEMPORTATION   122

Paris Qui Dort   Johnny Mnemonic   Frequency
He Loves Me, He Loves Me Not   Donnie Darko
The Thirteenth Floor
Eternal Sunshine of the Spotless Mind
The Butterfly Effect   2001: A Space Odyssey
Being John Malkovich

5 **VR FROM CIMNEMONICS TO DIGITIME** 164

The Forgotten   City of Lost Children
Bad Education   The Final Cut
Caché   Syriana

6 **MEDIA ARCHAEOLOGY, HERMENEUTICS, NARRATOGRAPHY** 206

Minority Report   The Lake House
Happy Accidents   Brokeback Mountain
The Jacket   Irreversible

BACKWORDS { *Appendix: Precinematics; or, Reading the Narratogram* 249
Notes 267   *Terms* 283   *Index* 287

ACKNOWLEDGMENTS

Unlike movie credits, which often parade a phalanx of technical experts while burying behind-the-scenes debts to so-called script doctors, I want to give evenhanded credit to interventions both verbal and visual. My gratitude, first, to the initial and anonymous Press reader whose probing diagnostic report, along with an equally subtle and comprehensive workup for the Press by James Chandler, each of them luckily judging the manuscript viable, helped make it considerably more so. I owe a similar paired debt to two doctoral assistants—and gifted surgical interns in matters of argument—who also, like the Press referees, elevated underpaid labor to intellectual collaboration. With his deft theoretical touch, Joshua Gooch was on call when my pages were first coming to light—almost before they were quite chapters. After intervening months of cutting, reconstruction, and new screenings, Glenn Brewer's precision and wit, as well as his wide and canny acquaintance with contemporary film, were available for consultation in numerous office visits during the final weeks of postoperative mending.

On the technical side of things right from the start—and as if with the swell of exit music cuing their recognition at last—my intense thanks to Linda Edge-Dunlap and Greyson Purcell at the University of Iowa for making possible the book's frame captures. Their keen digital skills—and remarkable geniality, even through many a retake session—allowed me to illustrate in some detail the narrative fallout from a proliferating electronic technology to whose secondary search-and-seizure operations I had no access of my own. Nor was I left to my lone devices in the remaining stages of manuscript preparation. For their exacting, fast, and affable work in proofreading, Ryan Clark, Bridget Draxler, and Matthew Low have my serious gratitude.

Effortless as the advice may have seemed at the time, and no doubt long forgotten by these friends, my further thanks to Elisabeth Gitter and William Galperin for having sent me to crucial films I might otherwise have missed—even when, in Betsy's unguarded case, the merest hint could well precipitate my sudden weekend visit to catch a limited New York run. Even before the first stirrings of this project, I also owe more than either of us could have suspected at the time to passing remarks by Ed Folsom that sent me back to *One Hour Photo* for what turned out to be a book-instigating second look. And as always in acknowledgments of mine, debts go too briefly itemized to Nataša Ďurovičová, who long ago helped guide me to the unmarked and risky intersection where film studies meets cinema studies and has, ever since, kept me looking both ways before I cross.

INTRODUCTION: *On Optical Allusion*

There's a great throwaway joke—implicitly about motion pictures as entertainment versus "the cinema" as art—in Alan Bennett's 2003 play *The History Boys*. A laggard jock, tutored for his Oxbridge entrance interview to say that he's "interested in film," thus avoiding the lowbrow taint of the word *movies*, has misunderstood the tip. Announcing at first that he's very much taken with "a film," the cornered movie buff, when pressed by an examiner to say which one, is forced to admit that he means "lots of them." *The History Boys* is itself historical, however, set back in the 1980s. The present book exists because two decades later, identified as cinema or not, certain movies, lots of them, are at base not film at all—or at least no longer all film.

And it doesn't take a cineaste to notice this encroachment of the digital. Screen narratives often note it for themselves in progress. How so is the main concern of these Forewords. But "forward" is just as much the point. The coming chapters are notes "toward" in an entirely historical rather than partisan sense. From word one of this book's subtitle, then, the preposition is meant not to signal the prolegomenon to some new screen poetics but simply to suggest an undeniable trajectory of the medium in its second century. What follows outlines no case for the aesthetic possibilities of the electronic image, a case amply made elsewhere. Nor, for that matter, does it lodge any argument against computerized mediation or its experimental artistries. It concerns not what can be done with the digital, but what the digital has already done to cinema. And more than this, how cinema has seen fit to picture its own transition.

Such a question, once generalized, is meant to get under the medium's skin, its projected surface. In the continuous *illusion* of screen optics, how can, or do, cinematic effects allude to their own means of visibility—to the technologies that either generate or imprint, then copy and project, their images? Allusion is essentially a form of reference, regularly intertextual. But what takes shape when an image is found alluding to its own optical conditions? How would one need to conceive of the cinematic system in order to see it performing this signifying work? And of what further use is such allusion in looking back on the whole history of the medium from the crest of its current electronic revolution? At least since the 1995 centenary of motion pictures, and more spottily before, we have watched a filmic medium's original serial imprint yielding to computerized adjustments at every level, from the generation to the editing of projected images. Increasingly, the temporal transit (mechanical) of the image, frame by frame, gives way to its temporal transformation (electronic) within the frame. This is obvious enough. What isn't, or not without some further reflection, is the frequency with which the later phenomenon is not only facilitated but inscribed by certain film plots of fantastic time travel. Less obvious still is the relation of these fantasy scenarios of Hollywood plotting to the uncanny temporal twists of recent European cinema. Hence the intended compass of my title. Framed time is a narrative inflection as well as a psychic topography operating across various genres. Its effect draws on the new cultural dispensation of virtual space and time as much as on any specific digital instrumentation, and is thus made available for an experimental and contemplative aesthetics of memory, premonition, and often transnational second sight.

Before electronic imaging, frame time—passing by at a punctual twenty-four per second—often brought with it a whole range of textual implications that tended to emerge on-screen at points of technical extrusion and narrative intensity. These involved the rhythms of consciousness itself, both serial and recursive, the bracketing and layering of memory, the motor impact of psychic shock, and so forth. But what was once the universal principle of such frame time in cinema no longer holds. In the fantastic turns and reversals of many recent narratives, whether openly digitized or merely contextualized in the cultural surround of electronic transmission and interactivity, frame time gives way, on several fronts at once, to that flashpoint of mediation I am calling framed time. This is the spatialized configuration of time itself as in its own right a malleable *medium*. How the on-screen results are different, more specific, more thematically explicit, and often a good deal more narratively implicated, than the generalized "time-image" famously advanced by Gilles Deleuze will grow gradually clear.

The films at issue do not need in any direct or considered way to address their culture of mediation, and certainly not to redress any aspect of it, in order for plot and style to test for tendencies—in the way a symptom can discover a syndrome. In regard to cinematography giving way to electronic representation as image paradigm, and even as model of temporality itself, we are concerned not with narrative intentions but with certain ad-hoc conventions that seem to be growing up around the digital unconscious of not just a wired globe but a now only residually mechanical medium. Beyond the obvious thematic of cybernetic dystopias in certain films, one eventual question would be why, in this international climate, plots are so often drawn to *non*electronic modes of virtuality, ghostly or neurotic, rather than to the going—and still coming—thing itself. And a follow-up question: drawn, not by what strategic design, necessarily, but by what free-associational slippage of fascination or anxiety? On the one hand, B-grade mass-market thrillers are certainly not rescued in these chapters for the A-list of auteurist reflection and critique. Nor, on the other hand, is the demanding philosophical and political work of a filmmaker like Michael Haneke given privileged access to the psychic drives—and medial phobias—of an epoch. Narrative originality is not the only return route to cultural origination. Some films think through what others leave merely intuited or scurry to displace. Films that don't even seem to know what they're doing may nonetheless do what they too well know—and thus act out their own repressions for all to see.

COMING TO MATTER: STRIP VS. SCAN, TRACK VS. ARRAY

Filmic cinema: temporal change indexed by segments, then remobilized frame by frame. Digital cinema: time seeming to stand still for internal mutation. The main purpose of this book is to note where and how this historical shift in the cinematic medium's own material constitution comes to matter on the narrative screen. From there, we turn to the more pointed question about why the gradual erosion of cinema's photographic base should so often coincide with plots obsessed by the psychosomatic contours of human temporality and human memory, including the exponential means by which those grounding conditions of identity and desire can be evaded, falsified, erased, or remade. Why this coincidence of the postfilmic image and the postrealist narrative, and where specifically? What I am suggesting by the title of the fifth chapter, with the conflated term *cimnemonics*, is that the original photogrammar of film tends to operate, sometimes at more than one level, like a kind of ocular portmanteau—by overlap and interpenetration. To reverse the sentimental cliché to which "the virtual" in the work of Deleuze can too easily be reduced, it isn't that filmic cinema, like lived time, is memory in the making. Under the rule of cinematographic speed, of retentive evanescence, or in other words under the optical lure of movement's own virtuality in projection, it is rather that the memory trace is time in the making: time coming forth as image, where what you see is what you lose. How might postfilmic cinema operate otherwise, or attempt to?

The question rapidly gathers momentum. Where, that is, would the form taken by this convergence of medium and theme make itself available to what I will be calling narratographic perception? My *premise* is that nothing rules out this availability: that there is no necessary chasm—despite rumors to the contrary—between local narrative response (including interpretation) and the assessment of a postphotographic cinema's reconceived place within either film history or, more broadly, media archaeology. Quite the contrary. My more specific *hypothesis* is that screen plots of the last decade continue to remember and address cinema's partial derivation from—and formerly total constitution by—the photomechanical imprint. This happens even while (and often quite directly because) the photogram—the single imprinted rectangle of the celluloid strip—is increasingly invaded or replaced in screen imaging by the digital photon.

A surprising retrospective example, from as late as the year 2000, holds our attention in the first chapter. It is the Merchant-Ivory production of *The Golden Bowl*. With what amounts to a lingering, almost clinging emphasis, its graphic treatment indulges in those material conditions of the film medium that had so frequently been foregrounded, for at least a century before, as stylistic markers in the articulation of screen stories. Every chance it gets—and not just by showing photographs in the hands or on the desks of its characters—the film feasts the eye (or sometimes besets it more aggressively) with the exposed photomechanical sequencing on which its own narrative realization necessarily feeds. As a heritage film, it thus exhaustively repictures the photographic lineage of its own mediation. With that 2000 release as summa of the filmic tradition, the rest of this book goes on to wonder, in some detail, how a digitally intermixed cinema might realign the fit between photogrammatic seriality and the psychic sequencing of plot. Answers, though, are less important than the five-part, plot-based questionnaire (chapters 2 through 6) that the inquiry generates. Rather than any prognosis for screen media, the resulting diagnosis offers instead a cultural symptomatology of a recent strain in both European and Hollywood narrative. I would say *symptomatography*, but I reserve that suffix, with its emphasis on the graphic texture of the image per se, for the more specific determinations—and refocusing—of narrative theory to come in that first chapter.

As stressed by the title of my 1999 book *Between Film and Screen: Modernism's Photo Synthesis*, filmic cinema is produced by the light thrown *between* spooling strip and screen rectangle.[1] But this is only because its mechanism activates the difference on that strip (and bridges the gap) *between* those single cellular frames that file by in a succession as unseen as their synthesis is visible. The photogram is thus taken up from moment to moment by filmic sequencing in the way that photography was incorporated into cinema in the first place. Thinking of the much-cited title of a book by Bolter and Grusin appearing, alas, only in the same year as mine, I have several times wished since that I had been able to paraphrase this process—in historical and material terms both—simply as the "remediation" of photography by cinema.[2]

A particular series of overt optical allusions in recent art practice is uniquely revealing on this score. Over the same decades that have seen the growing concessions of a filmic cinema to newer electronic technologies, thereby effecting an increasing historical distance on the commercial hegemony of the moving-picture medium itself, one still photographer's ongoing work insistently recalls the material fundament of his medium's longtime rival—and not only in its derivative imprint basis but in its fragile conditions of projection. These smiting photographic experiments do so by distilling the innumerable luminous frames of a projected film into a single moment of supersaturated blank light. It is as if all the flickers and flashes of a full-length feature were to attack the eye at once. Working in high-contrast black and white, when Hiroshi Sugimoto reshoots—in empty movie palaces or on deserted drive-in screens—the projection of an entire film in one prolonged exposure time, every shifting detail of the screen image is washed away in the sustained rectangular glow, the viewer blinded to all telltale shadows by the continual overlay of light. The result wryly effaces the very process it allusively retraces.

These site-specific temporal composites by Sugimoto, begun in the late 1970s, are at the same time genealogical parables.[3] Photography, formerly assimilated by filmic animation, in this case feeds on the medium's sequential alterations—now wholesale and impacted, and thus newly invisible—and returns cinema to its precarious support in projected rather than trapped light. In this way, the photographic has trumped the cinematographic from within the latter's own premises—and screen perimeters. By a medial recession in Sugimoto's work that is also a double remediation, the visibility of the on-screen image has not just reverted from shimmering motion to the stasis of a thousandfold beamed palimpsest. It has also been converted from one medium back to its precedent other, from countless photomechanical transparencies to a single opaque imprint. There is nothing digital here. But there is nothing potently cinematic anymore either. Encroachment operates by a reverse historical path. Whereas the stilled image was once overtaken by the filmic cinema it empowered, the exposure time of photography in Sugimoto's screen pictures now claims again the upper hand by absorbing all serial flicker into the fixed image of framed time itself.

A vestigial nomenclature is telling here in its very obsolescence. Apart from the use of archived production stills, what illustrated publications on film used to call "frame enlargements," a term still commonly used, has in most cases faded to misnomer. When based on DVD images, these reproduced shots are instead frame *reductions* of the on-screen image. More rightly called "frame grabs" or "frame captures," they are now electronically seized on the run—usually with the aid of a pause button—from a given configuration of the digital array on the monitor, after which they are printed on the page at a smaller scale. Only in the purely filmic moment of cinema did publication about the medium return to the celluloid strip for the painstaking rephotographing of the miniature photographic frame. This quiet phasing-out of technical vocabulary aside, and apart from such fresh coinages as *remediation*, common parlance, too, often spells out more than it is taken superficially to say. As a popular successor for the term *cinematograph* or *animated pictures* (among film's earliest nineteenth-century designations), the nickname "movies" came into widespread circulation for at least one good reason. The term aptly (even if subliminally) names, like "the flickers" in continuing British usage, or "a flick" in American slang, something that cinema *does*—does behind the scenes, so to speak, of what it images. In order to make the picture move, cinema moves pictures. I have previously called this photogrammatical motion the "lost cause of cinema," gone and forgotten in the visible goings-on it forges.[4]

Yet in some cases it is gone altogether now—or going. Digital imaging on the narrative screen has no nickname yet. And "timies" isn't likely. But you never know. For time can now be framed in its change rather than just derived from a change in moving frame. In filmic cinema, the real time of rotary motion (frames plural) is converted to the secondary or inferential time that follows from the image of on-screen movement. In postfilmic cinema, no image precedes the one we see—or follows from it in sequence. All is determined by internal flux (frame singular). Between whisked-away imprint and the whiplash action of convertible pixels, then, lies the difference between electrical and electronic cinema: segmental transit versus fragmental transformation.[5] Photograms graph motion, inscribe it by succession itself. By contrast, the computed picture, timed by binary (com)mutations, is more like a weightless easel for pixel tessellations, bit by microbit. The question remains: where do such differences leave their narrative mark upon screen stories? In particular, what consequences follow for the temporality of narrative within divergent modes of the fantastic that quite directly worry the relation between time and its (s)pace?

Certainly, no inherent logic invites us to extrapolate automatically from constituent optic module to screen shot. But correlations often impinge by optical allusion as the plot unfolds. They are pertinent only because they are *not* inevitable. Indeed, so far from necessary are such scalar leaps from material base to image that work on the new media in relation to narrative cinema often treats digitization as if its predominant function were to disappear into a simulation of the photomechanical index. The computed optic field becomes a stunt double for the cinematographic image within a substitute credibility of *mise-en-scène*. Discussion of this sort tends to suggest that the revolutionary technical fact of the newly compromised photogram should make no fundamental difference to the phenomenology of spatiotemporal coordinates either for agents within narrative or for spectators without.[6] But if a given film *does* choose to recruit its material base for its own narrative texture, what then? Especially when this foregrounding of either photomechanical *procession* or electronic *process* is found to concern rather directly an organizing thematic of time, memory, and their uncanny migrations, what are we likely to find?

We might, for instance, guess that narratives intent on exploiting the filmic substrate for figural leverage would be drawn to a metaphorics of succession and evanescence, whereas movies relying on computer electronics for their graphic tropes, even in the new mongrel status of a partly digitized cinema, would tend to gravitate toward themes of a more radical transmutation, even transmogrification. I think we'd be right in this supposition. Certainly, Dr. Jekyll will from now on morph into Mr. Hyde in a way unknown to any strictly filmic special effect, as composed of mere dissolves and double exposures (the celluloid trick par excellence). Those days in the chemical laboratory are gone, as much for cinema as for a mad scientist. Plot magic may once have been in this case just an intensification of ordinary scene change in the exaggerated spectrality of dissolve. Now, however, such magic happens to the image unit (no longer unitary picture) from within rather than from without.

One result, among others, is frequently thematized in recent fantasy plots. Narratives of preternatural temporality, whether telekinetic, recursive, reversible, proleptic, or any combination thereof—plots where the present is melted or dismembered before the eyes of its own occupant—make for spectacles in which morphing, not so much of human agents but of whole captured spaces, tends to replace superimposition as the reigning time-lapse function. Not a sprocketed drop of frames spooling past the aperture but a coded phasing-in and -out of the graphic grid; not an ocular rhythm at the threshold of perception but an algorithm beneath it: such are the newly divergent means that may elude visibility even in allusions to them by plot. As succession yields to intrinsic oscillation, transformation thus outstrips, as it were, the routine frameline of mere spatiotemporal transfer or transit(ion). From the ocular ground up, narratography often goes to work where, in just this way, frame time yields to framed time. As defined here for our response to on-screen events, both in the attached glossary and by example in the pages to come, the work of such narratography is, in brief, the reading of an image and its transitions for their own plot charge.

Our eyes held by the instantaneously rescanned frame of electronic mediation, we often see time itself imaged—with whatever degree of violence—as a process malleable, even reversible, rather than an incremental procession. In many such plots, the very feel of time, having been ratcheted up from material to narrative level in this way, takes shape as a portable interface rather than an organic interval: a phantasmal zone internally convertible without being traversable from one integrated moment to the next. Time is reduced, in the plastic bases of its own thematic, to a permutable optic field—differential from the inside out rather than, as in filmic cinema, troped for the most part by way of deferral, succession, and return. The surprise, perhaps, is that such visual options in high-tech screen imaging, maximized not only in generic sci-fi but also in Hollywood fantasy plots and their sometimes electronic rather than cinematographic execution, should find their mostly low-tech equivalent during these same years in the eroticized magic realism, virtual borderlands, and parapsychic rapport of European screen narratives.

At this stage, one may be guardedly intrigued by an easily deceptive correlation. The difference between segmental transit and internal transformation may well seem an overly schematic version, more like a travesty, of Gilles Deleuze's distinction, within classic cinema, between the movement-image and the time-image. For it is under the spell of the second, of postwar temporal imaging, that the spectator, according to Deleuze, waits to see what will happen across the effect of a given image. This is a matter of what will occur *in* the image itself—or occur *to* its attentive viewing—rather than the strictly event-based question of what will be pictured as happening next.[7] We of course need to beware. Any facile parallel between the stratified and elusive mutations of the time-image in Deleuze and the inherent flux of electronic registration is not only a superficial but a potentially misleading one. Nothing could be more foreign to Deleuze's actual claims for cinema through the early 1980s, or more confounding in any attempt to chart the fate of the time-image since, than a hasty alignment between those temporal figurations intrinsic to the film image and the work of digital implementation at large.

But while curtailing such premature parallels, there is still this much to say about temporal figuration when confronted with a prevalent strain of the Hollywood fantastic. American cinema's recent time-warp plots, with whatever degree of computerized editing or enhancement, often seem to be operating according to an implicit *digitime*. Further, despite the "uneven development" of European versus American cinema of the same last decade or so, the former's less aggressively fantastic plots of transnational identification, uncanny boundary crossing, psychosomatic empathy, and the rest involve a comparable visual stress—though often more cinematrographic in cast—on the effects of unsettled presence and spectral temporalities. To what degree comparable, and on what grounds?

IN MEDIAL TERMS

There's a new adjective all over the Web, if not yet in Webster's. Copyeditors have queried me on it, and I sometimes hold firm. At other times it is easier simply to say, "of the medium." But sometimes *medial* seems just right, a term that has come into circulation, one guesses, by back-formation from the more common *intermedial*. Just right, I might add, for reasons all there in the title *Between Film and Screen*. For it is exactly the betweenness of the moving track in respect to strip and screen, to material base and its projected materialization as image, that makes film a privileged site for the *medial* in both its overtones—pertaining to a medium, but also navigating a mean between separate terms: a midpoint, a median. It goes without saying that the phenomena we call media do not just link reality to its secondary perception. Their own mediation introduces a third and more or less material term, whose tertiary perception and estimate is, in certain cases, the work of narratography as these chapters will be practicing it. In the medial operations of filmic cinema, the mean(s) between strip and projected image is indeed the speeding track of a technically and historically prior photography in its now transparent and automated mode of advance.

Insisting on—and looking out for—this medial function is what put my earlier book directly at odds, in one respect, with the most challenging discourse on film at the time of its writing. In order to preserve the immanence and malleability of Bergsonian *durée* as a paradigm for cinematic fluidity, Gilles Deleuze thought he had to reject the logic of Bergson's essential sticking point as regards the then new medium: namely, the way projection falsifies in process its dependence on the discrete photographic moment by *simulating* the flow of movement. With that earlier French philosopher, however, I couldn't have agreed more—especially when the fact of the strip, with the resulting activity of the track, makes itself felt in, and against, the conveyed on-screen action.

What has happened lately in a postfilmic cinema of fantastic temporalities, however, has returned me to Deleuze's second volume (on the time-image rather than the Bergsonian continuum of the movement-image) with a new sense of urgency. Yet this entails no recantation in the least of my claims for photography's former, and still residual, role (its on-screen role) in obtruding the temporality of past (t)race into the ongoing time of a screened event, where stillness is already "remediated" by motion and refigured as movement. Furthermore, between Bolter and Grusin's broad-gauged survey of the new media in 1999 and the present writing, other recent books on cinema, especially in relation to temporality and its filmic base, should help clarify what I had in mind with my own 1999 study—if only by contrast in one case—as well as what one might now see at work in cinema's digital turn. After canvassing their salient points, with a promise of returning to them in subsequent chapters, we will move toward the medium-specific method proposed here for the analysis of narrative inscriptions in either photomechanical or electronic modes.

To begin with the case that most directly contrasts with my own sense of the medium, Sean Cubitt's *The Cinema Effect* (2004) combines a sweeping Deleuzian periodization of cinema with an unspoken reworking of suture theory.[8] Cinema for Cubitt is divided into three facets, with the regime of "the cut" and "the vector" being, counterintuitively, *preceded* by that of the "pixel," by which Cubitt simply means the pinpoint glisten of luminosity on the first film screens. In describing Lumière's inaugural film of workers leaving a factory, Cubitt does come close to acknowledging the photogram, if I'm decoding his remark correctly, in noting a kind of subperceptual translation from spatial displacement to temporal synthesis. As he puts it, "The motion inherent in the instability of the frameline acts as a given, as something that, since it sums all movement as equilibrium, is perpetually now" (71). Strip—manifested as frameline—races past the aperture as the always destabilized basis of the projected track, even in a fixed shot. As opposed to this timeless "now" of motor projection, however, screen narrative's articulated sequence requires for Cubitt first the cut, to objectify the "given" of the screen picture as a selective imaging *of* and *by* some agency (personified or not), and then the vector, to induce the expectation of narratively focalized change over time.

Slightly askew to his main triad, or extending beyond it to include the pixel in a quite different (and, in our terms, postfilmic) sense, another of Cubitt's threefold distinctions is certainly clear enough on its own terms. There are three predominant—and often overlapping—manifestations of the cinema effect. To compress a summary appearing late in his discussion: "One cinema represents. . . . A second cinema reproduces. . . . The third mode of cinema generates" (360). With only pixels of projection, no motion of the shot, the shimmering image plane represents the world in fixed views. The disjunctive cinema of the cut, and its further vectorization, reconstructs a narrative world rather than conveying an immediately recorded space. Digitization arrives to produce rather than simply reproduce its world. Since my interest begins with the film effect, so to say, underlying the cinema effect—that is, with classic cinema as a film effect—I would be inclined to add that vectorization at the macro level (everything from rapid montage to the Steadicam) should be understood lately in connection, however invisible, with the rudimentary dynamics of the new image field itself in the electronic scan of pixilation's instantaneous array.

In Cubitt's historical sequence, before disjunctive cuts and vectoral drive there is only the first facet or phase of sheer scintillating image—the minimal form of what Tom Gunning would call the "cinema of attractions" and what suture theory would see as the bliss of spectation before the fall into editing, into system, into plot, or, in other words, into the scission that makes for variable temporality and narrative direction.[9] In choosing to term this founding moment the time of the "pixel," however, Cubitt inevitably obscures the "flicker effect" of advancing photo frames of which early audiences were so well aware. If this rush of frames is the "given" in his mentioned disclaimer, the taken-for-granted, it is also what is magically—or better, spectacularly—overcome. Part of the "attraction" value of early cinema as spectacle was indeed—a fact generally submerged in too much film-historical discussion—that audiences could be counted on to remember the previous medium from which animated photography springs so captivatingly to life.[10] To put it by contrast, where we might want to say that Sugimoto's photographs of glowing screen frames reflect the art of the luminous pixel, motion pictures depend as well on the mechanics of the photogram. For all the genuine achievements of his schemata, Cubitt's seeming anachronism in characterizing the pioneer (and prenarrative) moment of cinema as one of pixilation also inhibits the long view he attempts, for instance, in accounting for the electronic successors to the "neobaroque" cinema of the Steadicam.

And if Cubitt has oddly squandered (or at least diluted) the vocabulary of digitization in advance, his stress on pixel (rather than flicker) distracts us, first and foremost, from the way the advent of the cut does its definitive work. For this transitional development en route to full-scale screen narrative depends on exactly the temporal as well as spatial molecularization (seriality rather than pixilation now) of the sequential frameline. Inherent to photographically based film projection, this is where the fixed frames of visual imprint are motorized across their own discrepant gaps, or where (in Cubitt's own passing concession, if that's what it was) the "instability" of motion in the apparatus is overcome by being "summed" as present movement on-screen. It is only the clean edge of the photogram that makes possible the "cut." Yes, film appears before us bit by lit bit, but—unlike the physics of the digit—not all within a single composite image. Rather, as we have reviewed its mechanism, filmic imaging is strung out, rung by cellular rung, along the vertical spool of its own incremental passage. It is in this sense that filmic cinema stems at base, at the lower limit of functional discreteness, not from the collective stippled gleams of apparition across the enlarged screen rectangle but from the unseen sequential drop and roll of the photogram: strip projected as image track in its manifestation of screened scene.

Carol Reed's 1948 film *The Fallen Idol* makes the foundational relation between cut and vector, with their codependence on the serial photogram, unusually clear—and at its plot's dramatic crux. With Graham Greene's script revising his own story to introduce these visual ironies into the staging of the screen climax, the young boy whose POV has organized the film is seen fleeing down an outdoor fire escape from a traumatic moment of violence: a bitter wrestle between his idolized butler and the man's shrewish housekeeper wife, who has just discovered her husband's affair with a younger woman through the boy's unwitting disclosures. Disillusioned and in shock, the child-hero descends from outside a huge upstairs window that frames the wife's enraged thrashing of her husband, near an ominous interior banister, to the parallel lower window that (just a few precipitant seconds later) reveals her body at the foot of the grand staircase. More than the boy's idol has fallen in this vectoral tracking shot. And more repressed facts have been disclosed than just those of illicit desire in this plot twist. In the downward elision between rectangular framed views, the focalizing child, in his momentarily occluded vision, suspects a murder where we know there has been only an accident, not a push but a slip. In this blank between windowed apertures has transpired as well, in encoded graphic (or narratographic) pantomime, not only the photogrammatic exchange of vertical frames at the strip level but also the profilmic disjuncture that allows the actress to move, invulnerably, from upper railing to the floor below across separate camera placements. With rare clarity from within a given plot, the constitutive manipulations of photographic film and narrative cinema alike—the intermittent image and the thereby permitted cut—are together laid bare, under misapprehension, to the innocent eye view.

In contrast with the terminological backflip of Cubitt's history, with its emphasis on the originary pixel rather than the photo frame, Lev Manovich, in *The Language of New Media*, looks back to the discrete nature of celluloid imprint from the vantage of postmodern pixilation. This hindsight is often eye-opening. Unlike computer-generated visuals, it is not the framed image which is piecemeal and composite in filmic cinema, but rather film's projected duration. This is a frame time no less artificial for being automatic—as Manovich is able to show more clearly than ever in his millennial overview of mediation, where cinema is merely a brief historical interlude, or detour, in the arts of temporal representation. This is to say that Manovich's emphasis, unlike Cubitt's in its stress on the inaugural pixel, doesn't feel anachronistic (or impressionistic) in its paradigm, but rather archaeological. Cinema's stratum in this evolutionary process is as clear as it is unique. Before cinema, time was always constructed by media, manually inscribed, whether through pigment or alphabetic words. By contrast, cinema arrived to mechanize succession. But technological time hasn't stood still off the cinema screen either. Once again now, with the coming of computer graphics and digital editing procedures, sequence can be artificially constructed by hands-on keyboard input. This is certainly the case at one level, even though such somatic input may be further disembodied on the way to visualization by speeding photon cells. In this sense, cinema is merely a blip on the historical screen of ocular representation, with mimetic production coming full circle, after a century of industrial practice, from the former pointillist jot to the electronic blip itself.

Stressing input rather than reception, we might say that the image under *digital manufacture* returns us to the anatomical etymology of both those terms in the sense of manual articulation—as well as harking back further yet to the prefilmic arts of time. For a century-long interregnum in such manufacture, however, time was for the first time ever, in Manovich's apt term, actually "sampled" by art rather than fabricated. Film projection is the result. Indexing the real even in the single camera shot, time on-screen is a continuous automatic sectioning-off of duration for the purposes of illusionist spectation. By contrast, in the stop-action photo sequences of Étienne-Jules Marey and Eadweard Muybridge that preceded cinema, temporal process was isolated for the opposite purpose of motor analysis. But cinema's stake was in simulating rather than merely dissecting the time-space ratio. The other modes of representation indicate or picture time, imitate it. Film is actually based in and on it. Cinema digests time in both senses, chewing it up in useful mechanized bits for redistribution across the functional ellipses of narrative. In contrast with a strict machinic capture of filmic time in serial record, however, it is the subsequent art rather than the raw technology of cinema that returns imaging to the local adjustments of handicraft. In part laboratory based, and hence a matter of editing as much as of recording, such is the labor that summons back the manual over and above the automatic.

Here we verge on a third recent account of cinema's medial coordinates. In *The Emergence of Cinematic Time* (2002), Mary Ann Doane emphasizes how it is not just the narrative craft of cinema, but its very conception of time, that requires the laboratory intervention known as the cut.[11] Her study goes back to the experiments in chronophotography (Marey and Muybridge, again) that were designed to parse time rather than resynthesize it: to seize on and record what Walter Benjamin calls the "optical unconscious" of ordinary perception.[12] That's chronometric photography in a nutshell. By contrast, film puts the pieces of the real together again in the synthesis of screen movement. As Doane explores the technological allergies of that emergent historical moment, this is exactly why chronophotographers like Marey resisted cinema as much as Bergson did, if according to opposite agendas. The science of vision, as practiced by Marey and Muybridge, was newly wedded to the increment rather than the flow, to motor stages of motion rather than its continuum. Cinema overrode that commitment by rendering invisible again the phases of succession. The philosophy of duration in Bergson, by contrast, saw cinema misrepresenting time in an opposite way—by having any traffic whatever with discrete imaging. Time cannot be sampled for Bergson, only lived. Its vitality is indivisible. Oppositely, motor duration is never for Marey a seamless continuum, but only a chain of positions. On Doane's showing, Freud is a third figure whose resistance to cinema only works to clarify its true temporal dimensions. According to psychoanalysis, time is entirely a production of the unconscious. To render it in art as anything like a natural and unbroken continuum thus counts as a mystification.

Though this is not where Doane places stress in the long run, it should be clear that all three experimenters with the consciousness of lived time—Marey, Bergson, and Freud—might have viewed cinema differently if only they had perceived it like Eisenstein did: as an actively dialectical rather than a loosely synthetic or homogenizing medium.[13] In Eisenstein's aesthetics of collision, there need be nothing passive or inert in the sequence of frames. The overriding of photogrammatic traces in the viewing of screen narrative—their typical relegation to the unseen and the unconscious, both as regards the framed image itself and the discursive intervention of its cuts—can easily be reversed under certain narrative pressures, and even thematized in the process. Materiality can in this way be made to bear narrative yield through its own deviant exposure, becoming figurative on the spot.

For Doane's turn-of-the-century trio of conscientious objectors, moments of this sort might, or could, have been sufficient to disrupt the complacent acceptance of screen temporality and thus credit the apparatus with genuine investigative and aesthetic potential. Freud, especially, might have been delighted with a full-blown narrative cinema of irrational cuts, lacunae, and symptomatic ligatures. Be that as it may, it was not to be. His early objections, along with those of Marey and Bergson, unanswered by movies as they knew them, have passed into history—along with cinema itself in its solely photomechanical constitution. In looking ahead to the medial implications of recent hybrid cinema, suffice it to say for now that digital imaging and electronic editing (their aesthetic and thematic explorations aside) at the very least require a new theory of the technological repressed. We cannot see (or ordinarily even think of) the pixel in seeing the picture, any more than we can see (or remember) the photogram, the serial transparency, in the spooled movement of cinematographic projection.

But if there is a digital unconscious in recent cinema, what can a narratographic reading of it make us productively conscious of? Here is where new work of the last half decade brings us back round to the open-ended conclusion of *Cinema 2: The Time-Image*. Sudden electronic possibilities of the screen field—"untheorized" (266) by the early 1980s—induce for Deleuze not a forecast but instead a review of cinema's postwar transformations: "We do not claim to be producing an analysis of the new images . . . but only to indicate certain effects whose relation to the cinematographic image remains to be determined" (265). Despite this disclaimer, Deleuze actually goes far toward laying the groundwork for just such an analysis. And certainly, as it turns out, he is never more forthright about the photographic basis of film than when articulating its contrast with the digital. Unlike binary imaging, cinematography depends on representational units that he now stresses as "superimposable" (265), separable and successive, whereas the video or computer image is instead "reversible" (267). Intervals and interstices have been replaced by the flux and reflux of internal conversions. In prewar cinema, intervallic movement was not simply measured by elapsed time in an empirical sense. More abstractly, for Deleuze, time became the very "measure of movement" (271). Motion was thus the indirect image of time. As the movement-image slipped from prominence in the stasis or introversion of postwar cinema, time came to be imaged more directly. What, then, about time—and its tempi, no longer its (s)pacing—in postcinematographic screen practice? That is the question Deleuze leaves readers to voice for themselves.

The digital image can be altered in any direction across its own frame, but not by accessing off-frame space, whether inferring it or moving to include it next. There is "no outside" to electronic images, "any more than they are internalized in a whole" (265)—as for instance in a conception of the strip as the vanishing frameline of a sequential narrative projection. As his striking comparative analogy succinctly puts it, "the shot itself is less like an eye than an overloaded brain endlessly absorbing information" (267). Less, say, like *Man with a Movie Camera* than like Hal the computer cyclops in *2001: A Space Odyssey*. The brain itself does not see; it processes sightings. In this way the meeting ground of being and time surrenders the necessity of any ocular focal point, any Albertian myth of the "window" (265). (Deleuze is writing, of course, before the ideological misnomer of the Windows operating system.) Human memory needs no recourse to a figurative mind's eye rather than a mere cognitive scan. So the open question remains, for Deleuze, whether electronic transmission or generation thus "spoils" the time-image—as image—or instead "relaunches" it (267).

Thus does the fading hegemony of the cinematograph—as gradually eclipsed by what one might term the electrograph—press new questions in turn, even about films with no interest in brandishing their computer-graphic (presti)digitations. To think of the screen image as brain pattern rather than eye view might well revise the entire apparatus-based monocular logistics of the POV shot—might, and in recent cinema often does. The advent of the electrograph cannot help but lead, in connection with the time-image, to an inquiry into the "psychological automata" (266) that such images often serve to narrate, for instance, in Hollywood cinema. How is plot thus correlated with the "electronic automatism" (265) of its own manifestation?

Inquiry would look, for instance, to the digitally implanted (or otherwise virtual) memories of cyborgs, out-of-body agents, posthumous self-specters, autonomous laser holograms, computer simulants, schizophrenic time travelers—the whole cast of noncharacters in the postmodern fantastic. At the same time, it would survey their counterparts, for European cinema, in transnational doppelgängers, erotic revenants, the blurred embodied thresholds of past and present, the whole magicalized horizon of the real. In both filmmaking practices, and even when the means of recording is predominantly photomechanical, recent plots often tap a new latitude in screen mediation for their own most striking optics and metamorphoses. The main further inquiry left hanging in Deleuze—about how to wonder productively whether the time-image is obliterated or somehow renovated by digitization—inevitably concerns one's particular method of approach. In this respect, the present effort via narratography has a specific academic genealogy that should be useful to spell out.

SEQUEL FOR AN AFTERMATH: THE CHAPTER PLAN

Most critical projects leave in the dust their detours and jump starts in barreling toward results. In the present case, however, accounting briefly for the genesis of this book may help offer the best ground plan of its chapters' linked claims. Writing began with a paper for an Italian symposium on liminal or threshold moments in film at the University of Udine, a conference very much in the mold of Italian structuralist semiotics. Once inflected by a symptomatology of cultural production, however, the convening rubric offered an opportunity to look more closely into the thing that had struck me again and again about screen narratives of the fantastic since my 1999 book on filmic cinema. Even before plot has taken hold, both European and American films in this vein tend to plant an image—unreadable at the time in its full implications—that turns out, in the often surprise (and frequently supernatural) ending, to have been an instigation and a clue.[14] And more than this: to have installed the first trigger of a curious homology between the collapsible spacings of narrative temporality and the psychosomatics of time-consciousness within plot. Other screen releases followed in the wake of that early paper, of course, as well as further conferences on old and new media alike, including the next year's symposium at Canterbury under the title "Stillness and Time: Photography and the Moving Image." Through it all, in my own investigations at least, the liminal moment kept coming to mind, and to view: the moment where photograms are first sent into motion—or photons into coruscation.

Structural functions, that is, grew to seem more and more materially rooted and medium specific. For these inaugural gambits are exactly where the ensuing moves of plot may leave a residual image in their wake, overlooked, forgotten, repressed, or at least deflected. Moments of ocular onset—whether in cinematographic or electrographic registers, and often quite *pointedly* in one rather than the other—may well disclose a precipitating image for subsequent narratographic uptake. Laid bare in images of this sort is often a pure graphism before story. Such a visual limen—by conditioning rather than exactly constituting the first thrust of plot—can seem exclusively material, tapping the medial support before the momentum of narrative purport. Structural semiotics, or its subdiscipline of narratology, might readily be satisfied on its own terms (Michael Riffaterre's, for instance) with an account of suppressed matrices and the superstructures derived from them. Everything fits, even when it arrives silenced and well in advance of any revealed context. But a close-grained narratography can seize on a further yield in film's liminal manifestations. Hovering on the threshold of story, picturing is often a medium-deep rudiment as well as a plot spur. Again and again, whether photomechanical or digital, the sponsoring first image graphs its own optic means into the open before being assimilated by the story's drive toward closure. And when that point of closure returns in echo to the launching image, it is likely to effect this return, yet again, by reverting to medial assumptions and their optic coordinates. At which point more than a narrative circle is closed. In league with the deviances of plot and psychology, cinema as projective system rounds on its own instance as providing a latent archaeology of differential visuality: a mode of picturing once strictly mechanical, now variously electronic.

After an orienting glance at applied narratography in the analysis of prose fiction, the first chapter will offer a review of the temporal photogrammar of cinema's first century from the vantage of a single period film. This is a densely metafilmic heritage production from the year 2000 that in fact begins with its own liminal moment in a close-up of "artificial light." *The Golden Bowl* opens on a fluttering pre-electric torchlight answered to in the last frame by the iris fade-out on a primitive and flickering film-within-the-film. The contemporary screen narrative is bookended by two stages of its own visual prehistory. Turning next to recent narratives in the mode of the Eurofantastic, the second chapter will move to follow out in detail the semiotics of trick beginnings and trick endings in these plots. On view there (in repeatedly binational coproductions) are mysterious narrative events falling not only under the potent and widespread influence of Krzysztof Kieślowski's humanist fables but under the yet longer shadow of Proustian spectral mnemonics and, darker still, of a traumatic historical memory. These are often films that silently participate in the psycholinguistic disorientation of the New Europe as well as its geopolitical mutations—even when their plots may have nothing explicitly to do with the Realpolitik of reciprocal postnationalist accommodation. As such, they work to envisage the virtual, rather than the actual, as a more operable route to the real.

From there forward, emphasis will fall on the structural obverse of these European psychodramas in the new Hollywood fantastic, with its own more resolute flight from politics. The third chapter takes up the supernatural versus digital plot twists in alternating subtypes of American genre production. The fourth moves to films either of recursive temporality (the second generation of time-loop plotting after the likes of *The Terminator* [1984], *Back to the Future* [1985], and their respective sequels) or of downright mnemonic psychosis. These are films, in either case, whose escapist logic is anchored, often covertly, in a culture of digital interface that is for the most part backdated to (and masked by) precedent technologies. In many such narratives, the paradoxes of *temportation* are floated, even flaunted, within a technics rather than a plausible logic of ocular mutation. After submitting some of the most recent examples of such temporal paradox in the fifth chapter, and via Deleuze primarily, to a more thorough account of the virtual in the perverse collisions of the new *timespace*-image, the way is then prepared for a fuller reflection on method in the last chapter, concerned with the debated place of narrative analysis in media theory. In pursuing such metafilmic and postfilmic effects under the sign of the virtual, we take our cue from Deleuze's own glimpse into the potential temporal ironies of the "reversible" rather than "superimposable" image, the optical signal "numerical" (that is, binary) rather than additive and sequential. Doing so keeps this study in touch with those layered historical strata of mediation that lead on—in and through genealogy, and through narrative interpretation as well—to the wider cultural and technological purview of visual archaeology.

The closing chapter is tuned in this way to the analytic repercussions of all that precedes. It refuses, in particular, to accept the terms of a divorce widely filed in many recent venues between visual or media studies, whether of a genealogical or an archaeological stamp, and the hermeneutic energies of narrative analysis and cultural critique. If, choosing not to perform appendectomy, the reader elects to read on about the prehistory of cinematic narrative in the protofilmic syntheses of Dickensian lexical and syntactic dexterity, then cinematic archaeology has claimed a yet broader field of cultural production as its own. This book's experiment in narratography—as begun in the coming chapter with the syncopated "intercutting" of Henry James's prose—will thereby have taken its widest bearings in the history of differential narrative forms, from skidding letters in a shuttling syntax to flickering photograms to oscillating photonic shifts.

That these are all legible effects is no dead metaphor: legible in their difference and in their local deployment. In the case of filmic, then digital, cinema, what kind of thing is it that grows thereby readable? And how—to repeat the opening question—can screen illusionism thicken its signals in optical allusion to its own modes of visuality? Only if it does so can a given film become part of visual studies as evidence instead of just instance, providing investigative testimony as well as merely example. Only then can a film's narrative effects be closely deposed by an archaeological inquiry, say, rather than itemized in passing by taxonomy. Across the historical transition from filmic to digital screen imaging, it remains the case that—within narrative cinema—the visual can *make reference* to its own history and conditions only by passing through its inherent materialization as *text*. This opening discussion, as expanded by a whole range of examples in chapter 2, proposes one method for intercepting that materialization at its point of emergence. In the liminal instant, optical allusion is often all there is. As this book proceeds, it will be the work of analysis to appreciate how, and under what kinds of narrative stress, the technical means of such narrative are delivered up for inspection as the image of plot's own logic. In a continuous reversal of the truism, it is by such paths of attention that the message delivers the medium.

LEXEME TO PIXEL: AN EXPERIMENT IN NARRATOGRAPHY

In terms begin determinations. Yet in locating its distinction from traditional narratology, the suffix of my proposed method doesn't quite suffice without further explanation.[1] As with the term *medial*, then, it's back to the dictionary, though this time with an overload of models and associations. It is one of those slippery and ultimately conceptual ambivalences inherent to English nominalization that *graphy* regularly faces two ways at once—and, in this case, with far from unwelcome results. For that suffix points to writing *about* something as well as to writing *by means of* it—as for instance, in the case of orthography, to both handwriting analysis and the script itself: designating, in that case, a potential discourse both on and in writing. The distinction applies as well in examples less strictly linguistic. Often, too, the bivalve reference is equally perfect and complete. A mapped region of the globe, for instance, may both have a geography and enlist its discipline, just as topography comprises both the features of such a locale and their description. Iconography, for another example, is both a system of images and their study as such.

The case may seem more complicated with the two syllables of *narrat(e)* as stem, perhaps because of the currency of that middle term *narrativity*, which is abstracted from any given inscription in order to evoke the *structure* of a telling. Here easy parallels desert us. Organic beings do boast an inherent biology even as they submit to its scientific precisions. But the narrativity endemic to any story doesn't *have* a narratology; it only invites one—just as thanatology is the study of death and dying rather than some mortal core an autopsy might reveal. Further, what if one wants to specify the procedures of that narrativity for a given text in a given medium? If this isolates the labor of narratography as inscriptive process—so the rules of word formation would readily allow—it also, and more pertinently here, indicates the field of narratography as the study of such mediation. In this book, then, it is the writing *on* narrative's graphic effects, either lexical or filmic or now electronic, their category of study (rather than the writing in and by them of screen effects), that the term *narratography* is meant to help focus.

This is not in the least to deny, however, that every gratification is taken in the closeness—at such a medial node—between theory and its object, between analysis and visual praxis. The method of narratography proposed and developed here for cinema is meant to evoke less the first-order operations of a term like *cinematography* than the plottings of *cartography*, even when the latter term is thought of more in connection with the making than the studying of maps. This is, simply enough, because the ground covered preexists the analysis—although it is certainly reimagined in the charting. Beyond all dictionary quibbles, we can phrase the difference in the most familiar of terms. However much style it may brandish, narrative doesn't manifest a stylistics. It only prompts one. Narratography is just such a medial specification of poetics for a given story text.[2]

It offers, in turn, a mode of analysis inseparable from the material innovations of media history. In its concern with material inscription, such analysis is everywhere alert to the evolution (if that's the word) of one reigning graphic condition as it succeeds to another: letters through photograms to pixels, to name but three, each as constituent structural subunits of narrative representation. Yet each stage in this genealogy alters the ground of representation entirely. Narratology knows all this, of course. Narratography shows it in operation. Concerned in prose texts with the subtending enchainment—and potential surface tension—among letters, words, and syntax, or in other words the contributory give between signifiers and the larger patterns that render them the increments of narration, literary narratography is lexical and grammatical at base.

When turning to cinema, this level of attention turns instead, though comparably, on the filmic modules of advance and transition—and lately on the play of pixilation across the optic sector formerly reserved for photochemistry. After a shift in focus from literary prose to cinema, it is therefore the further shift, within screen projection, from the strictly filmic to the presumptively digital that the rest of this chapter will follow out across the metacinematic texture—and closing twist—of a filmed novel. But that's not the main connection between Henry James's prose and James Ivory's film practice that I am hoping to draw out here. Rather, concentrating on a "heritage" adaptation in the lately prevailing mode of self-conscious optical mediation allows us, from the retrospective vantage of the year 2000, not just to appreciate the ways in which optical recursions of the filmstrip within the image track have their equivalent effects in the sprocket noise and clutched advance of syntax itself. It also allows us to name narratography as a means for that appreciation.

Debate comes, of course, from the commentators who do not find any such subliminal microunits, either photograms or photon bundles, to be in fact signifying traces at all. Deleuze's emphasis on the "plane of immanence" rather than the generative track may stand in for a widespread discounting of the intermittent strip in any but the most aberrant screen practice (in Deleuze's case, the work of Dziga Vertov). But before the plane of immanence, there is the strip of imprint. One way or the other, whether we incline to a distant linguistic model or not, an important fact remains. By definition, the propellant cells of cinematic movement are processed in the frameline and disappear (but not always) into the projected image that they are recognized (and sometimes actually seen) to animate or transform.[3] In even a loosely applied parallel to grammatical communication, syntax is to the syllabic and subsyllabic components of diction rather as dissolves or cuts are to the unbroken continuum of the shifting framed image. While attending to each separately, its interests now lexical, now cinematographic, still the work of narratography bears just this ratio in mind whenever crossmedial comparisons are invited.

That such comparisons are possible shouldn't deter conviction, I hope, in regard to my claim that the narratographic method is in fact more medium specific than traditional narratologies, whose "ology" is by nature global, transtextual, intermedial, pansemiotic. Take the leading film exemplar of historical narratology. In good formalist fashion, David Bordwell puts structure (plot sequence, or *syuzhet*) before content (story events, or *fabula*) by discovering the latter as in fact manufactured and assembled (rather than just reshaped or interpreted) by the former. In fine-tuning the "discourse" versus "story" duality that colors the thinking of everyone from Gérard Genette to Seymour Chatman, Bordwell's viewer-response approach suggests that narration stands forth as *"the process whereby the film's syuzhet and style interact in the course of cueing and channeling the spectator's construction of the fabula."*[4] This is the most general and content neutral of definitions, and it seems loosely complemented by Tzvetan Todorov's earlier stress, in *The Poetics of Prose* (1972), on the way narrative begins in a disequilibrium that must be set right in the end, when story events have fully conformed to the structuring dynamic of plotting.[5] Hence the prototypical role, for Todorov, of detection plots and their spur of mystery, where there is a presumed but occluded story that can be found out by plot only by being (re)constructed. Such instigating disequilibrium is specified in turn, for a single related genre, by Todorov's own separate narratological study of the fantastic a year later, where the advent of the inexplicable, either uncanny or supernatural, is the most salient mode of disruption.[6] The initial and structuring discomfiture, at the launch of plot itself, is one with which story is always trying to catch up, until at the end stasis and explanation coincide. Or, say that *syuzhet* constructs from the start what style alone (more below) may help us decide upon in the end as the real cause of such an "unsettling" *fabula*, whether uncanny or marvelous.

Following upon Todorov's prose poetics, Michael Riffaterre's *Fictional Truth* goes on to subdivide the maneuvers by which such disequilibrium is set right. Textual progress is mapped from an initial model of instability or semiotic deviance that he calls "ungrammaticality"—though its markers are mostly figurative in their signifying slips, rather than strictly grammatical. This founding detour proceeds across a plot-long subtext on the way to resolution.[7] Thus is fashioned an atemporal semiotic pattern (or transformative infrastructure) beneath the flow of plot: a pattern made present to us in various signals (Riffaterre is at his most psychoanalytic in comparing them to neurotic symptoms) that surface at intervals on the way to resolution. One readily notes that the place of such a subtext (not so called) in Peter Brooks's *Reading for the Plot* is taken by an even more explicitly Freudian sense of the textual unconscious, one that troubles events at the first stage of disturbance in the psychoeconomy of narrative—and that comes round to explain it in the end. Casting Brooks's terms in the most schematic fashion, Eros triggers the disequilibrium that the death drive, or Thanatos, will foreclose.

Bordwell's highly generalized account aside, these other narratologies are deeply teleological. As modes of an engineering science, they seem concerned in good part with the light at the end of the tunnels girded by plot. In Bordwell's terms, though, and whether triggered by the *inexplicable* (as in the genre of fantasy this book will concentrate on) or some lesser cognitive upset, the onset of *syuzhet* always offers the first disequilibrating model—the first uncertain thrust—in the working up of a *fabula*. Other narratologies converge to explain the rest: how, for instance, narrative is organized from there on by repeated incursions of structure into story, or subtext into event, until petering out into a new stability at the end. This is the destined end of every plot tendency, where all is rendered quiescent again by marriage, by death, or by some other closural trope. In the genre of fantasy, this involves the prolonged (and only at the last minute resolved) tension between events amenable to psychological explanation and those that retain the weight—or weightlessness—of the supernatural.

From the most general to the most specific of the narratological analyses (say, from Todorov's foundational prose poetics to his particularized genre demarcations), one tendency stands out. What is consistently downplayed in the overall formulations of these approaches, if not always in their analytic practice, is the matter of style or technique. For Bordwell, one recalls, narrative takes shape, and definition, from the way *"syuzhet and style interact in the . . . construction of the fabula."* Despite the correlative grammar of his formulation, style (acting style, editing style, *caméro-stylo*) is for Bordwell mostly a bonus, an add-on, an adjunct to structure. Even if intrinsic to narration, style is not internal to its shaping logic. For Bordwell, two distinct levels or aspects of form in film (structure on the one hand, as interacting with its local visual or dramaturgical cues on the other) converge to generate on-screen content—and do so by guiding cognition. But think again—or differently. What if style in the cinematographic sense were not secondary or complementary so much as constitutive, more an internal supplement than a "plus"? What if the surface of the medium, with all its textual inflections (and deflections), were the only way in which the abstraction of *syuzhet* could make itself fully felt in the first place? Who, in fact, can really doubt this? If style or technique, broadly speaking, were acknowledged to be all we see of abstract structure on-screen, even in its formation of narrative content, the picture would look a little different, would need perhaps a tighter focus in analysis. And if, in turn, that stylistic stratum—tracing, so to speak, the very stylus of imprint and impression—were reconceived to be something like the interface between form and content, or, say, the structuring *mode* of content, then the need for a more closely graphic (rather than a broadly schematic) mapping would be all the more obvious.

In that case, one might wish to back up Bordwell's definition to the textual level so as to say that "technique is the *manifestation* of *syuzhet* in its structuring of *fabula*." This would not, in any sense, be to collapse all of Bordwell's pertinent distinctions into the dichotomy of discourse versus story, losing all detectable sense of structur*ation* in the process. It would only be to appreciate discursive technique (or style in this sense) as the mode of apprehension for a plot's operational work. This, then, would certainly still allow "reading for the plot" in Brooks's fashion, reading for *syuzhet*, where the chain of metonymy across narrative time is regularly "bound" by its conversion into metaphor, into symmetries and equivalences. Moreover, this view of technique as the legible face of structure in the latter's mapping of story over time would obviously align itself with "the discourse of the fantastic" in Todorov. This would include his sense of a distended periodic syntax of events slung suspended between undecidables. Such a revised stylistic paradigm also lines up neatly with the variables of grammar and agrammaticality in Riffaterre, as these are paced incrementally, in the form of a subtext, toward a return in closure to the matrix of unspoken thematic generation.

To recenter cinematographic style or technique in this way, as more like a definitive *medial* function, is thus to suggest the narratographic purchase within narratology—and, once again, to tap the medium-specific precisions associated with a given disciplinary object. Cinematography, inscription, textuality: these are what manifest in process the work of discourse as storytelling, with a narrativity that must—it can be said without theoretical tendentiousness or fanfare—be read even when viewed. These graphic practices are what narratography must engage. The need is mutually reinforcing. Narratography without narratology may well be "mere" stylistics. Narratology without narratography is barely reading at all.

Let me head off at least one potential misperception. In its terminological ambitions, narratography is not specific to pho*tography* and its mobilizations. The term derives, as we've seen, more from the generalized graphics of representation (etymologically: *graphein*, "to write") than from any inferences of photochemical imprint. Be it conveyed by the writer's pencil, by William Henry Fox Talbot's "pencil of nature" as the first master trope of photography, by Alexandre Astruc's *caméra-stylo* (camera-pen) as a definition of cinematic art, or by the letterpress form of the literary artifact, at one level it hardly matters.[8] Narrative process is, for the purposes of narratography, as graphic in fiction as it is in film. By contrast, narratology is about the transtextual function of narrative, its abstractable superstructures: how it is and moves—and, in the process, moves from one medium to another. Further, narratology is usually about the being-toward-death of narrative, its closural impetus.

If narratology appears in this sense a matter of vectoral engineering, narratography emerges, by way of distinction, as a seismic metering in process. It graphs the tensions and contradictory force fields along the armatures of plot in a specific mode of textuality, whether visual or verbal: the seams and junctures, the rifts and transferred pressures, the folds, overlaps, and undermines of sequencing itself. With narratology equipped to analyze the underlying (or overarching) structure of storytelling, narratography is then given to engage the surface tension of its materials. Narratology tells us what narrative does. In a given medial appearance, narratography shows us where. And that "where" may well hover at play in the border region between textual wrinkle—or warp—and its estimated intertextual, cultural, or intermedial resonance. The second half of this chapter will, for instance, track moments of optical distortion and photogrammatic impaction in a given film—twisted mirroring, phantasmal match cuts, superimpositions, hallucinatory animation effects performed upon diegetic photographs, and so forth. It will do so by way of noting how a single filmic text about aesthetic collecting and the photographic dossiers that abet it—in its own fetishizing optical dynamic on-screen—gets referred away to the very history of modern mediation and its commodified visual pleasures.

What is sketched by such narratographic apprehension in this first chapter is thus the method of the rest. For again and again, as we move ahead into narratives of virtuality either psychic or electronic, actual digital invasions of editing and even of the single image will come into clarified relation with moments of ocular distortion and mnemonic mirage not linked directly to technological manipulations any more recent than photography or film. But in each case, the passive succession of frame time may find itself converted to the nervous distortions or psychic violence of framed time. To put it somewhat differently, narratography is the mode of approach that grasps most clearly the sliding points of articulation between tempo and structure. With three influential volumes by Paul Ricoeur alone on the inextricable link between time and narrative, one is disposed to admit insinuated thematic links between temporality and storyline.[9] The two go hand in glove. Plot is the very timer of narrative, calibrating story's uneven pacing of event and significance. Narratography clocks just that unevenness, that mismatch. Where narratology concerns plot types and broad dynamic patterns, narratography is caught up in the local mechanics of interval and transition. Where cinematic narratology concerns the visual discourse of plot, filmic (or digital) narratography plots out the textualization of the image itself. In this, a major precedent comes to mind.

WRITING WITH MOVEMENT/S

Well before Deleuze on film, another French philosopher approached the question of the medium's layered temporalities—now mechanical, now representational, now narrational—in a way that closely approximates (with one elided step) the narratographic attention to storylines via framelines proposed here. I refer to a too-little-discussed essay called "Acinema" by Jean-François Lyotard, from the untranslated *Des Dispositifs Pulsionnels*, published the year before the same author's the *Libidinal Economy* (1974). It is an essay anthologized in English but never really given its due either in film studies or in narrative theory. In it, Lyotard shows how discrepant pulsions of the image track can be reinvested, at a higher level, in the internal economy (not yet designated "libidinal") of narrative drive.[10] It is there that no detour ever really goes astray, that no glitch can fail to be stitched back into meaning. His lone example, detailed in a single paragraph, is nonetheless broadly revealing. From "among thousands" of potential illustrations, he chooses the minor American film *Joe* (John G. Avildsen, 1970). In its story of generational and social anxiety over the youth cult of the late 1960s, it is "a film built entirely upon the impression of reality" (174). Yet this is a standard cinematic realism whose "movement is drastically altered twice" by scenes of fatal violence: an "accelerated murder finding its resolution in the second immobilized murder" (174).

According to Lyotard's sketchy plot summary, the first alternation occurs early in the narrative, when a jealous father beats to death the boyfriend of his daughter, as conveyed in a sped-up "hail" of rushed, blurred fists. The second disruption in recorded movement comes at the end, when the girl herself is accidentally shot to death in a massacre at her commune, the break with normal action conveyed by "a freeze-frame shot of the bust and face of the daughter" (174). Each "perversion of the realistic rhythm" is "obtained by waiving the rules of 24 frames per second" (174), Lyotard notes, but this is the closest he comes to acknowledging the frameline as such. His analysis is not materialist in this way, at least not as regards this lone example. Emphasis falls instead on the "arrhythmies" that serve to wrench time itself out of shape, even while the deviance at the level of image is restored to coherence at that of plot. "So while they may upset the representational order, clouding for a few seconds the celluloid's necessary transparency (which is that order's condition), these two affective charges do not fail to suit the narrative order" (174). Note, there, how the momentary ambiguity of antecedent for the parenthetical "that order's condition" is quickly cleared away. Celluloid is for Lyotard not a third order but a disappeared condition of the first.

Yet there is more to notice in just this respect. In that last sentence, "transparency" is itself a metaphor lifted (unwittingly—or perhaps not) from the plastic condition of the photogram on the strip. What Lyotard really means is the *invisibility* of the imprint, a unit or image cell that ordinarily renews itself fast enough to project an image of motion. But this aside, the interplay of cognitive levels remains clear enough. Spatially imagined, these dents in representation are smoothed out and recuperated at the plot level. Temporally imagined, such discordant syncopations are reharmonized by the ironic intervals of parallelism itself, becoming part of what Riffaterre would call a "subtext." What seems to be operating in Lyotard's account is a systemic adjustment, one by which narrative *discourse* is rescued from the perversions of antirealist *figure* (to borrow the terms of an earlier work by Lyotard that go unused by him here, *Discours, Figure* from 1971). This rescue action takes place when paired optical distortions resonate off each other in a discordant but shared *figuration*—a textualized trope—for the violations of lived time. In this sense the discursivity of the visual would seem to bear out the important (and subsequently unpursued) opening claim of the "Acinema" essay itself—to wit, that the apparent figural rendering of the moving world is actually a writing effect, a marking or tracing:

"Cinematography is the inscription of movement, a writing with movement, a writing with movements" (167).[11] What is that last prose movement in its own right? An overscrupulous apposition? A material specification? Could that odd shift from singular to plural be meant to relocate motion not exclusively on the screen but on the multiplied segmentations of the strip as well? The strip: whose frames are rotated past the aperture in a constrained vertical rush so that the framed images projected onto the screen can achieve the full illusory latitude of their embodied and largely horizontal motion.

Whether one uses movement and/or movements with which to implement discourse (as in writing *with* pen), or as the very material in and through which that discourse emerges (as in writing *with* or *in* ink), or as the object itself of the discursive act (as in the writing *of* books), the idea of inscription carries one back to the medium. Instrumentality cannot reasonably disappear from view in any full account of the textual product. As to movement singular or plural, movement in general or its generation in staged or blocked increments, the rest of Lyotard's essay at least keeps the issue open. In his linked examples from *Joe*, an early accelerated lurch finds its dissonant chime later in a climactic freeze-frame. What is unmistakably "written" by the marked departures of tempo—or overwritten by such anomalies upon the norms of represented action—is narration per se. How could we think of "style" as merely a complement here, when it structures the death scenes whose temporal ruptures it also glosses? Within the general practice of a writing "of," "in," or "with" movement/s, such aberrations in the normative speed of figured motion—of action figuratively represented, for instance, by moving bodies—supply discursive turns whereby image is converted on the spot to figure in the other sense: metaphor rather than representation. What is therefore figured together, by acceleration over against precipitous arrest, is the suddenness of violence and the finality of death—as tied back by plot in either case, as Lyotard in a single phrase does suggest, to a filial possessiveness that has become life denying. The subtext of paired murder, that is, works in the end only to reconfigure the very structure of patriarchal incest. For such is the psychic structure seen by this plot—or, more accurately, plotted by its visuals—to erase, with both sides of the same gesture, the future of its erotic object as well as that of its defied and vanquished rival.

To move aside from the elusive dichotomy between figure and discourse that Lyotard might have evoked in this case, let me put it this way, at least in methodological prospect. Concerning the two deaths, as punctuated in their thematic echoes by such deviations as sped-up or arrested motion, narratology would have no trouble interrelating these departures from a Deleuzian movement-image. They would emerge "at the culminating points in the tragedy of the impossible father/daughter incest underlying the scenario" (Lyotard, 174). And for this storyline, of course, intertextual analogies would abound. Yet once plot is *located* by narratology as the stages of an incest scenario with classic prototypes and intertexts, it is narratography that would *read* the tension, the spatiotemporal interference, generated by such an "impossible" and blocked desire: the way *timing* itself is tortured out of joint. Its legibility would be discerned across the local frictional mismatch (in specifically photomechanical terms) between representational and narrational succession—so that the resulting "time-image" would figure the abrogated future for an erotic body, first male, then female, by the onset of a patriarchal aggression. Borrowing Deleuze's terms again, we may say that a default in the movement-image *is* the time-image. And the time is now—now and no longer. Happening in an instant, death takes an eternity.

One should be quick to note that there is nothing inherently poststructuralist in that reiterated Lyotardian distinction between representation and narration—except for his willingness to see in a given narrative instance, or half see, a "writing" effect that deconstructs the celluloid transparency on which it depends. Roman Jakobson's structuralism itself would readily notice the artifice (read, "poetic function") of the specific devices in question, which map an "equivalence" (altered image speed) onto a succession (chain of events), with metaphor overlying metonymy.[12] Later analytic perspectives can only draw out the implications of such a patterning. In the psychoanalytic terms of a narratology like that of Peter Brooks, himself distantly influenced by Jakobson, one would still be operating in line with Lyotard in accounting for the way those twin veerings from normative "screen time" enforce the temporal ironies and elisions of an incest plot.[13] Certainly, these anomalous tempi of projected images, each concussive in its way, impose on normal duration the narrative recurrence of doubling and dead end. In Brooks's analysis of the incest function, just as miscegenation is one narrative name for the structural problem of too much metonymy and free association, so does incest result from too much metaphor, too much likeness: the structural double bind of the oversame. This is as much the case in a picaresque novel as in a countercultural screen melodrama, in the dodged incest of *Joseph Andrews* as much as in *Joe*. Narratography would pinpoint instead the medial tension (again in both senses of *medial*: material and mediatory) around which is pivoted—and renegotiated—the dyad of representation versus narration. This is where a narratography of the filmic within the cinematic fastens on the photogrammatic actual rather than the projected virtual. Thus can it track, more definitively than otherwise, such eruptions of medium into imaged plot. And definitive they are. For at such moments the constituents of the medium—rather than the epiphenomena of stylistic enhancement—surface from within the image precisely as the narrative form of that image's representational work.

So this is where we can press harder on Lyotard's central insight—or at least push down one level from the projected screen motion that concerns him. Narratology, as we've seen, can well plot the collision courses of desire in a story of displaced incest. More in a Lyotardian spirit, narratography looks to the points of tension—often of frayage—that result from exactly the fissure (and hence slip) between the two primary and ordinarily collaborative "orders" of scene and meaning. Lyotard's paradigm is one of internal economy, since these orders of representation versus narrative are ultimately coordinated "following the cyclical organization of capital" (174). Narratography would understand this by glimpsing that third term after all: the celluloid order itself. Such is the "rule"—and its "waivers"—evoked by the arrhythmies and syncopations of the track. These anomalies emerge not as the "clouding" of a transparent image but as the obtrusion of its constitutive photomechanical fixities. This paradoxical third order is both between and beneath. It is an order at times sprung loose by representation only to be recontained by narrative. It thereby stands forth as the *material base* of any such economy, any such "capitalization," libidinal, cognitive, or thematic. Realist transparency, then, is not occluded by fast or frozen motion; it is clarified—and precisely as to the causes of its normal narrative effects. In the kind of example brought forward by Lyotard, the economy of affect that subsumes representation to narrative would thus be seen to do so, in short, by eliciting the normally unconscious awareness of the strip on the viewer's part, in all its own somatic investments and disavowals. This is an awareness, or call it a preconscious apprehension, that typically gets suppressed by the viewer's ocular attachment to the projected movement-image. Suddenly surfaced and refigured in Lyotard's example, recognition of the frameline is buckled back into a twofold temporal order of irreversible fatality. What has happened is that text has indeed capitalized on its own deviance, cashing it out at the level of meaning.

Such an impact in reception can be induced, as in Lyotard's example, even by technical or stylistic cliché. For the film *Joe* is drawing directly on postwar modernist and New Wave devices of "irrational" editing (the Deleuzian phrase) that infiltrated even Hollywood technique in the late sixties. At the level of deviant editing, such reconfigurations of time itself can happen only because what the tampered track reveals first of all is the track itself: in other words, the underlying fact (*pace* Deleuze's sense of "immanent" sensorimotor pacing) that filmic cinema is movement inscribed and intermittently reconstructed rather than movement given to us whole. Which is only to say that in these heavily marked cases of material disclosure within screen story, the differential boundary between a "transparent" representational order and a structured narrative order is a shifting one. Between image train and systemic synthesis lies exactly that inscriptive stratum where cinema is revealed, for Lyotard, as a writing with/in movement/s in the skid of filmic difference.

THE TEXTUAL RETURN

Preoccupied as the remaining chapters are to be with other violent temporalities of screen narrative three decades later than that 1970 "late-modernist" exemplum of Lyotard's, one large question takes shape: What happens to the interplay of representational and narrational systems when the third (or celluloid) order, the inscriptive regime, if you will, is no longer entirely filmic in its composition or its succession, but finds itself instead either digitally composed or digitally edited? And following hard on the heels of this question: Why should the new "arrhythmies" that result, especially in Hollywood cinema, be bound up so closely with plots drawn repeatedly to represent lives that are themselves lived achronically, almost (one might say) nonnarratively? These are lives passed, and passed before us, not in the mode of succession and transience but in the mode of a different libidinal economy of time altogether. What I will be calling the logic of temportation, with its paradoxical spatialization of time, can in these ways subdue duration itself under the alternate force of Eros or death wish. Why now? And how—by what means, exactly? Site-specific, narratography knows it can only get to the why through the how.

Even as an engineering science, then, narratology often remains impervious to the medium in which it works. This huge fact is easily overlooked in the grip of its many and different precisions. Whereas narratology may be intermedial, it is not really interdisciplinary in the strict sense. It does not anchor itself in the disciplined understanding of any one material form or format, whether verbal, sculptural, painterly, filmic. That is its utility and its power. Narratography, however, must continually cross between alternate disciplinary grasps of text in order to prosecute its closer notice—even if only by contrast. It does so, for instance, when registering what is unique, let us say, to a filmic rather than a lexical effect, as for instance the lap dissolve versus ellipsis in cinema (transition by superimposition versus cut)—where each may serve to figure time in process.

Same with the three "orders" of filmic cinema as we have been working to distinguish them. Frame time is a function of the celluloid order, screen time of the representational order, and a specialized sense of framed time derived from the narrational order. Aberrations of the first system, having become defaults of the second, might be converted to figures in the third (as, for instance, in certain heavily marked forms of the Deleuzian time-image). That third level of narrative regulation is imposed of course by design—or, say, by the interaction of *syuzhet* and style. In recent films of the fantastic, such regulatory narration may nonetheless seem to arise from within—across a mismatch of the other two orders—so that normal temporal intervals are replaced by ocular gaps or distortions, disjunct spatial layerings, visual disjoins. When representation doesn't run smoothly, one looks to narrative for a motivation of the distressed view. That looking is the start of narratography. And where the fantastic anomalies of framed time are concerned, the object of scrutiny may well take shape more as a perversion of the Deleuzian time-image than as a searching new version of its figured temporality.

Narratographic attention, however, is a looking that always includes reading—and is often manifested, on the literary side of the matter, as the familiar thing itself: the reading of a typeset page. Either way, though, in response to verbal or visual narrative, what is invited is a return, via mediation, to the *textual* model in analysis. Ushered in long ago by the so-called linguistic turn, this is a model that can be revived in these pages, despite a faltering reputation, with the usual confidence in its analytic yield: in the present experiment, for instance, its close-grained narratographic returns. To move immediately to filmic and digitized textuality with no attention to its literary equivalent, however, even though justified by the "linguistic turn" as conceptual premise, would do less than full justice to narratography as method. Yet before passing to the book's main cinematic topic via the intricate and constricted prose of Henry James, and from there to a fiercely claustrophobic film of his work—with each narrative *textual* in the extreme, inscriptively fretted and overwrought, and deeply invested in a *hermeneutic* response—we had best pause for a moment over those two increasingly suspect adjectives.

In doing so here, and then again in the concluding chapter, we are, in effect, cross-examining their so roundly discredited implications in certain quarters. These seemingly harmless adjectives, and the concepts of a work and its interpretation lying behind them, are not everywhere or equally under suspicion, to be sure. Yet they do send up red flags, or white, in many a book bent on engaging with the newer archaeologies of media with visual (or visuality) studies that cast both these mainstays of formal and cultural attention into irrelevance if not disrepute. No need for textual interpretation to consider surrendering, though, without first marshaling its forces in defense of critical analysis. In these pages at least, wherever narrative remediates its own signifying conditions—especially by the marked devices I am calling optical allusion—attention tries keeping its eyes peeled. To be willing, when the plot thickens, to call that attentiveness interpretation is only candor, not debate. Not yet. But the time (known here as chapter 6) will come.

If the textual turn across disciplines can be credited with some of the most important breakthroughs in postwar critical writing—and who would deny this?—then "suture theory" comes to mind as perhaps the most prominent instance of this former textualist mandate in the film-theoretical vanguard of an earlier decade. At the seam (or interface) between character and viewer, suture is the construction of narrative space by the internal relay of sightlines. Under its description, the narrative uptake of mediation's founding look—the look at the screen—is reconceived on-screen as signifier: the image of imaging itself (by an embodied eye) as it enters into narrative "discourse." But even to evoke such a quasi-linguistic paradigm is now often deemed to overlook (rather than continually to rethink) the medium in which its semiotic functions operate. By interpretation, material means are often thought sacrificed to meaning. Not necessarily, and certainly not here.

True, resisting a suspected retreat by hermeneutics from visuality's broader contexts has helped energize the turn to media archaeology. Results are in, many of them—and impressive. Less convincing is the tacit sense that this line of study renders irrelevant, rather than stands to reorient in progress, the entwined concerns of textuality and its interpretation both. Why have so many past gains been forfeited in the name of a single line of advance? What happened, for instance, to the sense of complacent viewings productively undone by the recognized psychomechanics of imaging and identification channeled by the filmic apparatus? And, anyway, what is to keep the work of reading from historicizing the medium in which it steeps itself? Such questions cannot seem rhetorical for long, since these half-dozen chapters work more or less steadily to offer one, if only one, answer. As soon as the analytic yield of these narratographic exactions has been logged in—their yield, that is, for a cultural account of postfilmic cinema and its symptomatic narrative refigurations—we can return to a methodological justification in further detail.

For now, let's just see whether reading Henry James and James Ivory next to each other, reading their inscriptive devices in action, does or does not put us quite directly in touch with their respective media—and, in the case of the film version, with an archaeological sense of visual culture more broadly. From here on, this book doesn't count on any particular acquiescence in any one of the dozens of interpretations, film by film, it proposes. What it does count on, though, and makes bold to promise, is that with all of these readings behind us, the case for interpretation as the only *felt* way, the only somatically engaged way, to a genuine medial apprehension of a given narrative and its grounding conditions will, having cumulatively been made, be easy to summarize.

SUTURE 1904

The question isn't how James Ivory might have adapted Henry James's last paragraph in his film of *The Golden Bowl* (2000)—or why he didn't try. Narratography's concern is with what James's represented exchange of stances and stares sounds like on syntax's own terms. In a "shot analysis" of the novelist's mutedly disjunctive prose, one might imagine that the three Bergsonian facets of the movement-image in Deleuze—perception, (re)action, and affection—narrow to a single exchange of glances in a recognition of the changeless. Even more than in the standard marital closure of novelistic fiction, what results, across the unhinged rigors of signification, is an image of time struck still. An attempt at a cross-medial definition might therefore go this way. Suture names that articulation of narrative space whereby reader or viewer identifies (aligns) by turns with the verbal or visual axes of character interaction. Narratology could well find in the byplay of such exchange a switch point for the major drives of plot. Narratography would go more closely to work there on the disjunctures of the shunt, the friction of the interface, the mismatches as well as alignments of answering frames as they plot their own contributory ironies.

The James Ivory film of *The Golden Bowl* gives us in closure, like the novel, a forced marriage desperately recoupled. In the film, though, unlike the sutured interchange of the book, each partner is balanced precariously against the other in a "two-shot" within the same harmonizing (or imprisoning) frame, after which we cut mercifully away to a metafilmic coda unauthorized in any way by the novel. What James gives us, instead, end-stopped and finally zeroed out, is the most rigorously minimal and most complexly intercut dialogue sequence in English-language fiction, one searing line each for his two principals, circling round in echo like a noose. Even the crudest outline of the story shows trouble coming from the first. Maggie Verver is so attached to her billionaire father that when she marries an Italian nobleman in part for his title, she also arranges for her best friend, a penniless expatriate, to hold (without really filling) her place in her father's affection. Yet they are often still together anyway, father and daughter—and not least because the patriarch's bride, much his junior, is also, as fate would have it, the former mistress and occasional current lover of the heroine's new husband. I say "heroine," because the second half of the book is called "The Princess," the name of Maggie Verver being subsumed to the title she has bought. The first half, filtered in good part through her husband's point of view, is called "The Prince." Narratology would readily appreciate this bifurcation without synthesis, the duality without intimate bonding, the class-based power play of the structural shift itself. Narratography moves in to graph the marital confrontation of the final sentences, once the incestuous bond has been loosened and her father packed off to America with his powerless wife.

Powerless, but "splendid" (Maggie's proposed term for her mother-in-law, Charlotte)—and thus at once too captivating on the one hand, too proud on the other, to let her marriage to Maggie's father come crashing. "That's our help, you see," James has Maggie add, "to point further her moral."[14] It is this "emphasis" that triggers the deadlock of closure:

It kept him before her therefore, taking in—or trying to—what she so wonderfully gave. He tried, too clearly, to please her—to meet her in her own way; but with the result only that, close to her, her face kept before him, his hands holding her shoulders, his whole act enclosing her, he presently echoed: "'See'? I see nothing but *you*." (580)

Beyond the unpunctuated viselike clang of "before"/"therefore," the ensuing syntax is like a coiled spring that takes on a force of its own. With the Prince "trying to" accept her forgiveness and her solution, yet not quite succeeding, even the sentence cannot extricate itself from this same impasse, lexically figured. Hence the slant rhyme of its subsequent grammatical span: "*He tried, too* clearly, *to please* her—*to meet* her in her own way"—or, in other words, to meet her "halfway," in the idiom that James seems to have tweaked in the direction of her priority ("in her own way"). In narratographic terms, the drag of echoism and distracting recurrence, rather than working to effect some arc of achieved symmetry, unsettles and finally shatters all sense of negotiated marital parity.

I can find no better way to illustrate this narratographically registered tension than in fact to graph it, to render it diagrammatic: to generate from it, in fact, the prose equivalent of a screen storyboard. In blocking out this decoupage, I have pulled to the left margin (and into italics) those intractable echoes and stumbling parallels of construction that oscillate across prose's own aural recoils. Figured here, by rhythm alone, are tacit recognitions on the couple's own part that can be divided into imagined shot breakdowns to capture *the look of the Other* in both senses, eyed and eyeing. It all transpires in the asphyxiating chiasm—and visual metaphor just barely literalized—between "you see" and "see . . . *you*" (with the castrating slant of James's own italics slicing across the Prince's last word).

```
                    SHOT OF MAGGIE
     "That's our help,
you see,"
     she added—to point further her moral.
     -------------------------------------------------------
               SHOT OF PRINCE FROM HER POV
     (the reverse camera position gradually confirmed by free indirect discourse)
It kept him before her,
     therefore, taking in—or
trying to—what she so wonderfully gave. He
tried, too
     clearly, to please her—to meet her in her own way; but with the result only
     that, close to her,
     -------------------------------------------------------
                    SHOT OF MAGGIE
                    (FROM HIS POV)
her face kept before him,
```

```
his hands holding her shoulders, his whole act enclosing her,
```

REVERSE SHOT OF PRINCE

```
    he presently echoed:
"'See'? I
see
    nothing but
you."
```

After the rutted grammar of "trying to. . . . tried, too," another echo cuts across the syntactic grain: namely, the dismantled reprise of "kept him before her" in "her face kept before him." The second time around, *kept* is not a verb in a freestanding clause but an adjectival past participle, its passive grammar seeming to drain him of agency on the spot. Her shoulders fixed in place before him, he is ready to level his parroting reply: "'*See*'? I see nothing but *you*." The cage is closed. It could have gone otherwise—verbally, that is. With her Prince kept in view before our heroine, prose might have equilibrated their fresh acceptance of each other in some happy, appeasing reciprocation, might have suggested how "her face kept before him," say, a prospect of all that they still meant and might mean to each other. If that alternate sense ripples across the gathering grammar even now, it does so in the form of a sardonic false lead. Instead, the so-called absolute rather than independent clause ("her face kept before him") gives us the fact of her enwrapping gaze and then its upshot in peripheral blindness (nothing seen but. . . .). Grammar has thus splayed open in a nonsynchronous syntax of dissonant rather than symmetrical concord. The contortions and overexertions of closure, its stray repeats, its unruly tractions, are registered in a syntax of difference within repetition, squirming to make space for meaning within sameness, with all possibilities ultimately closed down.

In the light of his gaze she sees how completely she fills it, how utterly she occludes his view. Beyond suture, it is now as if she were seeing her own reflected image in his eyes, eclipsing all else. For, in the paragraph's and the novel's next and last sentence, with reference to the Prince's last words, "the truth of it had with this force after a moment so strangely lighted his eyes that as for pity and dread of them she buried her own in his breast" (580). With the close-range demonstrative, "this" rather than "that," the withering counterforce is all the more immediate. It seems already accomplished, even before described, in the oddly inverted Jamesian period. Narratology would of course recognize the ingredients of catharsis, "pity and fear," borrowed from another genre and come ironically to invade the closure of a marital comedy. Narratography charts the lexical demotion of the latter monosyllable from all sense of sublimity to the low-level domestic trepidation of emotional "dread"—closer to the mundane *dreadful* than to any self-transcending awe. Such closural effects are available to attention in precisely the way they graph the whiplash ricochets of exchange in this abdication from all plot momentum into irreversible stasis.

Just as narratology has no particular allegiance to translation studies of the novel-into-film variety, narratography bears no intrinsic relation to comparative stylistics. To bring together a novel by Henry James and a film (of it) by James Ivory is a convenience of this methodological chapter rather than an inevitability. In each medium independently, narratography probes beneath the ostensible to the tensile forces that compose it. No narrator before James puts the incidents of language so much in the way of event. No recent film has put the incidents of photomechanical registration (even at one point digital) so much in the way of period recreation. The result for *The Golden Bowl* is that the constructedness of the filmic image infiltrates—seditiously, like a double agent—both the labor and the thematic work of period reconstruction.

PHOTO/SLIDE/FILM//PIXEL

Concerning the ironies of cinematic rather than syntactic texture, the slant of this subheading—like the angle of method it advances—stresses at first a three-stage slide: photograms slipping past on the strip, the track in action, and the resulting artifice of moving image. This last is where spinning vertical motion is transferred to a mostly lateral action, moving pictures to screen movement, real speed to virtual duration. Not just tapping such a conversion from photomechanical motion to imaged movement, but replaying its mechanics, the screen version of *The Golden Bowl* goes on to turn the material interrelation of photo/slide/film into a structured sequence of technological evolution on the way to its covert incorporation of the digital. The movie's own plot begins with a photographic archive, pivots on a photographic slide lecture, climaxes in a subjective and hysterical "filmatization" of still images, and then closes with a documentary film-within-the-film whose backdated historical authenticity, at least as regards its graphic surface, is partly the result of digitally tampered evidence.

Hence the double slash after the first triad of this section heading. The technical as well as historical slide has become a constitutive slippage of medium within the image frame, no longer mechanical but now electronic. The fourth term of the sequence thus inserts a more extreme difference, a continuity within projected imaging but also a supplanting of its automated imprint basis. The quantum leap to pixel marks an irreversible threshold crossed—and crossed for the most part, as at the end of *The Golden Bowl*, imperceptibly. It is there, in the closing newsreel, that a specifically cinematic genealogy of instantaneous transcript and automatic projection must widen to include both an archaeology and a videology of image manipulation, where keyboard doctoring returns some notable distance toward a precinematic mode of manual handicraft. That such a visual deviance can be historically recontextualized in this manner—not despite, but by dint of, hermeneutic pressure—is the point of what follows in the rest of this chapter, as it will be the theoretical claim of the last. In the form of narratography, interpretation does not bracket the largest issues of mediality but reframes them for a more sustained look.

For filmic cinema, this is no doubt clearest when a film character is held riveted by a photograph even while his or her on-screen image is composed of merely hundreds of its transparent equivalents per minute. The film version of *The Golden Bowl* takes this kind of optical congruence—such optical allusion—a step further. For at one point the mechanical as well as historical slide from photo through still projection into moving-image film is enacted before our eyes, via the subjective gaze of a character. This happens in a nightmarish hallucination of photographic documents come entrappingly to life. In a narratographic grasp of the sequence, where photos skid and vibrate into film, a species of representation seeds itself in embryo within the gestation of a single complex instance. And when the photomontage of America—to which the traumatized viewer, the "splendid" Charlotte, will soon repatriate—has become at the end of the film a newsreel of her return, its digital enhancement (in the form of simulated age and degradation of the print) only escalates, as we will find, the media-historical irony that has pervaded the film.

The first instance of such technovisual irony goes back beyond projected spectacle to the archaeological prehistory of still photography. Tapped here is the predecessor of the optical trace in the embossed or imprinted image: the coin or medallion (here the relief of a noble ancestor given by the Prince as an engagement gift—and insignia of his pedigree—during his first scene with the Ververs).[15] The keynote is struck by this struck-off image, the very badge of rank. Right from the first, *objets d'art,* imprinted, sculpted, or otherwise, are coins of exchange and tokens of power. Associated in the same scene with Verver's soon-contemplated photographic dossier of sculptures and paintings for sale, the medallion, in its status as material imprint, initiates a subtext of mechanical reproduction that keeps the power of photography—and, by association, of film—contrapuntally before us in relation to other modes of illusory presence, from low relief in precious metal to painting to preserved corpses to wax dummies and their uncanny movements. Here, within a single film, are the archaeological strata of optical impress and simulated animation that lay the fertile cultural ground of cinema's own emergence.

Yet it is the photomechanical genealogy of cinema that remains most prominent in Ivory's *The Golden Bowl.* Whether in a dealer's portfolio of print samples, in framed family photographs, in a series of backlit and projected slide transparencies, or finally in the flickeringly mobilized splutter and jumpiness of early documentary film, the fixed cellular imprint of photomechanical record secures its place as the ultimate measure of simulated life. Photographic capture offers a technology pitted against both the signature effects of high art and other modes of representation, from impressed heraldic profiles to embodied waxwork monstrosities. But there is an extra medial twist at the very first turn of this sequence. Concerning the initial episode of linked artisanal and mechanical reproductions, with Verver studying his new medallion before turning to his dealer's inventory of photos, art history combines with commerce to offer a latent metahistorical gloss on the evolutionary motif of primitive mechanical impress leading to automatic record. The unspoken irony still cuts deep. The very medium of photographic print, which aided the disciplinary rise of art history in its capacity for stylistic comparisons at long distance, reverts in Adam Verver's case toward photography's first mercantile uses, half a century before, in the off-site display of saleable real-estate (and other nonportable) properties.[16] Comparatist scholarship has degenerated into competitive advertising when the same technology that has elevated art history also facilitates the exponential commercialism of the modern art market.

Machines of reproduction and authentication, modes of effigy and replication, form a close weave of association in the film narrative from here out. Exceeding anything even hinted at in the novel, these simulations come into eventual connection with the photomechanical ironies of the newsreel closure. But this is, as suggested, only one strand of the braid. Image reproduction is tightly attached as well to other forms of mechanical mediation and material simulation in Ivory's version of the story. These include the un-Jamesian segue, bridging space in real time, that accompanies—and refigures—the film's first long-distance scene change: Charlotte accepting Maggie's invitation to the wedding in telephonic voice-over. Greater distances are later spanned—through yet more layers of mediation and surrendered presence—by another communicative instrument in the film: the latest upgrade of both telegraphic and imprint technologies as commercial tools. For when Adam Verver ends up staring in private rage at the Prince's medallion, contemplating a violent revenge for his sexual betrayal of Maggie, his desktop tickertape machine waits in the background, like a spool of the film's own inexorable plot (fig. 1.1), to keep him in touch with the publicly sanctioned field of his power.

So it goes: impress and coinage, sculpture and painting, photography, telephony, telegraphy, and finally not only slide projection but silent cinema. The film has inscribed its own place in the evolution of both mediating form and the technologies of transmission. In its picture of a craving for both graven and two-dimensional images, a collector's lust inflamed by cash and monomania, the film aligns its own serial gallery (or strip) of framed pictures (photograms) with the twin origin of their recreated images in realist art and industrial reproduction. Only

1.1

in recognizing this can we begin to locate more precisely the film's carefully positioned effect within a recent tendency in the Victorian and Edwardian ranks of period cinema. This is the tendency, often in complete detour from the source text, to encode the medial technology of film's own process as a filter of its retrospect. Again and again in heritage films of the last two decades, cinema is the insinuated future tense of a retrospective plot. Often this modern paradigm of image generation is glimpsed through the local folds or torn seams of a film's visual grammar: jump cuts, skid editing, freeze-frames, any such unexpected wrinkles or rips in the transparent continuity of historical presencing, any such reversions to the filmic strip itself.[17] Within the general narrative debt of screen melodrama to Victorian plot formats, here is where narratography responds more immediately—or, rather, more closely—to the thickness of mediation. At other times the cinematic intertext of Victorian or Edwardian narrative is stitched in wholesale, whether as an isolated cinematographic screening or in some predecessor forms of visual exhibition: a film-within-the-film, say, or its magic lantern or slide show prototypes. In every sense of the phrase at once: historical projections.

One format of such projection in *The Golden Bowl*, a precinematic rather than actual film experience, comes well before the newsreel closure in an abruptly introduced and undermotivated scene that derives from a negligible offstage event in the novel. This is the slide lecture on despotic Italian history to which we unaccountably cut as the adultery plot thickens (fig. 1.2). In the novel, Maggie looks into the Prince's history at the British Museum. Here that archive is theatricalized by projection. By the time it is introduced into the film, however, the slide lecture has become not only a medial upgrade of such paper researches but a structural irony about visual narration per se. In its blatant plot disjunction, denied any transitional scene on either side of it, the slide show is forced into a strained continuity with the film that embeds it. Its backstory about the Prince's ancestral lineage is a pallid and euphemized version of the Renaissance bloodbath (adultery caught in the act and punished by twin beheadings) that Ivory has added to the film by way of an un-Jamesian prologue. The slide show both resumes and euphemizes this same subtext of violence. Intercut with shots of a rapt, genteel audience, such a sedate illustrated lecture—as highbrow institutional precursor to the "cinema of attractions" (and its own frequent vocal commentators)—refigures the film's own audience for a Renaissance,

1.2

then an Edwardian, retrospect as lurid as it is curatorial.[18] Such, in heritage cinema, is the premium placed on faded grandeur as narrative instigation. Here also is the typical place of precinematic optical pleasures in helping to orient our own rearview fascination. As audience to a projected historical reconstruction heavily invested in the erotic intrigue that brings it to life, and to collective appeal, suddenly we watch ourselves watching by proxy.

The slide show—as moving-image spectacle—has certainly been slotted into place, as we've begun to see, amid an already taut network of artificial ocular and technological mediations. Instances of such visual transcription and simulation alternate between photographic display, in its genetic bond to cinema, and more far-flung modes of mechanical and illusionist replication, each taking its place in a steady run-up to the closing film-within-the-film. Urged to marry by Maggie, to ease her own guilt over "forsaking" him, Adam sits alone one night shedding a tear over a photo of what should be the lamented and now displaceable wife but is in fact a portrait of Maggie herself (fig. 1.3). As such, this is the one photographed thing of beauty in the film that Adam, rather than consider purchasing, must strategically relinquish. Yet the relation of this mechanical artifact to the whole ethos of imitative art breathed by the film is made unmistakable when the photograph fades out in transition to the newlyweds in Italy by passing—in momentary dissolve—through a decontextualized and unmotivated painting on the way to a materialization of the same hilltop scene in real (cinematographic) space (figs. 1.4.–1.5). Unmotivated, because here is a painting that is eyed, or owned, neither by Adam, as far as we know, nor by anyone else in particular. It evidences merely the painterly itself as the very motor of the story's own filmic spectacle: the painterly, the aesthetic, the objectified, and hence the always implicitly acquisitive.

This transition from the space of image to the place it pictures shows just how deeply the free-floating aesthetic of mediated presence, of self-luxuriating mimesis, has infiltrated the very chinks and linkages of narrative sequence. Along this trajectory of simulation and trace in the film, another scene not in the novel includes an actual photo session at a costume ball, with its echoing wordplay on pose and disclosure in the photographer's "Composed, madame? Then we shall expose!" The irony is scarcely light handed. The magnesium-flashed photographs taken of Charlotte and Amerigo as a Renaissance couple are in their own way a laying bare of the liaison

1.3–1.5

being repressed in time present. And it is just this double image of the couple, in the open secret of their undue intimacy, upon which the plot's main melodramatic disclosure will later depend. When the titular golden bowl is delivered C.O.D. to Maggie's parlor, the antique dealer, chancing to see on a tabletop this unidentified costume photo of her husband and mother-in-law, unwittingly mentions recognizing them from their visit to his shop long before, a "couple" one could not readily forget.

The past will not die. Photographic tracing is time's memory work, in this case the return of a suppressed event. The historicized contexts of imprint media in the film are related as well to a favored site of disciplinary archaeology itself—the work of mummification, whose relation to photography has a major theoretical lineage, from Bazin forward.[19] Here is a preservative function whose gothic variant is alluded to in the Prince's early anecdote about his ancestor stuffing the corpses of his enemies in order to display them as monitory totems in his own private museum. Life converted to image stands forth as primordial power, taxidermy as threat, preservation as an armature not only of death but of sustainable violence. And this motif is extended to other lifeless reproductions in the film. There is the present-day mausoleum of effigies at Madame Tussaud's wax museum, to whose lower-class tourist precincts Charlotte drags the Prince for a bout of confidential exchange. Among other simulacra on display, we first see a masked executioner and his guillotine, symbol of the life-arresting violence associated with so many of the film's two- as well as three-dimensional images, images embalmed in their own stasis. To round out the circuit of fixed image and fixating gaze, this summarizing tableau is soon followed, in the waxen lineup at the museum, by a weird

1.6

1.7

stationary figure who, like some external embodiment of the couple's paranoia, actually follows their whispering movements with living eyes. The effigy turns once again emblematic. A plastic image come to artificial animation, as is of course the case with all human images on film, operates in this instance as a double for the scene's own prying omniscient camera, anthropomorphic and impersonal at once.

To escape the glare of spousal surveillance, the lovers must separate, must ultimately put an ocean between each other. Indeed, their unnerved flight from the wax museum is visually reprised in the scene of their final parting—in which they are severed before our eyes not only from each other but from their unfleshed mirror doubles. Once the panning camera has edged past Charlotte to lock her reflection into the CinemaScope-shaped middle frame of a dusky triptych mirror, her ghostly image is left calling out to the on-screen Prince, who, avoiding her gaze, is trying to turn away from her craving once and for all (fig. 1.6). Only his reflection in the mirror's third panel, fused with a bug-eyed gargoyle at right, seems still drawn toward her passion in the lingering murk of their intrigue. Such graphic subdivisions of a framing-within-the-frame quite directly recall the series of fun-house mirrors where the on-screen body of the Prince, also with his back to us in the far right frame, was matched by the distorted reflection of Charlotte in writhing isolation (fig. 1.7). In the compositional grid works of these separate scenes, dramatic, ocular, and medial estrangements are impossible to segregate. As the later crisis schematically replays that carnivalesque allusion to a precinematic optical toy in its commercial emplacement, the fun house has become a torture chamber. And in the later serial strip of images on the mirrored mantle, though their framed sequence is rotated into the horizontal axis, it is as if their oblique sightlines have disarticulated the cinematic suture from within its climactic scene, marking the gaze of reciprocated desire at its spatial as well as temporal point of no return.

We know the full measure of Charlotte's desperation in this scene, because we have re-

1.8–1.9

cently seen her own recoiling vision of what awaits her in the return to America with her husband. She has been expected to enthuse over a clutch of photographs including shots of the American construction site of his monumental museum. Instead, she is left alone with these photos and her own anxiety, shuffling through them in mounting panic (figs. 1.8–1.9). In a subjective sequence framed by her own mind's-eye view, the forbidding images of an industrial landscape are overlapped and slammed together, generating by impacted association the hammering soundtrack of their own internal screening. Here we have, in the terms of Deleuzian semiotics, the *sonsigns* of a virtual future—and mediated in this case, synchronized, by a run of photogrammatized images in the precursive form of opaque glossy imprints.[20] Single visual documents are swallowed up as cellular increments in the unrolling of an inexorable destiny.

One upon the other, these congested shots—of photographs convulsed into action—are envisaged from Charlotte's point of view as the gruesome industrial truth beneath her perfumed leisure. All the grimy engines of commercial progress churn forward at once, from clanking urban trams (fig. 1.9 again), screeching railroads, and hissing steel furnaces to mine shafts, underground elevators, and groaning coal wagons, including the mass labor that sustains such productivity. In this channeled pandemonium, the black-and-white documentation is gradually overtaken, the more she looks, and looks forward, by the sulfuric yellow and molten red filters of commercial modernity's own volcanic inferno—everything the aggressively cultured Charlotte has fled America to avoid. Exacerbating to a kind of apocalyptic violence embodied in the serial model of the preceding slide show, these swiftly interchanged photos, atmospheric soundtrack included, are montaged as a nightmare color film of the all-too-near future. This is a self-generated filmic abyss that at one point even replays its initial jump from squarish photo to full-screen format by tracking the widening bleak prospect of a coal train out from a railroad tunnel (figs. 1.10–1.11). In the metafilmic process, it thereby rehearses as well the historical evolution of screen ratios

1.10–1.11

on the way to Ivory's own favored wide-screen format.

Charlotte's garishly colorized hallucination is soon after tamed again by the closing newsreel footage, silently removed from all subjective participation. Plunged into this epilogue with a violent shift in register all its own, we watch the relocation of Adam and Charlotte in a documentary rather than feverishly internal view of their new land. It is, nonetheless, an equally tumultuous and industrially gutted American cityscape into whose commercialized publicity machine they are instantly swept up, the stars of the new social season and its tabloids. When Ivory's film funnels down to a supplementary film of its own—one screened before us from an unidentified source, a film annexed to omniscient narrative as both documentary extension and publicity blitz—the narrative's concerted self-scrutiny as mediated artifact seems complete and inescapable. The airless chambers of the preceding story, dislodged from private satieties and violations to be thrust upon public history for the first time, are of course the springboard for this metafilmic epilogue and comeuppance. In the process, all the productive labor of machine engineering, of the wheel itself, from coal cart to railroad, gives way, as it did in Charlotte's cinematized vision of their optical documentation, to the new industrial use for rotary motion: the reeling film mechanism itself.

At one level of mass-cultural poetic justice, the fledgling medium of cinema arrives in the coda as the democratized future of the connoisseured artifact. But such an implied technological history turns sociological at the same time. Thus, media exposure becomes the very future of mass wealth. Here again, then, is the increasingly familiar double helix of self-referential media involution in Victorian and Edwardian period film. In screen narratives set either on the eve or during the actual dawn of the cinematic apparatus as a public institution, the very condition of cinema is repeatedly wrenched into the open at pressure points where the crises of plot are intersected by stressed gestures of visual technique.[21] The common denominator: a recursive disclosure of cinema's photographic undertext, its racing cellular makeup. This is most often marked by

1.12

1.13

either a dislocating stutter or a fixating seizure of the track, erupted against the normal streaming of the screen image. Such "modernist" devices often have an anachronistic feel in these period treatments. In Ivory's tampering with the image at the end, the effect is instead historiographic. In all this, the stage directions of the screenplay depart entirely from the novel. On the verge of their departure for America, Verver tries talking Charlotte down from her hysteria: "They've never seen the likes of you. You'll make their eyes pop right out of their heads, as they say over there." As her crying jag subsides—or at least stiffens—under the icy comfort of this promised future, assuring her only of the desexed public spectacle she will make of herself (as the commercial documentary will immediately confirm), we cut to primitive, decaying, weather-beaten footage of their ocean liner plowing forward (fig. 1.12) and finally, in good newsreel fashion, passing the Statue of Liberty. This transported European artifact, in another gimmick of newsreel rhetoric, is spied through a binocular masking of the image (fig. 1.13) as the marital prison ship comes into port, with Verver's own sculptural tonnage on board. All is iconic, all show.

Yet not without critique. The private collector is captured by the machine of public exhibition—and not just captured, but arraigned. For this shrunken newsreel image soon includes headlines like that from the *New York Herald Tribune*, noting the return of "Croesus" with his stately bride alongside the protests of the labor unions against the unwanted largesse of his museum, which he is seen to be foisting on the populace at the expense of better wages and working conditions. As tracked by mass media, modern history is beginning its recoil against the enclaves of monopolist power and the self-aggrandizements of public beneficence. Cinema is there as part of the medium of transmission, agent in its own way of a certain democratizing of visual discourse. Not only does popular film footage, then as now, wallow in the star value of the arriving potentates in their triumphal publicity shots, but it reframes them in order to make room for the headlined stirrings of dissent as well. Cinema, again, is the plot's own future—complicit, but also potentially corrective. The proliferating visual medium of live-action footage, though inevitably falling into place as part of the network of mediated imagery in Ivory's one film, nonetheless breaks free of previously commodified portable artifacts and their circulation in the same narrative. It enters instead into a filmic system—and a representational network—not just of iconic transmission but of communication and exchange as well. This, in short, is the would-be utopian promise of recent heritage film into which the coda of *The Golden Bowl* seems plugged.

Familiar enough of late in screen Victoriana: cinema's last self-congratulatory gasp on the horizon of its own passing into history, its own outmoding by electronic images. But a negative checkpoint may be clarifying. In contrast with anything ameliorative or redemptive in such films, even at the level of cinematic conveyance, note the closing photochemical abyss of Terence Davies' *House of Mirth*, shot at roughly the same time as Ivory's film and equally influenced, it would seem, by the medial self-consciousness of recent turn-of-the century fictional adaptations. Once again a prose closure is revised by the techniques of visualization alone. Here is the devastating end of Edith Wharton's novel, with Lily Bart dead of an overdose of sleeping potion and the previously noncommittal Selwyn now kneeling beside her discarded body as the final shrine of her spirit, trying (we find, through indirect discourse) to rekindle some faith in their previously unconsummated bond: "It was this moment of love, this fleeting victory over themselves, which had kept them from atrophy and extinction. . . . He knelt by the bed and bent over her, draining their last moment to its lees: and in the silence there passed between them the word which made all clear."[22] All trace of this logocentric Liebestod is gone from the Davies film, its very metaphors turned bitter. On the film's reading of the story, the overdose is a clear suicide and the mourner yet again a leech feeding off the heroine's at last decisive energy: the voyeur as parasite. Even Wharton's saving conceit has been made pitiless by literalization, for the reservoir of life that has so recently "drained" from the Lily of the film is figured by the posthumous dripping of that now superfluous opiate from the bottle still clutched in her hand.

Davies has acknowledged how he originally intended to follow the closing deathbed title, "New York 1907," with a series of old Manhattan photographs—as if such an arbitrary picture archive (we can only speculate) would have appeared to swallow up this one transient narrative into the ongoing flux, and random interchangeability, of modern urban alienation. But this ending must have seemed too impersonal, too neutrally metropolitan. All Davies says is that "it didn't work."[23] Instead, he cuts to the window side of the bed, with the kneeling mourner's back to us, and freezes the shot into a full-screen mourning photograph in its own right (fig. 1.14), death

1.14–1.16

date inscribed. The revised idea still turns cinema back upon its predecessor in photography, as initially conceived—but without releasing the narrative into the documented wash of indifferent history itself. Slowly, excruciatingly, Davies drains the image of all color as well as all representational life, stopping just short of a reversion to the undeveloped photographic sheet itself (figs. 1.15–1.16). As the credits crawl past, we are left hovering over the ghostly dregs of the image, at the bare minimum of discernability—an image like some evanescent cloud pattern in which one could barely make out a corpse and her paralyzed nemesis. Davies has found a quite precise visual rebuttal to the indirect discourse of the survivor's last-ditch consolation in the novel. For here all is indeed, in Wharton's words, "atrophy" and virtual "extinction."

By extreme tonal contrast in *The Golden Bowl*, documented history is the only way out of the plot's hermetic bind. Hence the abrupt onset of the epilogue, and its rapid course to fade-out. Whereas the first image of Ivory's film is materialized, behind the main titles, in the fixed and long-held shot of fluttering torches—eventual illumination for the narrative's brutal Renaissance prologue—it ends, amid the muted crackle of sprocket noise, with the fading out altogether of the projector's light. From shots of the famous couple in the very moment of being translated into photographic image amid the bursts of magnesium flares that punctuate the final sequence, through the further debasement of their image to the mockery of caricature in a headline montage, editing then carries us to the ultimate fading away of Charlotte's oval-framed newsprint icon into the facelessness of the urban mob (fig. 1.17)—her worst fear come graphically true. From there the frenetic silent

1.17

1.18

1.19

newsreel, and the film with it, grinds rapidly down to another (and this time circular rather than binocular) mask in the final iris-in—and blackout—on a groaning horse-drawn cart laden with crates of Adam's European booty, moving lugubriously away across a city square (fig. 1.18).

Even before such telescopic remove has (to some extent) put acquisitiveness in its place, the newsreel has simulated the work of historical distancing in another way as well. Pocked and jittery with age, its footage is seen spluttering out at the end like the timeworn deus ex machina that it is. This is important as well as deliberate, and at more than one level. In contrast with the lusher canvas and color palette of the period recreation until now, the newsreel's print is not meant to be freshly struck. Rather, it appears retrieved as if from our own historical archive, credibly weathered, authenticated by the very flaws of time. In just that respect, though, this shaky newsreel montage could well come across as a minor tour de force of historical similitude within contemporary cinematic technique. But even the technical bravura of evocatively aged film is nothing new—except in regard to its execution. Another film-within-the-film about a power-hungry American magnate and European art collector, this at the beginning rather than the end of a larger screen narrative, has famously given us not just found and well-preserved photographs of the boy and his mother but also dated silent footage of Citizen Kane's public life (fig. 1.19)—newsreel images, of course, painstakingly hand defaced by the art crew of the 1941 film. Authentication by disrepair: that was Welles's gambit as well as Ivory's. But the medial progression of *The Golden Bowl*—as a vessel of heritage nostalgia filtered through technological advances both then and now—is yet more layered.

FROM *CINEMATÈQUE* TO CINEMATECH

Again: photo/slide/film//pixel—in this case with the digital incursion built into the cinematic erosion it simulates. For when the look of handmade mechanical stress appears, whether via Wellesian intertext or not, in the year 2000, rather than 1941, it has a quite different feel, a quite different archaeological orientation—and irony. Paolo Cherchi Usai's powerful meditation, *The Death of Cinema*, emphasizes the inherent terminal disease incurred by the fragility of film's material base as it is run through the machine time after time. He thereby identifies the filmic medium as bearing in progress the traces of its own wear, its own tears, its own steady deface-ment, screening by screening.[24] That this inherent degradation of celluloid exhibition could be simulated might seem itself an elegiac act—but with an unexpected twist here. Viewers may well respond instinctively to the crafty streaking and scraping that ages this footage as if it has already passed into that same history that will eclipse Verver's genteel and brutal monopolism. But if we go on to suspect that this inherent and definitive cinematic obsolescence, the entropy of exhibition itself according to Usai, the in-built mechanical abjection and "dating" of the im-age, is in fact imaged for us by a high-tech digital simulation in the new modes of CGI effects, then the metafilmic perspective is wholly (or at least materially) altered. This would be the at once distancing and clarifying work, yet again, of optical allusion.

We well know that the evocative timeworn stock descends, in all its inevitable damage, not from real time, real duration, real history. Still, we may think that it at least derives from the real duration of its manual simulation as laboratory doctoring. If not, the impact of this final "distressed" image is layered over with a further stress, a further metahistorical irony. If generated, that is, by a "motion graphics and compositing" tool such as the "aged film" filter of a Photoshopped "Aftereffect" (Adobe's trademarked name for this process), the historicism of the antiquing is further historicized. In that case, this 2000 film points as much forward as back. Its digital *trucage* of filmic aging, its showy electronic trick, may well remind us that only by special, rather than inherent, effects now—in the digital regime—does the instantaneously regenerated (rather than the photochemically imprinted and mechanically degradable) image ever genuinely erode. In Usai's sense, we might say that the filmic image now encoded and endlessly renewable is more like the vampirical undead of cinema. In any case, we would be readied by such technical deceptions as the digital mockup of an old-fangled medial death—the localized "framed time" of incremental wear—to look yet more closely into what the electronic dispensation of third-millennium cinema, in precisely its postfilmic devices, has done elsewhere to the work of time. Not just to the look of material aging, of history, of traced decay. But to the paradoxical look of time itself held in framed space—or closed out by it.

Beyond the narratological *mise-en-abyme* of a film-within-the-film at the end of *The Golden Bowl*, it is by the further texturing of these internal projections that a more specific narratography is called for—and called forth—in closure. That the film's ending should be no less susceptible to such narratographic response—even in its total abstention from the brilliantly precinematic logic of James's sutured syntax—shouldn't erode any methodological confidence as we move on. It offers yet one more demonstration of the way narratography hews to, and so highlights, the medium whose very mode of inscription it reads. Clearly, it is filmic cinema's ability to "sample time," to scoop it up intermittently, to collect and redistribute it, that—in the case of Ivory's *The Golden Bowl*—has put its photographic increments into such close fit with the plot's running exposé of thieved vitality and stockpiled goods, the whole commerce in reified images. A thematized digital simulation of the world would, of course, gravitate toward other plotlines in order to maximize its potential. Would—and has—and does. In this 2000 heritage throwback, however, in what might seem a late efflorescence of filmic and metafilmic cinema, a contrary thematic is erected upon the recording, seizing, and redisplaying function of cinema's predigital mechanism. As a literary adaptation, Ivory's film is not just a period piece. In the folds and overlays of its heavily filmic rather than just cinematic execution, it mounts (almost in the photographic sense) a nostalgic image of its own moving-image medium. Only if one guesses, or presumes, the advent of digitization in the final faked aging of the medium's own planned obsolescence is one given further metahistorical pause.

But such pause would only encourage us to linger over the irony as given. For when, in the last seconds of the closing newsreel—featuring the lumbering vehicle of Verver's collection—the iris-in narrows the field of motion to a spyglass vignette of tabloid-style incrimination (fig. 1.18 again), the very epoch of the movement-image has hit an impasse half a century before its time. All action, all movement, all historical motion, is in effect blanked out in a time-image that, almost by default, transacts within its telescoped optic field a valediction to all it has conjured. History in the making, absence in process. With a film like *The Golden Bowl*, narratography is elicited by the flickers and fades of the framed image per se: the optic field of the screen rectangle. Such are the moments when style is troped as import—and not just style at large but often, as we have seen, the photogrammatic (and later digitized) constituents of all such technique. It is in the intractable nature of the photographic base in filmic cinema—made visible at such times in what it otherwise makes visible—that it should refigure quite narrowly its own conditions as medial unit. In *The Golden Bowl*, the photogrammatic fixture turns figurative by an emphasis on record and accretion, on storage and its layered perusal. Which is why it can be so continuously subsumed to an ironic stance toward all things mediated, aesthetically distanced, aggregate, reified, and acquisitive.

By contrast, in the films of the digital gothic to be encountered (alongside the European uncanny) over the remaining chapters of analysis, the pixel tropes its own condition in the mode of dot and matrix, line and scan, all its electronic geometries contingent and relative, instantaneous, granular, and disintegrative. The contrast is absolute even when invisible. Instant by instant, the photogram captures whole—and comes to light in the mode of storage or its vanishing. Microsecond by microsecond, the pixel accumulates as array and comes to narrative light in the mode of fabrication and dissolution. That's why the reeling past of the past newsreel at the end of *The Golden Bowl* offers such a flashpoint of narratography *as archaeology*. In its laminated application of an "aging" filter, the digital can simulate the death of filmic cinema because it *is* that death. As Charlotte knew before she was the subject of one, but merely subjected to its sequencing (figs. 1.8–1.11 again), film is photography motorized, made an engine of serial motion. Instead, the digital image spills its grains across a single plane while racing to winnow and reseed itself before it is ever, even for the least moment, complete enough to be fleeting.

How an entire phase or subgenre of filmmaking—indeed a whole thematic turn in narrative tendencies across national cultures—is reconstellated in part around the spatiotemporal implications of this shift in mediation becomes the real and largest question (and the subdivided topic) of what follows. As discussion proceeds, though, what must be kept in view are further methodological questions as well, which might go far toward answering the primary one: What is the specific place of narratographic attention during a transitional period when the graphic basis itself of the medium in question is mutating beyond, as they say, recognition? Put differently, how does technique manifest structure in its fashioning of story when the technical basis of the image is so mixed, undecidable, and elusive? And when the very modes of visualization are thus invisible on-screen in a whole new way? What happens, in other words, when the imperceptible ingredient (and gradient) of motion is located not in the photogram's continual disappearance across the frame of the aperture but in an internal remaking of the digital frame itself? And how much cultural rather than technological context is necessary to approximate any kind of satisfaction in one's answers?

TWO

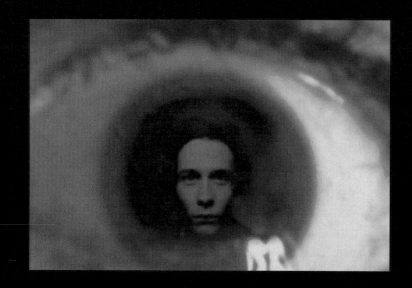

TRICK BEGINNINGS AND THE EUROPEAN UNCANNY

Typographically, the slippage of photo/slide/film has served to evoke the enchain-ment of the photogram within the racing frameline of projected motion. Cinemato-graphically, with Ivory's film version of James's *The Golden Bowl*, we have watched this same medial shift pace a sequence of mounting optical checkpoints across a single period plot. Such is the familiar narratographic transaction between medium and theme. In the process, that triadic formulation photo/slide/film also leapfrogs across a medial genealogy that, in its own right, is rapidly passing into history. Moving pic-tures in memoriam: 1895–1995. Since then, as the digital increasingly edges out the photomechanical, the instantaneous "sampling" and reassemblage of time is replaced, at least potentially, by its hands-on simulation. Marking a return to manual execu-tion (rather than automatic impress) in the representation of temporal event, this is the medial shift—and historical reversion—Lev Manovich characterizes as the move from "kino-eye" to "kino-brush." [1]

But the latter takes its underlying electronic shape, one might add, as a kind of automatized new pointillism associated with the triggered synapses of sensory cognition. This machine-gunning of the digit, with its spray of pixels—rather than the rotary advance of the photogram—certainly unbinds any inherent indexical link between the modules of record and their mobilization in the filmic synthesis. By definition, narratology would have every vested interest in probing the different mode of screen stories that may have emerged during this epochal transition, including any broad impact that electronic imaging or editing may have had on older paradigms of plotted temporality. At closer range, narratography would look to see where such difference makes its visual mark.

And one of the places such a narratography can look first is to a pronounced division of labor over the last decade between modes of a resuscitated fantasy in screen plotting, European versus Hollywood. In cinema's progressive encounters with the virtual, the elisions and slippages of time, space, and their points of contact serve to graph any number of fantastic displacements now psychosomatic, now supernatural, now explicitly digital. What grows clear is a sustained contrast that opposes the Eurofantastic of interpsychic trespass (the films of Kieślowski, Medem, Tykwer, and so forth) to a Hollywood ontological gothic veering between thrillers of the virtual afterlife (*The Sixth Sense*, *Vanilla Sky*, *The Others*) and what one might recognize as their sci-fi counterparts in the alternate realities of digitization.

These films of technological virtuality have given elusive new spin to the sci-fi genre's former interest in the *ontological* benchmarks of second-generation imaging. The photo of the scientist as he started turning into a fly, or of his charred shadow after disintegration by an alien ray, or indeed of the lethal saucer's first sighting, offer visual authentications of the marvelous from within a visual spectacle. Beyond photochemistry, too, the three-dimensional laser illusions of analog holography often testify at least to the volumetric presence (elsewhere—at point of recording) of the inhabited body.[2] What lately stands forth instead is a whole new region of virtuality in the paranoid digital ecosystems of a counterfeit three-dimensional real. In classic screen sci-fi, the photograph—or its analog offshoots—might well mark the outer limit of normative mimetic and evidentiary technology. It is against the expectations of photography that future simulacra and surveillance technologies are seen taking shape. Ever since cinema has in its own right grown no longer strictly mechanical, however, plots negotiating the engulfing force of electronic imaging have departed in two quite different directions from photography's indexical measure. Either these dystopian narratives have no (even residual) use for the photograph as touchstone (*The Matrix* trilogy, for instance)—or else they introduce it as a fabricated index of the real within a totally simulated space (*Dark City*, *The Thirteenth Floor*).

In addressing such virtualities of recent screen plotting, and their very different European counterparts, analysis starts from a question straightforward enough, at least in the phrasing. One wants simply to know why so many films of the last decade have taken the route of fantasy—and along such forked paths. Influenced in part by magic realism—as well as, one suspects, by the porous linguistic (and therefore psychocultural) boundaries of the "New Europe"—increasing numbers of screen plots in Spanish, French, German, and Central European cinema have been drawn to moments of the uncanny—to second sight, hypnosis, telepathy, psychic time travel, both amnesia and spectral anamnesis, reincarnation, out-of-body imaging, and the like.[3] The uncanny overtones of ethical connection and empathy in the films of Kieślowski are an originary influence—in just this respect—on the conventions of European narrative in the years at least since the Polish-French coproduction of *La Double Vie de Véronique* (1991). This is a cinema whose webs of psychology and destiny are repeatedly threaded by the fantastic.

On offer has been a wild variety of departures from normal perception into affective rather than technological virtuality. And through it all runs the suspicion that anxieties about historical memory and cultural identification, though attenuated since postwar modernism, are nonetheless being reconfigured around a certain bilateral thematic. Straddling linguistic and cultural boundaries (Warsaw/Paris, London/Provence, Spain/Finland), here are narratives of desire and its reciprocities in a confederated European landscape of permeable borders and receding horizons. At work in these frequent coproductions is a new "motivation of the device" where binational settings tend to be thematized by plots of reunion or escape, erotic relay or psychic retreat—and much of this under the shadow of an increasingly global traffic in electronic imaging. Benedict Anderson's sense of imagined communities seems so fully imaginary at times in these films as to be almost hallucinatory, a matter of extrasensory affiliations.[4] If such screen narratives remain potently Eurocentric, it is only because they are nationally, linguistically, and psychically decentered.

All the while, month by month over the same last decade, Hollywood has been turning out its digitized spectacles of teleportation, electronic time travel, and computerized virtuality as well as its thrillers of the undead, of posthumous agency and supernatural twist endings. What might, first of all, be the relevant common denominator between either mode, or genre cluster, of the Hollywood fantastic and their strikingly different European equivalents? And what, within Hollywood production, might be the tacit alliance between sci-fi formulas of electronic dystopia and the supernatural gothics of somatic virtuality, where the hero is only a ghost rather than a digital file? Put the other way round, what light might discovered links between American sci-fi and its gothic counterparts have to shed on the transatlantic operations of the fantastic? How does the sense of a shared narrative field help draw out the common strands of anxiety or desire, even phobias of networked connectedness, to which their narrative solutions are so differently addressed? And couched tentatively at the start, how far toward an answer do we get with a rudimentary distinction between an ontology of the occupied body in Hollywood and an epistemology of intersubjective consciousness in European cinema of the same years?

In setting out to explore such questions, another critic's motto was often in the back of my mind. This was Franco Moretti's recent prescription for a "centaur" criticism no more than half sociological, no more than half formalist.[5] He doesn't say which half is which, of course. The trick, I suppose, is not to let either drag its hooves. The trouble with any serious formalism, though, is that the details, even in the service of cultural explanation, can seem, in the rigor of their specification, somewhat top heavy. Be that as it may, I certainly wanted to keep such a bipartite (con)textualist agenda in view when turning to variant articulations of the fantastic across national film practices. Attention would need equal focus, in Moretti's simplest formulation, on the "how" as well as the "why" (6) of their making: on the formal devices that rendered graphic their implications and the sociocultural inferences that flowed from (as well as into) them. Given the quirks and recursions of time in both modes of fantasy, European and American alike, the different temporality of filmic and digital imaging would certainly remain one fact, or at least one question, to hold open.

On the formal end of the matter, there is an obvious place to go for a narratology of fantastic temporality and its articulation over plot time, an account at least provisionally relevant to these new avatars of the mode: namely, to Tzvetan Todorov's classic study of the fantastic. This is a genre that, having waned with existential and psychoanalytic modernism, seems to be remaking itself under the sway of an increasingly equivocated real.[6] We'll be returning to Todorov's influential proposals in some detail later. It is enough for now to say that this chapter is named for the uncanny end of his fantastic spectrum, with European plot explanations, however improbable, rooted in paranormal hypotheses that stop short of full-blown supernatural or technofuturist marvels. Still, for a medium-specific narratography that would cut deep enough to connect form to social function (or to social dysfunction, for that matter) and ultimately to the history of visuality—in which the uncanniness of the medium itself has programmed its own response—there is, as was apparent from the start, some seriously uncharted ground that needs staking out.

While narrative is always temporal, temporality has lately taken new narrative forms. Even virtual time, then, is of the essence in this book. And here is where the work of Gilles Deleuze is not only inevitable but invaluable. For the divided strands of contemporary cinema—divided in their approach, first of all, to the malleable textures of conscious perception and the sense of memory it generates—can, for all their internal differences, best be understood either as an extreme limit of or as a sudden break with the whole regime of the postwar "time-image." And this is not least the case in Hollywood films of a multilayered temporality that flashes before us as elastic, permeable, and ambivalent—a temporality whose mutable cast is itself conveyed by computer imaging as well as by digital editing. Plot's way of addressing this particular invasion of the cinematic frame by the digital array, so the latest evidence suggests, is to constellate all kinds of virtuality around a crisis of radical temporal negation. Such plots, in other words, turn escapist fantasies of time travel, transcended memory, or imaginary futures into ethically shaded allegories of the unreal.

THRESHOLD ACTION

Discussion will arrive at this more pointed emphasis on temporality, and what I call temportation, when enough film evidence has collected around the issue. For now we can turn back, just a few years, to the movies that first alerted me to a resurgence of the fantastic in contemporary film—and to my initial effort at accounting for them in structural terms within a broad range of "liminal" negotiations. In this respect, as in others, the pacing of narrative itself, its own structuring of story time as an artifice of temporal causation, comes into elusive and often suspect alignment with the retraced lifelines of the protagonists themselves. This will emerge. For now, the main assumption is as follows: Poised at the threshold of narrative, knocking only tentatively at the door of story, the free-floating liminal shot—once it has submitted retroactively to narrative drive—is often seen as the first real lead of plot.[7] Even though coming from nowhere, departing from nothing, it may induce, or in itself constitute, a moment of unrest. A classic account of narrative mechanics comes to mind at this point. For such liminal "disequilibrium" must eventually be set in balance again by plot. That's according to the structuralist narratology of Todorov in *The Poetics of Prose*.[8] And given the global nature of his formalist approach, the principle should hold not only for the epic, romance, or realist novel, say, but also for the "trick beginnings" that have come to characterize contemporary film narratives of both psychological and digital virtuality.

In comparing these thematic options as they become manifest in alternative narrative formats, we need, of course, a definition of fantasy as genre that would encompass, transatlantically, both tendencies of this contemporary "irreal": bizarre coincidence, spectral doubling, and related preternatural intuitions in Europe; ontological subterfuge in Hollywood. That genre designation comes immediately to hand, as we know, from the same author. According to Todorov, "the fantastic" is defined as the narrative span of undecidability—inaugurated with the first break from the normative into the deviant—during which a reader is held in suspension between incompatible explanatory options. If the strange events are resolved psychologically in the end, then the fantastic is cancelled, because settled, by the uncanny (*unheimlich*). If otherworldly rules of the "marvelous" are necessary for explanation, then fantasy is cancelled by the supernatural (or its sci-fi form of the "scientific marvelous").[9] Only in between, and for as long as that prolonged uncertainty can be sustained, does the genre of the fantastic exist. The European cinema of crisscrossed fates and permeable identities gravitates to the former (or uncanny) pole in resolution. In contrast, recent Hollywood thrillers approach the marvelous (or supernatural) pole, where lived reality is rewritten by the laws of electronic virtuality or ghostly afterlife.

Before its resolution in one direction or the other, the fantastic in film is clearly a genre that would have every use for the illusionism of *trucage* in Christian Metz's sense, whose effects can be either naturalized by the codes of screen storytelling or returned to spookiness.[10] *Trucage*, on Metz's showing, is a tampering with or "tricking" of the image track (superimposition, lap dissolves, fades, and ripples are his favored examples) that has moved, historically, from diegesis into syntax, or in other words from manifest special effect to sheer transitional device. But these laboratory manipulations can be returned again to motivated magic or spectrality in certain narrative contexts—or at least to uncanny figuration. And needless to say, digitization opens a whole new arena for the balance between "visible" and "imperceptible" *trucage* in Metz's primary division. Ultimately, Metz equates the artifice of montage all told with a "perpetual" *trucage* (672): the frameline itself disclosed as one extended trick, a prolonged optical illusion of embodied movement and temporal coherence alike. What would be the equivalent admission of deceit not only for postfilmic editing but for an entirely computer-generated image? In one kind of foreseen evolution of the screen image, pixilation would be the only trick left—once again inevitable, definitive, and engulfing.

Short of this, one of the most intriguing aspects of the parallel between the European uncanny and the Hollywood virtual, with their stress on epistemology versus ontology, respectively, is that each deploys a high incidence of filmic *trucage* in the localized and exposed sense, openly illusionistic. This involves the isolated and narratively motivated visual aberration, whether in traditional or digital form. By the frequent reflex action in a given film's self-conscious sense of its own optical genealogy, an otherwise low-profile syntactic device is tossed up by plot as a marker of ocular transgression or hallucination on the part of a character. A ready example would be the superimposition that lingers as ghostly memory. Such metafilmic reflexes are often associated with the moment of death (or its correlate in electronic disembodiment), where the immediate world gives way to the mind's split-second remediation of elapsed time—and often in the mood of psychic remedy, as for example in the montage flashback that we will find closing *Vanilla Sky* in the next chapter.

In light of Deleuze's sustained argument with Henri Bergson on cinema's "movement-image," the latter's interest in the "panoramic vision of the dying" comes directly into play with such narrative material. This is the moment, first discussed in Bergson's *Matter and Memory*, where all immediate perception turns inward to visual reprise.[11] Against the privileged nature of lived duration in Bergson, the moment of death is exactly the life-denying advent of time spatialized, compacted by juxtaposition rather than unrolled in natural succession—as in literary treatments of life's split-second replay in drowning. Of course, any retracing of elapsed life in narrative throws into relief plot's unfolding structure and its own retrieved point of departure. This is all the clearer when the beginning is the first lurch of a prolonged dissimulation.

If death is the only way back, or round, to clarification, then it is hardly remarkable that such an annihilation of temporality has become a touchstone moment in both European and Hollywood films of fantasy. The closest Deleuze comes to noting the equivalent enfolding of temporality in earlier screen modernism, though for once without any return to Bergson, is to quote Godard's parodic version of this trope in *Slow Motion* (1980), when Godard has his director-hero say, "I am not dead, because my life has not passed before me."[12] Plot after plot in the genre of the fantastic can be read, instead, as a nonironic mining of this metaphysical vein—and its potential ambiguities. Further, it is not surprising that the fixed image of photography should have its place in any such replotting of biographical trajectories. Often redefining subjectivity on the *apo-calyptic* (or etymologically veil-lifting) spot, death stands as the limit to which psychic transgress or digital illusion has been pushed. Anticipating such closural arrest, the place of the photograph en route—in its hovering status between the captured living and the frozen dead—has its recurrent diagnostic role, whether in plot or subtext or both.

Subtext: a loose-enough term, perhaps, unless taken in the rigorous sense mobilized by Michael Riffaterre.[13] Among the diverse theoretical positions that seem convened around a contemporary parting of the ways between fantastic subgenres in European versus Hollywood films, Riffaterre's semiotics is the one most alien to traditional film discussion. Yet its insights get at something not only structuring but ultimately filmic in the generation of screen narrative. For, better than anyone else, Riffaterre explains the hermeneutics of inauguration in any textual system: how first traces are swept from mind until they emerge again as variations on an always reticent but no less definitive theme. Riffaterre places no emphasis on the "liminal" per se, but his system, as we'll see, gives us alternate terms for it. It should be said, however, that no such sophisticated narratology is necessary to make sense of beginning's privileged status—or, in Todorov's terms, its functional disequilibrium even before a status quo has been fully established or stabilized. For there is a common sense in operation here—as well as a specialized semiotics. We all know that, even though few films end in a satisfying manner, most begin that way. This is because there is nothing to test the beginning against, nothing for it to fall short of. In David Bordwell's formalist distinction, it is all *syuzhet*, with no *fabula* yet constructed—or not quite yet: all shaping and plotting with no clear story showing through. Or perhaps, still in Bordwell's terms (as adjusted earlier), we encounter at such prompting moments a zone of pure style hovering on the verge of articulating the very structure it will then interact with as plot goes forward. At least for a liminal second or two: pure visual energy, ultimately a structuring energy, without yet a narrative to channel.

And structural latency without content is another name for a paradigm—after which the syntagmatic takes over, subordinated to the cause and effect chain of narrative linearity. With exceptions. These exceptions are both ubiquitous and crucial in the films I'll be taking up. One might call such moments, for short, the return of the repressed—if it weren't that Riffaterre has put into lucid circulation a shorter (though in the long run no less psychoanalytic) term yet in *subtext*. But how exactly do these exceptions to strictly linear or syntactic determination, these undue returns, effect their recurrence in film, a medium never once examined by Riffaterre? And what specific measure of the repressed emerges with them over narrative time?

Over time—or under it? That's half the answer already, in the inferences of terminological prefixes themselves. For the *sub*text bears in Riffaterre a definitive relation to narrative time: by constituting what amounts to its subterranean transcendence. Like the reversibility of the hermeneutic code in Roland Barthes's *S/Z*, the subtext reveals a kind of thematic iteration athwart the continuum of plot. It is as timeless, so Riffaterre puts it, as is the unconscious in Freud.[14] From the chronological succession of narrative's familiar bioplot, such unconscious or subtextual impulses alone are exempt. How film gives up its subtext to recognition while taking its own good narrative time: that is the question which the intermedial application of this semiotic narratology to recent structures of the fantastic invites us to pursue.

As often occurs in such visual specifications of narratological armatures, a more closely focused narratography of frame advance suggests a first clue, even in the abstract—even, that is, when no actual disruptions of the track impede the opening images. The material conditions of projection itself offer a deep orientation in their own terms. Like the already belated and always disappearing photogram, every plot increment comes only to pass, with the first instance being the most tentative and ephemeral of all. Story's first lurch is denied the preparatory context it can only posit in retrospect. In screen projection, as we've noted, the initial cinematic image emerges seemingly out of nowhere. Yet it often entails (or enchains) a hovering sense of source or origin, as yet undisclosed: a phantom undertone of first cause before the screen's first move. This amorphous sense of the formative—a seedbed of visual or narrative material that shapes and impels the initial image—lends a fleeting glimpse of pattern that recurs at unpredictable intervals across the course of plot's narrative syntax. In these recurrences, the vague sense of the germinal is always embedded and localized in its returns, even though no more than ghosted with its origin. So to understand the operations of this subtextual sequencing in recent films of the fantastic, we can begin by responding to the variant cast of its withholding and release in different national tendencies.

TRANSATLANTIC BEARINGS

Let me further clarify the poles, then, before populating them with examples. Again: epistemological slippage versus ontological outage. Against European plots of preternatural accident, temporal ambiguity, mnemonic mirage, and parapsychic empathy—in films that, as a rule, sustain their epistemological ambiguities through to the conclusion—Hollywood specializes lately in downright trick endings. These are final disclosures in the mode of ontological negation that require the total rethinking of a deceptive narrative line according to some revealed surprise of virtual or ghostly reality. Compared to the loops or short circuits of memory and desire in the European cinema of radical coincidence or mysterious fatality, the protagonist of the new Hollywood fantastic may turn out to have been dead, or merely digital, or only dreaming all along. Such films end up by upending their own premises. As a result, their "trick beginnings" are often linked—across a direct subtextual nexus—to some extreme closural reversal. In this sense, one might well expect these deceptive gambits to operate in ways pertinently different from the elusive leads or planted clues of the threshold moment in the European uncanny. The expectation is entirely justified. There is, however, a decisive common thread. One thing does bind form and content equally in the two modes of fantasy. And this is the thing we need to bear down on in this chapter. Repression, deferral, delayed release, retroactive clarification: these are manipulations of narrative plotting that in each case have direct repercussions for the experience of human chronology, or in other words of lived time, by the characters within a given plot. For the viewer, one of the best registers of such concussive returns across the span of duration remains the collision, if not intersection, of form and content monitored by Riffaterre's semiotics.

Under its auspices, a question persists. Why should the liminal shot or image of plot's initial thrust play a crucial role for film narrative? Before any further specification by genre, as with the fantasy plots to follow, one answer comes at once to mind: because in locating the shift from credits into plot's first move, the liminal moment, in any mode of filmmaking, establishes the implicit visual model of everything to follow. What Ivory's film of *The Golden Bowl* accomplishes, for instance, with the flickering torchlight inauguration of its violent Renaissance prologue is to model more than the adultery plot. Casting the scopic drive itself in a stark new light, in its links both to aestheticism and to sexual spectacle, these remediated images of primary illumination begin a film that shrinks its image in the end to the final containment action of a masked blackout on the flicker effect of primitive celluloid record. To say so is to compact by example the argument of Riffaterre's last three books: paraphrased (by the keywords of their titles in sequence) as the *semiotics* by which text *production* generates an ingrained *truth*. This is a "truth" (fixed rather than relative or differential) to be tracked beneath narrative fact or act. In more familiar terms, this third element, this permutable truth content, is what we might call a theme underlying plot. Together, then, these semiotic studies by Riffaterre stress the ways in which the iteration of an atemporal "subtext" beneath the forward movement of narrative is generated from the always unsaid "given" (or "matrix") of the text by the founding appearance of the "model"—often, in our terms for cinema, a liminal shot.

Before applying Riffaterre's threefold template of hidden matrix, manifest model, and recurrent subtext to transatlantic differences between modes of the fantastic in contemporary film, let me give two (I hope clarifying) examples from the narratively complex work of Christopher Nolan. Neither film is strictly fantastic, yet in each case its structural anomalies can be seen to derive in retrospect from a model shot. It may well be the attachment of these films to older plot formats of mystery and detection, even under extreme revision—and their distance from anything like the ontological fantastic (or digital gothic) of Hollywood films during the same period—that sends them back to the filmic track for their extracted photographic models. In *Memento* (2000), the medial "given" of Polaroid photography, as a nearly simultaneous memory trace, is inverted by the film's postcredit image: the reverse-action cinematography of a self-developing snapshot fading to blank as it slips back into the camera. This is the paradigm or model image for a film whose narrative episodes will themselves be told in reverse, inaugurating what Riffaterre might see as the unspoken matrix for an entire subtext of backward plot "development." One result is that this film does drift, after all, toward the current Hollywood mode of ontological subterfuge, if only in a limited sense—since the detective figure who follows the hero through the film to its last scene is revealed there to have been murdered by him at the start: in other words, a phantom presence all along.

The unspoken "matrix" called up by liminality in Nolan's next film, *Insomnia* (2002), is perhaps the cultural given (and Shakespearean intertext) that "murder will out," visualized on-screen in an evidentiary trace of blood. Backing the subtext up into the credit sequence itself, the title *Insomnia* takes shape like a stain behind the last listing of the cast. Subsequent credits then appear over increasing close-ups of blood seeping through the weft of fabric, as if through the revealed "weave" of the text—a case, no doubt, of "figuration" in advance of discourse.[15] Although without the wholesale hallucination of false realities we find in other Hollywood films of the last decade, the disorientations of traumatic memory loss and insomnia—Nolan's two encompassing pretexts—make for lives lived (and plots unraveled) like a bad dream. Then, too, whereas the original Swedish version of *Insomnia* (Erik Skjoldbjaerg, 1997), of which Nolan's film was a remake, begins with photographs of a murdered girl being taken by the killer in brief segments cut on the flash, the American version, like *Memento*, initiates its plot with an already processed photograph of a bloodied female corpse in the hand of the protagonist. A parallel forensic photo appears in the Swedish film too, but only after we have been exposed to the scene of the crime from which its image has been developed—not, in other words, as the film's first decontextualized image. In Nolan's version, we cut directly from bloodstains being rubbed into fabric behind the last phase of the credits (almost a punning clue to the detective's previous "fabrication" of evidence in another case) to photographic evidence shaking in the protagonist's hands during a bumpy plane ride to the scene of a new crime. In memory and premonition at once, the turbulence is both within and without.

In any comparison of the two Nolan films, one thing, one photogenetic thing, is obvious even before the crime plots set in. The quintessential photographic object according to a certain strain of image theory (from Bazin to Barthes)—the recorded corpse, frozen stiff in turn by photochemistry—offers the loaded first moment in both films. Each opens with the image of a slain body: the one film running backwards from obituary shot to its inexorable prelude, the other tracking out, first with camera, then with plot, from a single frame of photographic testimony to its subsequent investigation. And each film hides from us, for as long as possible, an act of murder *by the hero* as the eventual secret of its twist finish. In *Memento*, the trick comes via reverse plot chronology, when we find out who was already killed before the misleading homicidal search for him began. In *Insomnia*, the twist discovers our sleepless detective-hero as himself a nightmare killer. Investigative hero as villain: that's where each plot is heading when it leaves behind (but never completely) the opening mortuary image of photographic incrimination. Two paradigms collide under the eye of narratography, imploding the terms of agency and solution. The detective prototype of all fictional plotting, retracing the course of already elapsed events, finds its equivalent cinematographic model for time passing in the seriality of the single evidentiary imprint.

But that's only the genre-bound and site-specific, rather than more fundamental, pressure exerted on any such liminal photographs. Thematized in connection both with death and with narrative chronology, and whether stressing the reversibility of the latter or its options for forensic retrospect, Nolan's films brandish the photographic opening with typifying extravagance—typical, that is, not only of the director's self-consciously mediated style but, more to the point here, of the founding rather than contingent relation of photograph to screen narrative. For the photographic still deployed as inaugural image of any given film offers a threshold moment in metafilmic (or medial) terms as well. The "model shot" of such a repictured imprint (modeling if nothing else the photomechanical nature of the ensuing narrative chronology) must always mask, even if asking us to recall it then or later, cinema's own matrix at the level of the strip. Remediation is laid bare. The fixed photograph on camera, mediated by cinema's mobile ones (its photograms), especially when it has emerged in advance of all plot context, cannot help but double for the invisibility of its racing equivalent on the spooled celluloid reel. The opening photo is a picture of film's own pictorial capacity in dissection. It is the image of just what movies must put into movement. The ensuing rotary motion is manifested, of course, only in the projection that sweeps the fact of its stillness up into the mirage of action (or even of visual prolongation, as with a held shot of a photo). Whenever, in sum, an on-screen photograph offers the liminal impetus of a specific narrative development, whether as functional clue or derailing false lead, it is also a model shot in a deeper sense: a model for which the photogram is the unseen matrix.

As already noted, this insistence on the discrete fixity (under erasure) of the serial frame rather than on the gestalt of motion is, at least since Eisenstein, at classic odds with a major line of film theory from Bazin to Cavell to Deleuze. Not just time embalmed but "change mummified," in André Bazin's deathless phrase: that's cinema for the phenomenologist.[16] Temporal transformation is preserved in all its unfolding duration. Change itself is struck off as speeding imprint. The other way to take this, precisely because you can't see it, is to recognize the filmic spin beneath the cinematic scene. Along the serial strip of film, instead of standing still, each timed image only stands till erased by succession. That this isn't what cinema looks like on the screen doesn't mean that it isn't the way such an apprehension of filmic cause can help us probe the elusive temporality of screen projections. This is easiest to consider, because possible to glimpse, in their occasional syncopation—rather than synchronization—with the timing of plot. But especially when invisible, the photogram is definitive. For as in the psyche, so in the technology of projection: the unconscious is by definition counterintuitive.

Isolating the role of the photogrammatic imprint in recent instances of cinematic fantasy can, more particularly, help locate differences between the now psychological, now metaphysical disturbances to reality induced—across the polarized tendencies of European and Hollywood production—by their parting of generic ways. This is often because the moment of photography's remediation by film returns us—in quite different ways in the two modes of the fantastic—to film's own differential basis on the celluloid strip. It is there that stillness stands disclosed, or, better, stands exposed, as both the constituent and the antithesis of screen movement. Motorizing the serial strip, projection elides the rapid photo cell into the frame(d) time of screen motion, so that the whisked-away module reappears in its own momentum on-screen as a spectral phase of advance. All links transpire in the blinks of a mechanical eye. In a word, fantastic. Not all films of the fantasy genre take up this phenomenon as theme, of course, and least of all those concerned with distant electronic futures. But many do. And even avoidance is revealing. These chapters must close in, then, on the point of intersection between photographic temporality and narrative virtuality in the impacted photomechanical crux of numerous screen narratives on both sides of the Euro-American genre divide. These are films where desire often uncannily reanimates a fixed visual trace or, alternately, where the simulated kinetic image of living presence marks the death of the real.

A caveat may help at the same time to secure the historical bearings of my point. It should certainly be admitted that whenever the medium-specific rudiment of filmic motion in photographic arrest is somehow called up within the arena of the fantastic, the distancing effect may well feel residual in its very technique. It is likely to seem a leftover from modernism's habitual way of intervening in the classic transparency of the cinematic image. Before its consolidation as a narrative system, however, cinema in its earliest years was perceived as an overt photofilmic machination by its first spectators. As it happens, an indirect evocation of cinema's differential motion comes to us from these same years in the late-Victorian art criticism of a performance medium, dance, to which theorists of the mechanical image, Deleuze among others, have always known cinema's affinity—and to which Fernand Leger's experimental film title from 1924, *Ballet mécanique*, pays direct testament. In the slippery, self-enacting language of the Decadent critic Arthur Symonds, writing in 1898 within the first decade of cinema's absorption by cultural consciousness, the thrill of dance is that each discontinuous effect, truncated in material time, "lasts only long enough to have been there."[17] It is there as gone, a vanishing trace, which only attention can retain long enough to savor.

In cinema, the incremental effect, the frame itself, lasts only long enough to have been . . . to have been surrendered, that is, to the next in line, assimilated to it, shifting it as if from within, becoming its supplement in the oscillating slippage from split second to split second, moment to momentum. This is the dullest truism, perhaps, until renewed by a given screen narrative, where the filmic may obtrude with unnerving purpose from within the cinematic—often flagged by an on-screen photograph within the plot or a freeze-frame on the track. So how is this liminal reference—this optical allusion—to the medium's own material base likely to get staged in particular response to plots of the fantastic and their anomalous psychology? Nolan's films traffic in the uncanny without quite laying claim to the fantastic, or in other words without raising the genre dialectic of interpretive ambivalence. Explanations are delayed but not directly equivocated. Instead, almost like a counterfantastic, his films concern that deprivation of the unconscious brought on by amnesia or insomnia, where the inexorable momentum of time can be neither sorted by memory nor reconfigured by dream. As detective thrillers by any other name, seeking resolution in a stable solution, they steer clear of both the ontological gothic that characterizes recent Hollywood films (where clues are themselves tricks in an environment of hallucinated motion and presence) and the paranormal filiations of the European uncanny (where intuition and daydream have in themselves the force of action).

The closest thing in recent European cinema to Nolan's *Memento*, perhaps, at least with respect to entirely *un*supernatural violation of narrative temporality and cinematic sequencing, is Tom Tykwer's 1998 *Run Lola Run*. Skirting the fantastic in its nonetheless overt distortions of narrative temporality, Tykwer's film bears distant comparisons to Nolan's because of the way it summons the relation of the fixed photographic image to the arrest of biographic time. Equally to our point here, it does so by a multifold structural subtext that punctually interrupts the action generated by moving-image footage. What ensues briefly in each punctuating case is a barrage of freeze-frame sections of imaged space posing as quasi-photographic imprints or mementos. The logic unfolds in counterpoint to the already divided narrative. In a parallel-universe format, the frenzied sprint across town of Lola to save her boyfriend in the film's crime plot replays itself three times in entire alternate versions, fatal and otherwise. In the process, it interrupts its three main vectors of racing action with disjunctive flash-forwards to alternate lives of minor, unnamed characters whose paths Lola happens quite literally to cross. In all of these proleptic inserts, six in total, the indexical moment of photographic record is evoked (without being in any way narratively motivated) by the overlay of shutter sounds on the track, cut by quick cut. The woman pushing a baby carriage, for example, will eventually have her child taken from her on the grounds of neglect (click) until caught stealing another (click); and the young cyclist who later gets beaten by thugs will end up having his wedding photo taken with his former nurse (click). Either that, or the woman ends up winning the lottery and finding her apotheosis in a tabloid publicity shot; or the cyclist becomes—your quick guess is as good as mine—a homeless drug addict last seen as if in a police photo. Jump-cut editing and soundtrack collaborate in figuring time as a series of seized stills, thereby confirming Bergson's complaint against the medium, its participation in a widespread cognitive error that life is lived not as immersed duration but as a kind of mental photo album.[18] In none of this is there a clear baseline of reality established from which a departure into fantasy can be marked. In *Run Lola Run*, time is entirely contingent, up for grabs.

Six formal repetitions of this sort do indeed make for a pronounced—that is, an unmistakably articulated—subtext: the stilled photographic frame as emblem of time's conceptual dismemberment into moments, into discrete and arbitrarily segmented units of temporality. The emblem thus becomes in turn a model of plot's own trivalent structure in an alternate-worlds scenario of optional temporalities and avoided fates. Form and content have oscillated in so rapid an interchange of implied terms that the distinction has broken down between them over the triple tracking of plot and strip alike. Each titular run of Lola's is down a road not "really" taken. And the narratology of the provisional has found its perfect narratographic traction. For across these main trajectories of plot fall the minor characters and their miniature chronologies. The metafilmic pattern of all this runs deep. With the still frame operating as model in these elided serial lifelines, the photogram on the strip—as the liminal constituent of all motion—is again tacitly and belatedly divulged as the buried matrix of narrative temporality all told, arbitrary, manipulable, and elusive.

CROSS-EUROPEAN "INTERFACE"

But we turn now to clearer cases of the fantastic—and these, first, at the uncanny pole of co-incidence and parapsychic connection familiar from recent European production. Slavoj Žižek works overtime to guard the narratives of Krzysztof Kieślowski, for instance, against charges of new-age spiritualism. He does so by insisting on ethical parables of choice (rather than alternative universes) in some of the films' more strained coincidences and eerie doublings. The result of Žižek's reading is to refuse or neutralize the "fantastic" altogether, unduly normalizing the films in the process. Yet Žižek's important concept of the "interface effect" does often retain a trace of the uncanny that is not entirely subordinated to ethical deliberation. Interface occurs, for Žižek, as a breakdown in suture: and not just its local articulation in the shot/reverse shot format but its overall logic of immersion, identification, and signification, whereby the purity of image is subsumed to a logic of story, ocular pleasure to textual dynamic. What the "in-

2.1

terface effect" might be said to accomplish—might, because Žižek's remarks are entirely sketchy and passing—is to sustain an image longer than usual in the imaginary before its co-optation by the symbolic: to retain its visuality, its figurality, against the sway of its meaning.

With no intended invocation of digital screen formats, Žižek's use of the term *interface* occurs primarily when the normal pattern of shot and its answering shot is disrupted, the two instead overlaid upon each other in the same image, as in a figure seen against his own photo or shot through his own reflection in a glass pane.[19] In certain of these cases, one might say that the multifold cinematic given of superimposition—with one lit frame slammed upon another in process; projections layered in turn upon a screen; one shot overlapping and dismissing another in the grammar of dissolve—is laid bare even at the moment of its uncanny incorporation into narrative. But this is not the case with the more definitive and pivotal role of "interface" (in the more familiar currency of the term) as marker for the digital unconscious of Hollywood's recent ontological thrillers. Such plots of interactive virtuality offer a direct counterweight to the elegiac uncanny of the European fantastic. Once again we are concerned with the shifting valence of a common term, two variants of a comparable genre trajectory.

Žižek's clearest instance of the "interface effect" in the uncanny rather than marvelous scenography of Kieślowski's work comes in the second scene of *Three Colors: Blue* (1993), where the doctor is reflected in the eye of the heroine (fig. 2.1): a figure announcing the death of her husband and daughter in the car crash she has survived. At the ocular membrane dividing self from the world, his news hovers at the border of a traumatic internalization. But how has the film's actual *liminal* image, before any possibility of suture or its collapse, worked to anticipate the uncanny coincidings of self and other that the rest of the narrative plays out? As Riffaterre's system might well account for it, Kieślowski's opening shot gives us the car's revolving wheel just before the accident, from the very underside of fatality, the image emerging in the first place out of the full-screen tire's flat black surface. Plot's "model" is thus the vehicle of its own causal but arbitrary chain, summoning the unsaid matrix of narrative "drive" and thematic "momentum" under the double sign of *accident* and design.[20]

Commenting on Kieślowski's earlier and more overtly fantastic film, *La Double Vie de Véronique* (1991), Žižek, again ignoring the liminal impetus of the film, locates the "interface

2.2

2.3–2.4

effect" first in a Lacanian "anamorphosis"[21] through the refracting glass of a train window on the French heroine's trip to Poland, as well as in a glass ball's inverted estrangement of the same landscape a moment later (fig. 2.2). It is soon after this mediated looking that she snaps an unwitting picture of her Polish double from a revolving Warsaw tram, in a dizzying intercut sequence of circular pans (figs. 2.3–2.4). The actual print, processed unwittingly, is discovered only much later in Paris. But the uncanny subtext of optic interface has appeared as early as the twin liminal shots that "model" the whole narrative. These are the paired prologues from 1968 that reveal the psychically linked heroines, Weronika and Véronique, being shown in macrocosm and microcosm, respectively, with a maternal voice-over that directs the one at the stars seen through the reflective pane of a Christmas Eve window and the other at the veined underside of a springtime leaf viewed through a magnifying glass. Each optical partition—and resulting "interface"—serves to reflect or distort the perceiving subject at one remove. In addition, this seeing of the world for the first time through maternal eyes is interrupted for each character by the early death of the mother. As happens repeatedly in recent films of the fantastic, the loss of a female origin induces, in this particular case, an uncanny female twinning in compensation, confusing heterosexual desire in the process. Véronique's last love scene, long after the death of Weronika, takes place in psychic dissociation while she is glancing toward the posthumous photo from Poland shown to her by her lover a few moments before (fig. 2.5): the impossible photo of herself as the Other, brought into proximity once again with that inverting glass bauble at the foot of her bed.

2.5

Kieślowski's *Three Colors: Red* (1994) offers further evidence of the liminal moment as seedbed of the fantastic. By a deus ex machina in the last scene, the heroine and her destined lover are among the few saved from drowning in a Channel disaster, their rescue broadcast in an unlikely stop-action image on TV, then dovetailed with a prolonged cinematic freeze. The film's liminal shot has offered a "model" for exactly this climactic transmission. For the narrative opens, even before the credits, with a telephonic rather than video mediation across this same English Channel: a racing shot of cables passing underwater between her photograph on the desk of her former and never-seen British lover and the ringing of her telephone in Geneva. As figured in the crisscrossed wires of the model shot, parallel lines do in fact converge, spliced by the cross-national twists of fate.

A fixed image like that desktop photo, transferring the imprint of occupied space from one time frame to another, is in itself a mode of optical communication. Add to this image a voice and you have the detached rudiments of sound cinema itself, awaiting their consolidation as narrative. In contrast with the opening photographic anchor of *Red*, generating as it does no reverse shot of the lover picking up the phone, the place of a more complicated visual interface is delayed in the other two films until it seems more uncanny and unnerving: in *Blue* as the reflex of a cornea taking the imprint of reported disaster, in *The Double Life* as a reflex camera photographing the impossible—yet inevitable—otherness of self to itself. In each of these cases, the distance between model shot and deferred interface only defines the ongoing subtext into which the delayed effect is inserted.

The only strictly fantastic among these effects—namely, the heroine's accidentally captured image of her own double—locates *The Double Life of Véronique* as a disquisition on the photographic uncanny per se. In an ironic genetics of vision itself, mothers pass on to each heroine the lens through which they will see their worlds. But this includes the sense of being seen as well—in the eye of the Other. Any photograph may call to mind this extraction of the visual from the fact of visibility, of the eyed object from the subjectivity of vision. What makes the particular "interface" crisis of this one film's pivotal photograph a quintessential moment of the European uncanny (in its approach to the near verge of the supernatural)—a moment half fantastic, half allegorical—depends on something more. In the simultaneity of self and its own externalized self-image, Véronique's schizophrenic break is normalized by the preternatural nature of photography's own medium—to this extent at least: that Kieślowski's heroine is by the end quite literally beside herself.

One form in which the fantastic emerges for Todorov, and indeed as the special and intensified case of textuality in general, is that, at some triggering moment, it does indeed take "a figurative sense literally."[22] In *The Double Life*, a sensation like "I'm so plagued by self-consciousness that I feel like I'm always looking on at myself, or that someone else is" will materialize within plot as an actual double who must eventually be sent to death, her difference assimilated—but never completely. In our speculative convergence of critical paradigms, it comes to seem no accident that the foregone and unsayable matrix in Riffaterre is often just such a piece of received figurative wisdom: a metaphoric idiom that must be disguised by literalization across the plot and brought back halfway to recognition by the eruptions of the subtext. In the present case, the uncanniness of a photographic memorial bursts into Kieślowski's narrative as the residual emblem of unconscious connections in the past that are, though unrealized, never entirely past. Instead, in this case the traced shadow of displaced subjectivity is retained beyond the body as a trope for the spectator's (here the heroine's) own psychic "incorporation" of herself as other.

That's the way the film attempts resolution at the level of plot. But, yet again, we touch on a concern not only of narrative form but of medium—and of the filmic medium at that, over and above photography. This is not, I think, difficult to see. According to Todorov, fantasy—suspended as it is between the immanent and the unreal, the possible and the impossible—is the purest state of the literary (175). Emerging only very slowly from the deliberations of Todorov's book, and never quite fully explicit, such a metatextual understanding not only applies equally well to films of the fantastic but coincides with Metz's proposal that trick effects are only the special case of the cinematic illusion at large, since "Montage itself, at the base of all cinema, is already a perpetual *trucage*."[23] For Todorov, fiction is fantastic because its referents are strictly imagined. Similarly, all narrative cinema is fantastic because its presences and durations are strictly illusory.

At points of rupture or resistance, however, the photograph itself comes forward as the special case of fantastic temporality. In its cultural function, the indexical moment of photographic record is retrograde and prospective at once, putting a seal on the past and delivering it forward to a future not its own. Todorov, in fact, admittedly lifts the paradigm for his differential definition—fantasy as the dividing line between the uncanny and the marvelous—from philosophical definitions of time present. For him the "comparison is not gratuitous,"[24] since he means to evoke the instant, the *now*, understood as the definitive yet vanishing line between the accumulated past (of received understanding) and the nebulous future (of undreamt marvels). A yet stronger claim, however, may well seem invited. It is not just the accumulated knowledge of the past that would explain away the marvelous in settling for an uncanny resolution. For the uncanny, after all, is often the spectral *return* of the past, the lifted repression of banned desire or recurrent fear. So between a haunting by the past and a daunting future with unheard-of wonders, between preternatural disturbance and supernatural epiphany, the seesaw of fantastic uncertainty negotiates its plotline across the dubious present of its presentation.

And the photograph? What is it, too, but the invisible division between a past life or space or event, arrived before the camera, and a future image that will supplement or eventually supplant its presence? The image lives off the past in order to live on into its own future. In this sense of the image's fantastic balancing act, death always encroaches from the uncanny pole of photography, immortality from the supernatural. In the discursive function of its deixis, its pointing, a photograph announces that this here was there to someone then. Photography thereby inscribes the reciprocal vanishing of past and future into an objectified present transferable from taker to receiver, a present retained forever without change, a performative act of notice endlessly rehearsed as well as instantaneously preserved. What is deathlike about a human image under photochemistry is also, therefore, what is phantasmal about it: a corpse, but also a haunting.

A photograph has no memory. It is one: the material form of one. Photographed human images enact a perpetual displacement, from a subjectivity gone to another reconfigured by looking. The photographed person is thereby the prosthesis of a memory not belonging to it as agent, but only attached to it as object. When, moreover, such medial reverberations make their impact felt upon the doppelgänger motif of Kieślowski's *The Double Life of Véronique*, we recognize the bold, almost brash, simplicity of the photographic parable. An accidentally taken, processed, and subsequently unnoticed photograph of one's double, the Other within, is only a general condition of psychic life spelled out in uncanny vocabulary. For in mental life, doubleness is often *registered* without actually being recognized, psychically imprinted without being acknowledged. This is, indeed, one of the chief inferences behind the psychic trespass and sexual telepathy of so many fantastic episodes in recent European cinema. In many such cases, European screen poetics has passed beyond both classic linearity and modernist disjuncture, often residing now in the accidents and weird linkages of repetition and circularity. But part of my point here is that we lose the specific valence of the uncanny, and its complementary relation to Hollywood's fantastic cinema of the same years, by categorizing this as just another aspect of an already codified postmodernism. Something else is at work—a more localized

and consistent "writing with movement," an inscription in and across time—which is exactly what a narratographic analysis is designed to draw out.

Incorporating more obvious elements of magic realism than in Kieślowski, the principle of coincident plotlines and convergent fates is carried further yet in the fantastic cinema of Spanish director Julio Medem. And with an even more extreme pressure on the photographic index. In *The Red Squirrel* (1993), the unlikely final reunion of separated lovers is brought about through the mediation of photographic magic at the film's climax, in a shot that is literally pivotal. The hero fixes on a framed snapshot of himself with a former girlfriend, doubled by her own magnified image printed on a rock band T-shirt (fig. 2.6). Behind this image of the couple, we make out in the photographic distance, as the hero does for the first time too, a further female image: the once accidental and now uncanny figure of the new but missing lover (fig. 2.7): a woman whose unnoticed presence in the park that long-lost day was a sheer coincidence—and premonition. Under the force of wish fulfillment, the shot now magically *un*freezes and engulfs the hero's present space. Animated into lateral movement, the new lover and her former boyfriend first walk behind the couple in the foreground (fig 2.8) and then pass in a shot/reverse shot exchange with the on-site hero so that she may reveal the necessary clue to her present whereabouts as body rather than image (figs. 2.9–2.10). In this fantastic rescue of suture from within the two-dimensional plane of photography, a relationship previously threatened by death is submitted to the reanimation effect of desire.

The cinema of fantastic coincidence is even more obviously at work in Medem's next film, *Lovers of the Arctic Circle* (1999), whose liminal shot of a crashed airplane in a Nordic blizzard is intercut with its front page image in the newspaper, read

2.6–2.10

2.11–2.12

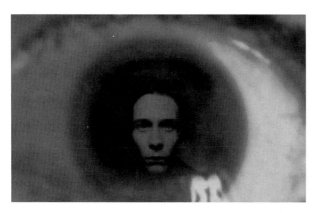

2.13

while crossing the street by a woman run down in the process by a passing bus. Her long-separated lover (also her stepbrother) rushes to her there, but only in time for a last fantasized embrace in the split second of her death. The doomed nature of this reunion taps directly into the plot's incest motif, involving the first divorced and then dead mother for whom all the hero's previous lovers have been a failed substitute. At the film's turning point, with the hero both boy and man in the same relived space, the adult son returns home to retrieve his camera and finds his mother's corpse amid the fly-infested debris of her kitchen. In a double wrench of separation, his absolute loss is backdated to a foundational lack that is marked by a primal match cut from this scene of adult trauma back to his former boyhood capture of the mother's living image (figs. 2.11–2.12): the taking of that very photograph the son keeps at bedside during his subsequent sexual affairs. By the logic of the "performative index," this photograph inscribes his mourning for himself, once present to her, as much as for her who was once there for him.[25] We are next translated to Finland from Madrid for an uncanny transnational reunion of the separated sibling lovers, whose names answer to each other as reciprocal as well as self-reversible palindromes, Ana and Otto. When the film returns in its closing moments to the reflection of the hero in his lover's dead eye (fig. 2.13), followed again by the downed plane, we realize that the whole narrative may have transpired in flashback from the moment of death—and of his death as much as hers, perhaps, since he was piloting the crashed plane. For in that lingeringly held image of reflected self in the death stare, what gets locked into place is the life-denying need to find one's adult identity mirrored in the eye of the ma-

2.14–2.15

ternalized female other. Tom Gunning's interest in the nineteenth-century idea that a murderer's quasi-photochemical image is left as so-called *optogramme* on the victim's eye finds a real-time but still fixed, suicidally transfixed, equivalent in this *Liebestod* variant.[26]

Time regained in a single photogrammatic reflex—regained and surrendered at once. Medem has here deployed some of the most insistently Proustian images in contemporary cinema. But not the only ones. There are others equally marked, if rather more forced, in an adaptation of Proust's work itself, *Time Regained* (*Le Temps Retrouvé*) by Raul Ruiz (1999), a coproduction not just of two countries (France and Spain), as in Medem's *Lovers*, but three (France, Italy, and Portugal this time). If horizontal borders are traversed at will by the generative energies of this film, so are the vertical divisions of times past and present. The plot is inaugurated by four all but simultaneous models, four liminal shots, rapidly succeeding each other in the passage not only from credits into text but into self-reflexive textuality. These liminal models are temporal, spatial, scriptive, and photomechanical, respectively, each a version of the other at the level of the film's ensuing subtext. Shot one: the audible tolling of time from the Combray belfry. Shot two: time's spatialization as flow in a surging river. Three: the frozen flow of ink across manuscript pages on the bed of the dying Marcel. Four: the marginal animation of family photographs in his palsied hand, and under a shaking magnifying glass, during his optically aided reverie, one after the other down through the generations from "Mama" (fig. 2.14) until his arrival at "et moi" (fig. 2.15)—his own photo as a boy.

The least Bergsonian of modern works, according to Georges Poulet, a novel that everywhere insists on spatializing time in the simultaneity of its overlaps, is therefore the most Deleuzian, where every "movement-image," especially through various modes of fantastic *trucage*, tends

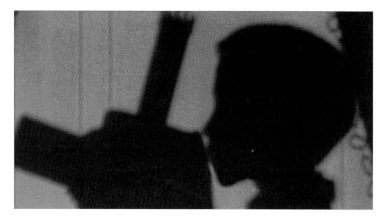

2.16

2.17

2.18

to reconfigure itself as a surreal "time-image."[27] In the process, the narrative evokes its own apparatus in a way that replays film's various routes of arrival as a narrative medium. Following those opening photographs, the initial party scene turns into a *tableau vivant* lit by the magic lantern of the young Marcel, seen at first only in silhouette (fig. 2.16) and paradoxically present there in the scene as an adult as well. Soon after his role as projector of superimposed illusion, the boy is told he is too young to see a stereographic slide of battle carnage. But when he takes up the binocular device anyway, what he glimpses is a briefly moving rather than still image of a cavalry horse dying on the battlefield. He sees it, in other words, with the eyes of an adult having passed through the war years—and through their cinematic documentation. Reading a letter from the scene of German invasion later, in a bizarre café screening room, Marcel floats surrealistically in an interspace between projection and the thrown image, accompanied now by his boyhood self at the helm of a lantern-turned-projector (fig. 2.17), throwing the shadows of history into relief even while imposing his own future form upon them. Young and adult selves now appear together in the same space of projective illusion, but still not communicating. This must wait for a final dialogue in an impossible pair of "two shots." The interface effect returns at this point as an uncanny culmination of the medial subtext. First, the younger self confronts his eventual mirror image as an adult male. Next, he faces his future shadow: two avatars of subjectivity brought together across the now dormant magic lantern of precinematic projection (fig. 2.18). Each shot returns, in effect, to the initial gap between self and its optically enlarged photographic double as "moi." Here is the very archaeology of ocular prosthesis as a metapsychology of subject formation and self-identification.

Moving forward again from the free-form adaptation of an elegiac modernist classic to the historical present of the contemporary trans-European fantastic, there is a particularly forced conceit at the core of a Hungarian-French coproduction by Ildiko Enyedi, *Simon the Magician* (1999), that in turn forces into the open a transnational poetics elsewhere implicit in such films of bilingual border crossing and erotic transference. So completely does the subtext of linguistic reciprocity and its mysterious barrier-leveling powers infiltrate the subject of such coproductions that it can become, as in the case of Enyedi's film, all but its only topic. Specializing in films of uncanny twinning and symbolic doubling (including *My Twentieth Century* [1989] and *The Magic Chase* [1991]), Enyedi constructs in this case the saga of a Hungarian mage imported by the Parisian police for his "paraphenomenal" powers. With the first postcredit scene of a transnational telephone call (as in Kieślowski's *Red*), he is brought in to solve a murder mystery. But he soon falls victim, on his own, to the mystery of love at first sight (the French malady) with a young woman whose job as a curbside pollster is to query people about their religious beliefs. A different kind of paranormal faith will eventually be tested, for spectator and heroine alike. In the meantime, his one-word answers are entirely contradictory, and she thinks that, despite his obvious interest in her, he is being annoyingly coy and oblique. In fact he can't speak French, and when asked by his former compatriot—the wealthy showman trickster who emigrated years before from Hungary to become the lionized new Houdini of the Parisian media—whether there are any tricks of the trade he might like to learn, all the hero wants is to be taught "a little French." Magic gifts are nothing compared with the empowerment of linguistic thought transference.

The hero's lack of that one power source has already been made clear—as the very backbone of the plot. By a telekinesis turned upon the technology of digital imaging itself, the hero can, astonishingly enough, move a computer-screen cursor without touch. His very sight is haptic. But his language cannot, as voiced desire, reach where he wants it to. The girl learns of all this only when the magician calls a police translator on a cell phone, passing the receiver over to the girl to have his passion for her decoded into her native tongue. My stress on the relation of linguistic decryption to the magician's libido is not selective. What I am reporting is virtually all there is of the film's whole plot—until a sudden public spectacle at the very end. After a hastily announced televised "duel" between the two magicians, each buried alive for three days, the hero is mistakenly thought to be dead. The video-screen flatline that shows his cancelled pulse may be the one monitor over which he has no remote control. Once exhumed and reburied, however, he is able to move the earth above him in the film's penultimate shot, the Paris skyline in the distance. Next, and without the magical contact provided by language, he now resorts for the first time, and almost by default, to telepathy. In this way he is able to communicate his "resurrection" to the girl across town, who has already assumed him dead.

Stopped dead in her own tracks, with a knowing smile breaking across her features, she gets the message. We have time to realize this, because the shock of his coming to life is registered by a long freeze-frame on her recognition, holding her in place, still waiting. Converted on the spot from a movement-image to a time-image (of expectancy), she appears transfixed as a receptive antenna for his renewed yearning. The moment is thus held suspended in a genre-deep equivocation. In this "fantasy" of contact, are we made privy to the uncanny power of empathy and affection crossing a sensorimotor barrier? Or is it instead a direct supernatural case of extrasensory perception? In any event, this optical figuration of expressive desire in the "trick" of a freeze-frame—this Lyotardian "acinema" of arrhythmic arrest—verges almost on a familiar sci-fi trope of realized *trucage*. Like Dr. Crase seven decades before in René Clair's Paris-set fantasy, as discussed in the fourth chapter, it is as if the man who could control the cursor without bodily contact can edit the world as well. And this later Parisian-based film, too, of the emigré trickster, has its own—almost requisite—trick beginning. For before the credits, we have seen jostled, canted, discontinuous images from a handheld video camera trying to record, without the least context for the film viewer, the arrival of the two magicians within the crush of press photographers—their much *later* arrival—at the scene of their competitive premature burial. Narrative structure anticipates its own climax in the subtext of the buried clue, advertising at the same time cinema's own paranormal powers of ocular premonition.

From the twist of premature burial in the gothic trick ending of *Simon the Magician*, with its fable of linguistic intersubjectivity and the white magic of romantic radar, let us look to another unmistakable instance of that "trick beginning" which many such films of fateful coincidence share, we will find, with Hollywood production in the same years. In a posthumously filmed script by Kieślowski brought to the screen by German director Tom Tykwer and set in Turin and Multepulciano, the film *Heaven* (2002) opens with a spooky luminescence of saturated, unearthly colors in a digitized haze of aerial landscape. The title scene? Without the viewer reaching for this protonarrative link, the view is left stranded for a moment, floating as pure image. In chromatic register and indistinct graphic definition both, the shot's opening effect suggests the liminal

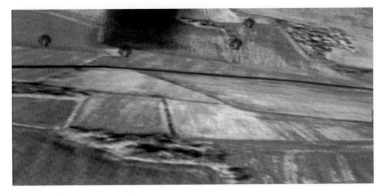

2.19

2.20

status of "figuration" in Philippe Dubois' sense (following Lyotard). Here is a visual display wavering at least momentarily between—between and before—meaning and representational imaging alike.[28] And further, in this case: between the medial orientations of celluloid and pixilation.

Not for long. Fragments of dialogue soon make clear that this is indeed an electronic and virtual landscape within the diegesis, if not quite within the plot: indeed, inside the cockpit of a flight simulator (fig. 2.19). Only as the camera pulls away from the eerie framed landscape do we realize that an unidentified woman, seen only from the back, is taking a pilot-training lesson. Uncanny terrain has thus been explained away as the reductive "interface effect" of sheer interactive technology. We quickly cut away to a narrative that has nothing to do, as narrative, with this visual prologue. No continuity permits us to contextualize, or helps us to remember at all, this throwaway opening. Only at the end, after the bitter complications of a backfiring terrorist plot, can this getaway story come to a head with the apotheosis of the flight-trained heroine (we may at last recall) escaping with her new lover from a barrage of police bullets. The couple takes its flight in a stolen helicopter at the site of their attempted arrest. In a perpendicular shot from beneath, this literalized deus ex machina lifts straight up from their certain death in the line, and roar, of automatic-rifle fire—out of narrative time altogether—into a silent sky above (fig. 2.20). And then vanishes as figure into its own boundless ground—a kind of corporealized fade-out ad infinitum. Heaven, again, and this time to be spelled out ironically in bold letters just before the end credits begin to roll past. Thus has the *Liebestod* ambush from a film like *Bonnie and Clyde* been translated to high technology and doomed liftoff. As the helicopter dims and shrinks in this way across the threshold of visibility, we are left not just with a symbolic vacuum of event but with a prolonged shot of sky that grows barely distinguishable from the freeze-frame hiatus of cinematographic rhetoric. Backing out of narrative, this spectral prolongation of the seen—echoing the virtuality of the prologue edging into it—offers a case of "figuration" degree zero.[29]

Visibility has in itself been rendered virtual rather than real at the start of Tykwer's film, then removed altogether at the end for the very vehicle of its plot: twin optical liminalities converged upon the mordant cross-purposes of the story in order to break free from its temporal deadlock. The subtextual development is familiar enough in the abstract. The inferred matrix of illusion per se, having found its liminal model in digital simulation, returns in the end as a lethal irony and a fantasized reprieve at once. Kieślowski's script, in Tykwer's hands, faces up to the very fact of its own dead end with an admittedly magic loophole rather than a full revisionary twist. It is a climax that Žižek would certainly be right to understand as giving full weight to the ethics of futility in the terrorist impasse of plot, even while conjuring its fantastic alternative. For, in the last shot, this doomed escape has been reconfigured in another axis as a quite literal flight from the real: a slow, gradually soundless, utterly incredible, and yet again wholly virtual (though no longer electronic) transfiguration of a death moment into a vision of continuance beyond the threshold of narration. The longer the shot is held, well beyond any last visible speck of the copter, the more its supernatural movement cedes all fantastic parameters to a terrestrial time-image: its duration wholly impersonal now, just barely discernible shifts of clouds in the course of diurnal motion. Precipitated out from event and meaning alike, time is yielded up to an all but fixed image of its own minimal passing.

If Medem and Ruiz and Tykwer, building more or less explicitly on Kieślowski, thus move far toward the realization of a pure "time-image," and hence an image at times of timelessness itself, American cinema takes a different tack in this same fin-de-siècle period. Its plots often serve, instead, to undermine the "movement-image" in its credible wedding to lived duration, as in the two prototypical instances to which the next chapter will soon turn: the digitized *Matrix* and the supernatural *Sixth Sense*. Before crossing to these examples, a last European exhibit. Beset neither by ghosts nor digital figments, the plot of François Ozon's *Swimming Pool* (2003) has nonetheless moved closer in spirit to the outright trick ending of Hollywood sensationalism rather than the recursive ambiguities of its European counterparts: closer, that is, to the radical ambivalences of the pure fantastic in its ultimate plot exposition (and final exposé). Then, too, the film begins by deploying, in two stages, a rather muted version of those trick beginnings that are as common to the thriller plot as are trick endings.

Swimming Pool opens behind the title with a presumed titular (that is, eponymous) shot of blue water filling the screen frame. An upward pan, however, reveals it to be a bracketed image of the Thames rather than of a contained pool. The film then cuts to the Underground, where a reader notices that the woman sitting opposite her is the author whose picture is on the cover of her mystery novel. Yet here, too, ocular mystery and deception seem to have set in. For this photographic evidence of the senses is immediately denied: "You must have mistaken me with someone else," says the presumed author dismissively. "I'm not the person you think I am." Index does not guarantee identity. A throwaway moment, one assumes—easily explained away by the author's revealed panic over her current writer's block. For the rest of the film, though, we submit to an openly voyeuristic study in voyeurism, only to find out that most of the characters aren't who *we* think they are either, but, rather, erotic projections of the writer as she hallucinates herself into the world of her new murder mystery. Derived ultimately from the play between virtual and real in Fellini's frustrated-filmmaker plot, *8 1/2*, and thus anticipating the fantastic overlaps of memory and refilming in Amodóvar's *Bad Education*, the premise of *Swimming Pool* is closer yet to the blocked-insomniac-screenwriter plot of Spike Jonze's *Adaptation* from the year before, where scenes are repeatedly virtualized as the film we're watching before—under entirely subjective refocalization—they are found being inscribed by sudden desperate inspiration in the hero's sketchy scenario. But the French film is actually closest yet, in its violent plot turn, to the stalled-author fantasies of Stephen King in their on-screen realizations, from *The Shining* (1980) through *Secret Window* (2004), released the year after *Swimming Pool*, where the nightmare obverse of writer's block is personified in a hallucinated double who levels a charge of plagiarism at the suspense writer while enacting for him a bloody revenge—based on their shared fictional fantasies, of course, but this time perpetrated in the real.

In *Swimming Pool*, after a violent homicide is covered up by the detective writer herself, the "switch" ending arrives in alternating match cuts of a real and a fantasy figure in the writer's mind's eye: the imagined sexpot Julie (fig. 2.21) and her editor's actual daughter Julia (fig. 2.22). With each reaching out in silhouette as if at the rear projection of an absent mother on the distant balcony (fig. 2.23), the shot evokes the classic opening titles of the child stretching his hand toward the imago of the mother in Bergman's 1966 *Persona* (fig. 2.24). Thrown by the disclosure at the end of *Swimming Pool*, we may be cast back to the film's trick beginning: that dodge of identity which now seems to have infiltrated the entire film from both the (destabilizing) establishing shot of rippling water and the subsequent photographic moment of indexical denial. In any case, the narrative game is finally up. We now see the forestalled truth—and seeing is indeed the mode of disclosure, as before it has been the means of deceit. In a bizarre overlap of wish fulfillment and revenge against her agent, we realize at last that the female creator (who began by denying her own authorship, as identified by photography) has mothered a lascivious spectral daughter

2.21–2.23

into projected presence, only for that figment of the unconscious to be revealed in the end as a libidinal phantasm. To fill out the imagined pages of her new novel, in other words, though the results are scarcely contained within them, the writer has brought this misbegotten fictional child into being out of a repressed desire for sexual contact with the father. Fictional conjuration thus stands revealed as halfway along the path to psychotic doubling. Filmic mirage is, in both senses, the *medial* function: both the median term and the apparitional form of its materialization. All one needs to do is factor in an electronic rather than wholly psychosomatic virtuality, in the Hollywood manner, and the problem of technofuturist hallucinations would take its familiar turn from epistemological mystery into ontological subterfuge.

What narratology would call a "garden-path narrative," the one fork provisionally taken being the road not really traveled, is what narratography would read in medial terms as a disjunction in the ocular unconscious itself.[30] In *Swimming Pool*, unlike in *Persona*, the filial images (daughters here rather than mothers) are not superimposed upon each other, let alone laterally collaged into each other. Their images alternate by alternation alone: the quick cut of montage point blank. Real and unreal girl (e)merge at the end, that is, by the false suture of hallucinatory reverse shots in the writer's reconsidered London memory of her self-invented Provençal intrigue. Virtuality redoubled finds its perfect technological articulation in filmic process: montage as pure *trucage* (Metz's dictum once again), or, in other words, editing as the primal filmic trick of screen narration. Virtuality is more likely to find its comparable articulation, in Hollywood cinema, through the interventions of the digital. Short of this, all it would take is for the personified figments of the female novelist's gestating brain to recognize

2.24

themselves *as such*, and we would tilt over from deferred epistemological irony into a prolonged ontological quandary more typical (though usually by last-ditch reversal) of the Hollywood fantastic, as with the hero's anxious plight in finding himself the invention of a writer he's never met—and hers in discovering him real—in *Stranger than Fiction* (Marc Forster, 2006).

I introduced the matter—and filmic materiality—of *trucage* early on in this chapter, alluding in the process to its updated digital equivalents as well. This is because its often roughened or eroded representational grain, or otherwise its picture-perfect illusionism, marks one of the leading places, especially in interpretations of the fantastic, where narratography catches hold. Yet the case for narratography must make room for itself within a larger methodological cross fire for which other aspects of the present chapter have also prepared terms. Merely to mention "interpretations of the fantastic" is to point to one of those terms. The plots that ask this of us do so right off the bat. A weird dislocation of normal reality sets in. What is happening? What does it *mean*? That essential hermeneutic question—raised (as Todorov saw) to metatextual prominence in the genre of fantasy—is submitted by his analysis to structuralist determinations. To these we have also added the micromanaged semiotic coordinates proposed by Michael Riffaterre. Yet exactly this motive of narrative interpretation per se is what more recent media study may often seem to find not only dated but distracting, prone to deflect emphasis from the medium itself to its broadcast meanings. Narratography prevents this deflection, or, better, resists it—resists it along the very line of resistance set up by the image frame itself, whether in mechanical transit or electronic transformation. And it does so in part by objecting to the complaint against interpretation in the first place. Why should we think that trying to analyze, with our eyes open, the often contradictory message of contemporary screen fantasies doesn't put us more directly in touch than otherwise with the medium they manipulate? And with its history. Rounding out the analysis so far, let the following chapters on postrealist Hollywood films cement by example a practical answer to this anti-interpretive stance before we round up its underlying assumptions for further cross-examination in the last.

THREE

OUT OF BODY IN HOLLYWOOD

In turning to the Hollywood fantastic from recent European experiments in the uncanny, discussion is drawn again toward the paradoxical temporality of retroactive initiation: the end that rewrites the beginning. This format grows definitive. It arises, in its vicious circling, both as a narratological irony and as a psychosomatic crisis for characters within the plot. And it is registered on the stylistic surface in moments of narratographic dissonance. Here are beginnings at times so thoroughly tricked that when their "switch" is finally pulled, and their ontological reversal pulled off, we realize that the protagonist whose quest they launch was never there at all—never anything to begin with.

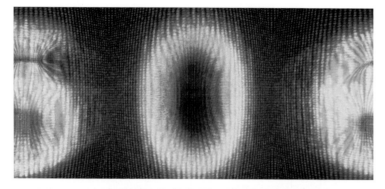

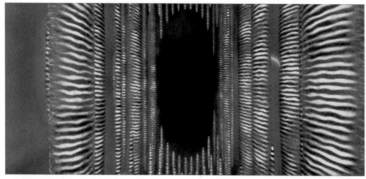

3.1–3.3

Swimming Pool is, as suggested, something of a maverick in the European context. And instructive as such. Despite the strictly filmic rather than digital *trucage* of its hallucinatory closing suture, the film feels closer in its contorted trick finish to recent Hollywood gimmicks, where digital or gothic paranoia can go so far as to reduce all human relations, in fact all biological origin, to the illusory. The renegade nymphomaniac daughter has all along been no more real than a digital hologram: that would suggest one Hollywood variant of such French hallucinatory eroticism. Certainly, in any attempt to adopt Riffaterre's vocabulary to such films of ontological subterfuge in the latest Hollywood mode, one among them comes first to mind. In testing the relation of suppressed matrix to instancing model in a film by the former title, we note how the first diegetic space of 1999's *The Matrix* (Andy and Larry Wachowski)—after an apocalyptic distortion of the Warner Brothers logo—emerges from the racing electronic array of a phone-trace system. The camera zeroes in, quite literally, on the zero integer between 5 and 6, burrowing through to a match cut on a policeman's circular flashlight beam (figs. 3.1–3.3). As model, this liminal shift from digital to visual works to keep under suppression the deferred matrix of digitization all told: the fact that extensive sectors of the human world in this film, including its habitable metropolitan space, are entirely a computer projection manipulated by robot overseers. The genre threat is ingrained: humanity under siege from a wholesale technological simulacrum. Including trick beginnings and endings alike, this genre subtype has proliferated for more than a decade.

MATRICES OF THE VIRTUAL

In an agenda-setting virtual reality film from 1997, *Dark City*, the liminal clue is seeded when a black screen cuts to the sudden projection of a star-thick cosmos over the transition in voice-over from "In the beginning there was nothing" to "And then came the strangers": namely, that alien race that has entirely projected the digitally simulated human environment of the *fabula*. In the revelation scene, a double-layered deployment of the "interface effect" has both the hero's body and its shadow, his empty silhouette, laminated upon the falsified slide image

of his own previous self. This is a projected still shot of his boy's body marked by the fire that supposedly killed his parents (fig. 3.4). The whole ontology of the photograph, in Bazin's sense, is dismantled, however, for the image is itself an orphaned signifier with no source or backing in the real, its icon of a wound bearing no index or scar on the present body. The hero of *Dark City* has seen this coming. For he has become, over the course of plot, the sleuth of his own elusive footing in the real. This narrative format has become more and more common in the years

3.4

since. Building on Todorov's acknowledgment that detective fiction is often thought to have replaced the fantastic in the popular imagination by shifting the hermeneutic of mysterious causality from marvelous to natural lines of investigation,[1] we can see a perverse reconsolidation of genre alternatives in recent trick plots. Heroes or heroines (*The Sixth Sense*, *Vanilla Sky*, *The Others*) hunt down the mystery of ghostliness, or other related anomalies of embodiment, only to find that they have exposed the secret of their own previous and until-now unrealized murders—in the latter two cases, with themselves as suicide victims.

In Hollywood cinema, the virtual self has therefore taken many inscrutable forms in the years since *Dark City*. In the process, as we began to see in the last chapter, a mainstay of the science fiction genre has been given either short shrift or a decisive new purchase in the epoch of digitization. For decades in this genre, the evidentiary photograph marked a fixed point-of-no-return in the dystopian technology of artificial bodies and environments. Several of the films taken up in this chapter are the structural flip side, and the cultural aftermath, of this generic trajectory. Traditionally, the photographed human body—having once been present to representation—is deployed on-screen to measure the fall from human agency to fabrication, as from organic body to cyborg in *Blade Runner* (Ridley Scott, 1982). But this particular significance of photography grows either vestigial or at best equivocal within a fully electronic dystopia, not to mention within a cinema increasingly enhanced—or commandeered—by the digital. There are no faked or ambiguous or nostalgic photographs, for instance, anywhere in the fully achieved electronic premise(s) of the *Matrix* trilogy—as there were, in its dystopian cybernetic precursor, the transitional *Dark City*. When the indexical image does occur in either digital sci-fi or its contemporary fantastic counterparts, photography may still operate to anchor (in however evacuated a fashion) the ontological gothic in either of its two common forms: sci-fi in the virtual-worlds mold (from *Johnny Mnemonic* through *The Thirteenth Floor*) or supernatural thrillers in the phantom-aftermath mode (from *Jacob's Ladder* through *The Sixth Sense* to *The Jacket*). In such narrative formats, where the hero is either cybernetic or ectoplasmic to begin with, a ghost either in or out of the machine, the photographic index of an inhabited body is often a touchstone of the crisis. This tradition finds its most famous precedent, no doubt, in the dubious maternal photograph of the cyborg in *Blade Runner*. Many a plot since then turns on false photographic witness: the record of a supposed past that is only part of an encompassing digital or hallucinatory simulacrum.

With the prominent exceptions of Spielberg's photographic nostalgia and its mortuary lyricism in both *A.I. Artificial Intelligence* and *Minority Report*, the photographic benchmark of presence within a surrounding digital *trucage* has certainly been marginalized by current sci-fi. Yet it has not disappeared. To some degree it has simply shifted genres: most obviously when it leaves behind a context in high-tech illusion for a narrative format of high-profile magic. In the *Harry Potter* films, for instance, every framed image in the land of wizardry, whether ancestral portrait or newsprint mug shot, can be found to writhe and speak within its curtailing rectangular frame—and to do so precisely as a self-referential marker of digital cinema's power to bestow such magical animation in the first place. Similar if muted effects punctuate those less strictly fantastic films, as well, where plot tends to put in doubt either the ontology of its protagonists or the epistemology of their vision, memory, and desire. As such, the current spectrum of the unreal or imaginary runs from the blatant ghost stories and digitized nightmares of Hollywood spectacles to the often elegiac uncanny of European practice.

A few more examples, from the Hollywood side, should then permit generalization. David Fincher's 1999 *Fight Club* is an unmistakable version of the *Jekyll and Hyde* story, of course, with Brad Pitt as the violent double, Tyler Durden, of an unnamed insomniac narrator, to whom his alter ego appears in flash insets (just a few photograms at a time) well in advance of the star's official "entrance." The film's liminal scene, before a plot-long flashback that circles round to it, has emerged, at high speed, from the electromagnetic circuits of the narrator's brain behind the credits—out through his nostrils and along a gun barrel stuck down his throat by (we much later learn) the materialized alter ego: the unsleeping dream self made flesh. Our next best clue to the displaced id embodied by this second self comes when the narrator turns for the first time toward the camera to explain more than we know he knows about Tyler, whose moonlighting as a projectionist allows him to splice subliminal footage of porno-film penises into standard family fare—just as the photogrammatic traces of his own extruded phallic presence have before been "flashed" at us below the threshold of full awareness: ours or the narrator's own, from whose unconscious this new potency emanates. Confirming a deep link between cinematic apparatus and psychic mechanism, therefore, is the repressed matrix as a founding play on "projection" itself. Focal point of this function in both the technical and the psychoanalytic sense, the film's star image will eventually be disclosed as just a figment of desire spawned in collusion—and what else is new?—between narrator and viewer. This ontological ruse has become so much a staple of recent Hollywood filmmaking that even the Academy Award–winning biopic *A Beautiful Mind* (Ron Howard, 2001), with its schizophrenic math genius as hero, participates in this new genre variant of fantasy by duping the viewer, for most of the film, into believing in the hero's dashing roommate as a fellow student rather than projected alter ego, to say nothing, at the level of Cold War paranoia, of the hallucinated CIA operative who motivates the ultimately undermined thriller plot.

More recently, the updated logic of brainwashing by communists during the Korean War in Jonathan Demme's remake of John Frankenheimer's *The Manchurian Candidate* (2004) has the cover story of martial heroism instead "wet-wired" by computer chip (in the language of cyberpunk) into the assassin's body. Time has been rewritten, memory fabricated. On just this score, the narrative takes a curious wry detour during a long expository sequence in which we learn, from a consulted scientist, that such technology, possible "in theory," is nonetheless "not supposed to exist" as yet: the near edge, in short, of the fantastic's "instrumental marvelous." In this downbeat exposition scene, the film goes on, almost recklessly, to raise the specter of an even more encompassing digital illusion—only to normalize the story from there on, and all the more firmly, in the mode of a genuine political thriller of military-industrial paranoia. For the cranky scientist asks the hero if he has considered the possibility that, instead of his suffering from actual bad dreams about his traumatic time in the Gulf War, he might in fact still be back in the desert, hallucinating the whole ongoing detective plot—including the present interview—by way of electronic implant (the digital version of a wartime death fantasy like that of *Jacob's Ladder* or *The Jacket*). The throwaway suggestion works by reverse psychology on an audience otherwise all too prone of late to harbor such suspicions on the way to postponed disclosure.

The film's actual political edge is thus sharpened by this bluntly raised alternative in the digital gothic. Most of the time it is otherwise. Audiences wait patiently for the unmasked deception they half anticipate, suspending not their blanket disbelief but their active skepticism, disavowing the subgeneric code of the trick ending in favor of a complacent investment in an *un*reality eventually disclosed as such. Then, too, beyond this extreme but overruled extension of the time-image as memory implant in the red herring of *The Manchurian Candidate*, there is another striking reversal in Demme's plot that connects with a more urgent dimension of the digital intertext in current American culture. By the closural resolutions of genre plotting, the film seems to be working to defuse certain anxieties over both invasion of privacy and unconstitutional violations of evidentiary protocols, all within the narrative's general ambience of antiterrorist demagoguery. This is the way the ending comes across when registered, in short, narratographically. For in the film's surprise finish, FBI technicians are seen electronically altering video-cam footage of the already digitally programmed but still embodied hero—before sending it out for TV broadcast. This is the surveillance video automatically captured when he entered the convention hall bent on assassination. What the FBI does is to substitute his image with the electronic scan of a known criminal employee of the global Manchurian Corporation, thus planting deliberately false and incriminating evidence of, in fact, the true culprits. With this twist on an ethics of simulation, the means are presumed to justify the end. From within a video transmission, that is, the further machination of the digital comes to plot's illusory rescue as the counterespionage image of a tricked real-time event. By graphic ingenuity alone, a broad-based societal paranoia is turned inside out to poetic justice.

But what if the hero himself of Demme's update—rather than either his whole simulated spatiotemporal environment since the war, as sardonically proposed early on, or only his digital body double at the end—were a disembodied figment of the imagination? This is the kind of further turn that the ontological ironies of recent Hollywood thrillers have taken. In contrast with those characters in European cinema who may seem burdened with some telepathic empathy or intuitive "sixth sense," the American film by that name turns this extrasensory radar on the protagonist himself. The liminal moment of diegetic image, emerging from black leader, is the preternaturally slow incandescence of a close-up lightbulb (fig. 3.5). Exactly what gets modeled by this slow-motion image in *The Sixth Sense* is the repressed matrix of temporal protraction, distension, or deferral that takes narrative shape in the plot-long dilation of the

3.5

3.6

3.7

hero's death moment. Bruce Willis is a famous child psychiatrist shot by an embittered former patient in the first scene. Assuming his recovery, we next follow his efforts with a troubled boy who gradually admits to seeing unacknowledged corpses all around him, specters desperately clinging to life. As discovered only in the film's closing moments, the hero is one of them. At which point his image bleaches slowly to a white screen (fig. 3.6): the syntax of fade-out as a *trucage* of supernatural removal, perhaps, or merely the hero effaced by the recognition of his own absence. In either case, out of this irradiated blank screen materializes, on a full-frame interior screen, his wedding video (fig. 3.7), which we now realize that his wife has been replaying at intervals, not in estrangement but in mourning. With no one there to watch this time, the memorial image lingers beyond suture, beyond even interface. Like a deathbed reprise. Modeled in the first instance by the decelerated ignition of the lightbulb in the emblematic basement space of the hero's own unconscious, this delayed fuse may by now be referred to the repressed matrix of "seeing the light at last."

But that light has taken a different form halfway through the plot. It has in fact taken on a specifically photogenic cast—and this in a strange episode of spectral record. Looking at a wall of her child's photos, the clairvoyant boy's mother notices for the first time that each image is streaked by a flare of light near the child's head—almost like a reflective glare on the lens, but hovering in free space (fig. 3.8). A three-di-

3.8

mensional flaw, a ghost, an aura, a cultist's orb. By some unexplained transference, the child's extrasensory vision thus seems displaced onto photography's own alert viewer. In the combined vocabulary of séance and image science, of sorcery and photochemistry, it is a case of the "medium" remediated. But so was that lightbulb at the start, whose slowly graded motion toward full glow did more than reflexively allude to the seepage of the projector's beams through the gradually more exposed areas of frames on the strip. The allusion was also historical, archaeological. Edison's remediation of nature's medium of visibility, light itself, by its electric recontainment (the form of the bulb embedding the content of a previous medium) offers indeed the opening example of Marshall McLuhan's *Understanding Media*, a moment from which Bolter and Grusin no doubt take their lead in *Remediation*.[2] And if the first of the material senses can be mediated, why not the immaterial sixth? Or reverse the question: if the parapsychic power of telepathy can conceivably be remediated (with its transferred affect and its virtualizing of an absent present), where is this more likely to happen—photographs of a psychic child aside—than in the transferential investments of everyday film watching? Examples? This chapter is full of them.

In reviewing complementary modes of the fantastic across national and cultural borders, it is no doubt irresistible to contrast a European virtual-reality film with its Hollywood remake: Alejandro Amenábar's *Abre Los Ojos* (*Open Your Eyes*) with Cameron Crowe's *Vanilla Sky*, 1997 and 2001, respectively. The Spanish original, for all its sci-fi overtones, is concerned more with sexual fantasy, erotic memory, existential choice. In the liminal cut from black leader, the hero dreams of awakening to the nightmare of a world without others—only for him to wake yet again into a reality of obsessive narcissism that amounts to the same thing. So it goes, until he is nearly killed in the suicidal car crash of a stalking lover. Nearly killed and largely disfigured. When reconstructive surgery fails, he finally kills *himself*. But—here is the trick—he discovers this suicide, along with the viewer, only in flashback. A century later. At this ontological turning point for the film as a whole, he learns that he had contracted for "cryogenic resuscitation" once the DNA technology was perfected. In the meantime, with his body lagging behind,

3.9

his reborn consciousness has been seamlessly "spliced"—that's the metacinematic word, in the remake too—into his own past, so that the life he thought he had been leading, rather than just dreaming, was as unreal as the stitched-together artifice of any film, our film in particular. Within this wish-fulfillment fantasy gone awry, he now jumps to his death from a skyscraper precisely in order to end the self-scripted dream, choosing a contingent reality instead. Blackout. Back to the liminal voice-over: "Open your eyes." Vicious circle or heroic redemption? Tone suggests the latter.

So too in the remake. But structure complicates, since the Hollywood version has indulged in a more ambiguous subtext of filmic evocation. The revelation scene that divulges the hero's artificially spliced resuscitation occurs this time in a glass elevator whose visual ascent is interrupted at pulsing intervals—as if by the bars of celluloid's vertical advance—by the regular occlusion of the view between floors (fig. 3.9). The scene involves not just a filmic disclosure of this sort, however, but a cinematic—and a cultural—one.[3] The life that the hero (played by Tom Cruise) has dreamed for himself has been (and who would doubt it?) artificially spliced not just in form but in content, modeled on his own identifications with image culture, including formative screen narratives flashed now before his mind's eye. These cultural intertexts include *To Kill a Mockingbird*, which has generated—through its image of a good father—the hero's imagined psychiatrist in the dream plot, and also *Jules and Jim*, whose poster we have previously seen in his apartment, and whose famous freeze-frame on Jeanne Moreau—in a scene rerun, in flashback, until its fixating arrest—has locked into place for him the fantasy of an exotic foreign lover realized by Penélope Cruz: the biotechnological version of a film-come-true.

Is Tom Cruise going to give up the dream and live his life in the real? "Your panel of observers," as his adviser gestures straight at the camera, "is waiting to see what you will do"—a suspense that figures in the film audience by proxy across the prevented suture of the screen's own barrier. When the hero opts for life over death by jumping free of the dream, the "leap of faith" (as repressed figurative matrix for his coming new existence) sends him plunging down the former trajectory of the elevator as if, with floor after floor flashing past, the mechanical increments of the vertical filmstrip were racing up to meet him. Intercut with this fall are dozens

3.10–3.11

of rapidly edited images in review of his past life, interspersed with the cultural life of his century, represented mostly by still photographic images, Sinatra, Dylan, and Martin Luther King included, all racing past toward the home-movie retrieval of mother and son together—and on, once again, to the negated projection of a white screen (rather than Amenábar's black). Yet once more, under pressure of death, the anti-Bergsonian spatialization of *durée* in compressed montage converts on the spot to something like a Deleuzian "time-image"—for the time that is up.

A further angle of approach suggests itself. In our transcontinental comparison, what happens when the Spanish director of a virtual-reality film partners with Hollywood for a gothic assignment in the new mode of ontological paranoia? The crossover is revealing. Again the black leader from *Abre Los Ojos* at the start of Amenábar's *The Others* (2001), but now with a children's story read into a vacuum until the tale's emergent illustrations then fade—by a match dissolve—from the lithograph of a rural mansion to its photographic realization in diegetic space. Erupting into this establishing shot, a further model of the hidden matrix: Nicole Kidman's agonized scream, her head upright in terror (fig. 3.10). But then the camera immediately straightens out the image toward the horizontal (fig. 3.11), so that we realize that she is waking from a nightmare. Or is she? This momentary *trucage* of artificial verticality, a mere trick of "style," is a true model shot. For it is picked up later by way of subtextual recurrence when one of the mysterious new servants in her household explains what Grace (Kidman) has found in the attic: a "book of the dead," with posthumous daguerreotypes of corpses propped up in chairs or beds—including (with shades of the opening unseen recitation) a transfixed reader. These photochemical records have been preserved out of a superstitious belief, as explained by the new servant, that photography could immortalize the soul.

3.12–3.13

That subtext of illusory vertical life returns soon again, just before the real trick ending, at what we temporarily assume to be the climactic revelation scene: the realization that all three new servants are in fact ghosts. For Grace stumbles upon a hidden daguerreotype of the servants, too, as trussed-up corpses (fig. 3.12). It is as if photography, in its evolved form as film, has indeed performed its supernatural magic by keeping them in artificial animation before our own eyes, a trio here panned across (as often before) while their images fill the screen (fig. 3.13). But this apparent resolution now passes through a further loop. (One hates to give away the plot—but what else are liminal shots there to do anyway?) What we soon discover is that Grace and the children, a fact unknown to themselves, as to the hero of *The Sixth Sense* before them, are ghosts too, made to look vertical and animate only by the film's own gothic premise. She has killed both them and herself. The whole plot is merely the negative "grace period" during which she and the audience come to realize this. Here we recognize further that the true model for the film's generic bearings lies back beyond the first scene of illusory vertical trauma, back beyond even the precredit launch of invisibly controlled storytelling, in the ambivalent title itself. For the phrase *The Others* is now recognized to have suppressed for most of two hours its inherent grammatical reversibility as a designation both relative and reciprocal.

Hence the finale, which confronts the new mortal tenants of the estate, the previously supposed ghosts, with those lingering presences who have made up our film's main cast. In the penultimate scene, Grace and the children burst in upon a séance held to contact them, where a witchlike medium and her assistant transmit via automatic writing the unheard screams of "We are not dead." What follows is the ultimate gothic match cut. By precise digital editing, the image

3.14–3.15

jumps from Grace, ripping up in midair the frenzied transcripts of her denial (fig. 3.14), to her body's instantaneous removal from the scene of shredding pages (fig. 3.15)—a violence, to living eyes, with no agent now in view. At work here, in a word, is cinema's own electronically aided "mediumistic" intervention in narrative sequence. Deleuzian categories have once again imploded upon each other in a single Metzian *trucage*. Existence—or, in other words, sensorimotor extension—has become in itself a trick played on its agent, whose duration as living person is entirely virtual.

So too in *A.I. Artificial Intelligence* (Steven Spielberg, 2001), where the robot-hero believes for most of the film in his own yearning humanity—and so sustains an eerie commentary on cinematic spectatorship. In his powerfully emotive performance, that is, this wholly credible robot child is no more real, but no less, than any other actor in the conveyance of an emotion not really his own. All of this develops according to the familiar threefold coordination of a liminal model, its repressed matrix, and the later recurrences of the subtext—the last taking the form in this case of an arresting and two-stage interface effect. With over two hours spent in a harrowing search for the Blue Fairy out of *Pinocchio*, who, it is hoped, will turn the mechanical boy real, what the robot child has forgotten from his dawning moment of artificial intelligence is exactly what we have forgotten from the film's first frames: the art-deco statue, just outside the robotics lab, that is the narrative's liminal shot after (at first superimposed upon) its oceanic prologue about global warming (fig. 3.16). This is the founding image to which the robot returns at the film's emotional climax. We have already realized in a much earlier panning shot that the artificial boy has been modeled on the lead scientist's dead son, who grows up before us in a gallery of photographic traces (fig. 3.17). A whole medial

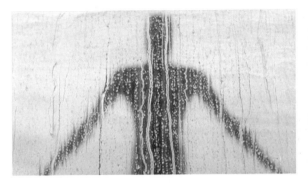

3.16

3.17

3.18–3.19

epoch seems to be edging past us in that leftward pan as it swallows up, one by one, those framed photos in its wake, and by association those photo frames: an epoch not just in the evolution of a species and its duplicates, and hence in the arsenal of a genre's tropes, but even in the photomechanical basis of its own narration. The robot himself only realizes his duplicate status in a comparable leftward pan across both these same photos and then his upright replicas in the discovery scene. From there he moves to recognize the lost female image of open-armed embrace. This happens in a densely layered "interface" sequence in which the constructed subject gradually approaches (fig. 3.18) and then moves to inhabit the mask of his own prosthetic existence—only to peel away from this technological double in a sutured but inhuman interchange with the industrial figurehead of deco modernity (fig. 3.19): that first image which he, in his unmothered laboratory origin, would once have seen and inevitably forgotten, repressed, displaced.

The film now moves into its supplemental coda, which replays this same logic at a new level of technological interface. Robotoid creatures two thousand years into the future have not only thawed out the primitive frozen robot but melted into rapid montage—in one visualized flashback segment after another—his whole unconscious memory bank: in sum, the last surviving record of the human race. Before the adoptive mother's actual DNA reconstruction

3.20

in the closing scene, electronic retrieval allows for a rapid montage of the boy's existence. This all flickers past as if in a classic death-moment reprise, channeled through the communal circuitry of the alien brains and projected onto their nearly faceless surfaces, one to another, first by touch (fig. 3.20), then by remote transmission. By such projections of the robot boy's own hardwired unconscious, he encounters the recovered image of the mother in the mirror of his own digital storage. Kubrick crossed with Spielberg—maybe *this* is the most Proustian moment in contemporary cinema: *le temps retrouvé* of the whole human species, screened by an apocalyptic interface with the Other.

GENRE, SUBJECTIVITY, TEMPORALITY

If *A.I.* gives us the time-image at the vanishing point of human time itself, many another film has posited, and struggled to figure, the far horizon of a single human duration. So where does this leave us so far in the comparative assessment of contemporary fantasy films and their tendency toward the aggressive reframing of time? In the narratographic terms attempted here—with emphasis on the liminal shot, itself based on classical film prototypes—we have seen how subsequent ramifications of an often forgotten model in the ensuing subtext will set the terms by which the contemporary equivocations of fantasy will eventually get resolved. Resolved not just over time, but in time's terms. These ambiguities will tend to sort themselves out either via the uncanny—as in the European cinema of fateful coincidence, erotic reverie, and mysterious second sight—or via the supernatural, concentrated of late in American cinema of the ontological gothic. At either pole, the subtext stands to its narrative materialization like latent to manifest content within a surface/depth model of consciousness. If we understand time itself to be produced in this slippage of topographic levels, as Freud did, then we can see at once—or at least with some further reflection—the special narrative purchase of a screen genre like the fantastic that so often reframes time in preternatural terms.

Within broad theoretical questions about temporality and its narrative variants, the matter of genre invites on its own terms some added thought—and precisely in regard to the place of the body as anchor of temporal consciousness: the body per se, its inner psychosomatic estrangement, and its outright carnal doubles. In tentative application of a major psychoanalytic dyad, one comparative hypothesis might go like this: On the one hand, if the resolution of fantastic material is achieved at the level of the uncanny, then the symptom returns from suppression in the form of a mostly *neurotic* disturbance, an obsession or sexual fixation or fateful anxiety manifested by some eerie knot of plot. If, on the other hand, the fantastic leans toward the marvelous for explanation, an abrupt departure from reality's cognitive paradigms, as in the latest Hollywood productions, then the returns of the repressed matrix betray the extremity (or at least its technological equivalent) of a *psychotic* break with the world—a matter of schizophrenia, paranoid delusion, or, more often, their digital equivalents explained away at the "supernatural" level of as yet unrealized electronic marvels, disabling the entire reality principle of plotted event.

Here it is necessary to recognize the way in which this distinction cuts at an unexpected diagonal across the related use of this Freudian dualism advanced by Todorov in the later chapters of *The Fantastic*, where he leaves behind the strict anatomy of a genre. His purview eventually widens beyond the fantastic, generically conceived, to a broader literary taxonomy of thematic constellations. We arrive thereby at a potentially confusing case of not wholly congruent dichotomies, as follows. Without intending to divide the issue between the uncanny and the marvelous per se in these closing stages of consideration, Todorov goes on, quite independent of this initial binarism, to isolate two main thematic networks not just in fictions of the fantastic but in literary narrative more generally: those of the self and those of the Other, inner- and outer-directed, respectively, concerned now with perception's relation to the object world, now with desire's.

Todorov's explicit turn to psychoanalysis at this point is indeed motivated in part by his sense of genre history and the transitory moment of the fantastic, which waned with the nineteenth century. For it is the arrival of psychoanalytic paradigms to account for all of human anxiety (in other words, the new century's global extension of the uncanny as a psychological field of explanation) that explains for him the dispersion of fantasy—and in turn the outmoding of its genre—in modernity. Until then, and as strictly literary rather than scientific material, "themes of the self" operate at the level of ego and cognition, even in such radical or psychotic metamorphoses as those in which a subject encounters its own double. By contrast, the "otherness" of desire and death incurs the operations of the unconscious, where it is typically neurosis, rather than psychosis, that harbors the ingredients of its perceptual unease (146–47).

For Todorov (though he neglects to say this in so many words), either theme might—when housed in a fantastic narrative—find its ensuing interpretive confusions ultimately resolved, as sheer plot mystery, at *either* pole of genre determination: by recourse to the uncanny on the one side, by the advent of supernatural clarification on the other.

One shape this process can take involves just the ambiguities we have been charting as they press toward interpretive closure. In the European cinema of the uncanny, for instance, it is usually some neurotic complication of psychic life, often an incomplete mourning trapped in melancholia, or an Eros undetached from Thanatos, that either dissipates the mystery in resolution or explains away the spectral coincidences. Oppositely, the trick endings of the Hollywood gothic or its technofuturist equivalents less often suggest a perturbation of unconscious desire than they mark a rupture in the perceptual apparatus of human consciousness. This, as we know, can be either induced by digital implant (if not englobing simulation) or brought about by death and its "belated" (posthumous) detection. In its oblique angle to Todorov's secondary Freudian dualism of thematic tendencies, then, fantasy emerges, in its own right, as the often prolonged equivocation of the very difference *between* self and Other, between agency and its manipulation from without, between identity and its perverse constructions, whether psychosomatic or machinic. In recent Hollywood films, such ontological uncertainty finds manifestation in the contested difference between the aberrations and reversals of desire on the one hand, the schizoid externalization of self by either ghost or digital simulacrum on the other. As the next chapter will examine in some detail for contemporary modes of the fantastic, all this can carry with it into plot an often unspoken cultural intertext: namely, the displaced agency of computer interface in the role-playing maneuvers of video games.

With either of those alternative Hollywood options, whether supernatural phantom or cybernetic double, a long-standing confidence in the modus operandi of character-driven plot has been eroded. A point of comparison in previous American cinema. In the political thrillers of 1970s Hollywood, even before the digital dystopias of 1980s sci-fi, paranoia plots operated to convert, as Fredric Jameson saw it, a former modernist totality to a closed system of unseen surveillance and control.[4] In the decade since he wrote, this syndrome has been transferred into a pervasive nervous uncertainty about narrative causality per se. One whole story repeatedly gets laid bare as the cover for another that undoes it. It isn't just that ghost stories have made a comeback. Current Hollywood production often feels haunted by the ghost of traditional story altogether, its narratives currently unfurled only under the sign of their own ultimate negation.

There is, moreover, a metageneric issue at stake in such plotting that extends beyond any pointed consideration of the fantastic. The old argument about classic genre narrative is well known: how it attempts resolving cultural contradictions that can readily surface, one might add in our context, as a visual instability in a given film's liminal shot. But this paradigmatic tension has evolved lately into a generalized contradiction at the level of plot itself, apart from its representation. Under the exposed artifice of trick structure in current Hollywood films, the *fabula* has on its own terms grown not just contingent or elusive but illusory. Openly marketed with a "don't-give-away-the-trick-ending" lure in mind, such self-identified "supernatural thrillers" can go so far as to narrate en route their own tacit promise to the viewer: the promise that a final plot surprise will turn everything inside out. When Bruce Willis in *The Sixth Sense* shows himself hopeless at spinning out a compelling bedtime story to his hospitalized patient, he is told that "a good story has got to have a twist." A whole subgenre now exists to keep in suspension this implied contract of plot in its most extreme forms.

Both aesthetic and commercial explanations for this are not hard to come by. The fantastic plots of retroactively nonexistent lifelines certainly do help compensate for any lack of psychological credibility in the protagonists. That's one reason they may appeal to the depleted fund of characterization in current Hollywood scriptwriting. Moreover, the accepted borders between reality and its alternatives have been repeatedly worn thin by the wish-fulfillment inserts of the MTV format: those prolonged imaginary digressions, intercut with performance video, that construct a dual subjectivity for the MTV "character." In a hypertrophic extension of classic film-musical formats, with their aesthetics of self-expression, video lyrics can generate an alternate reality before our eyes, complete with changes of costume and *mise-en-scène*. In cinema rather than in video production, narratives that fantastically reinscribe this difference between reality and desire, only to sweep it away again in the end by their decimating plot revelations, may have—through some diffuse cultural disposition—a preprogrammed audience waiting in a so-called crossover demographic. There's also another aspect of ticket sales to consider: the trick ending that invites a paid second look, to see where the film cheated a bit perhaps, or where you may have missed the clue.

But the itch for hindsight doesn't explain away the readiness to be duped in the first place. Here, then, apart from any given plot material, we may touch on the deepest cultural work of Hollywood's ontological gothic in an era of psychic retrenchment against the global—and its counterpart, the real. As one might well argue, by contrast, for European and often transnational productions of the uncanny, certainly these Hollywood films can't be felt to sensitize an audience to unforeseen connectedness. Equally certain, any question of a suspect totality is shifted from social or geopolitical to strictly cinematic structure, from complot to plot. In the process, these latest Hollywood variants of the fantastic often train viewers, from the first shot forward, to put aside all vestigial unease about logical contradiction or incoherence and accept, however passively, the given for the actual. In this, as it happens, they may well offer cinema's compensatory answer, as institution, to their apparent antithesis in "reality TV."

In any event, something like a revised pleasure principle stands exposed in such a cinema of virtual or ghostly agency: skepticism is now entirely contained as a form of repressed *narrational* (rather than political) paranoia. In this mutation of deceptive totality from the conspiracy thriller, the new thrill depends on letting the whole plotline conspire against you. So that liminality takes shape in these plots of ontological reversal much like it always does. It locates the *inevitable* caught in the already mobilized act of postponement or denial. What is new here is that when the supernatural facts do come to light, despite their encompassing shock to the narrative system, they turn out to have been quite comfortingly unreal to begin with.

PSYCHOPATHOGRAPHIES OF THE VIRTUAL

A particularly clear example of this transformation of conspiracy thriller to ontological gothic is to be found, and before the full onslaught of digitization, in *Jacob's Ladder* (Adrian Lyne, 1990). It is based openly on the film version of *An Occurrence at Owl Creek Bridge* (Robert Enrico, 1962), where the soldier being hung, whose escape we then watch for the rest of the film, has his neck snapped back in a final embrace with his wife in their fantasized reunion, at which point we realize that he has only fantasized his escape and has in fact died in the distended split second of his fall. The Vietnam veteran in *Jacob's Ladder* realizes slowly, in mounting flashbacks, that he has been bayoneted by one of his own squad after he and the rest have been made the subject of mind-altering chemical experiments. Administered by their own commanders, untested drugs were deliberately meant to send the men into regression as primordial killing machines. But the plot-long detection scenario that achieves this realization also runs alongside another story altogether—and this one in the genre of the fantastic. For early on the hero comes to suspect that he is in fact dead, and that the agonies he is suffering stateside in his New York life are indeed infernal. True to genre form, the dead metaphor "I'm going through hell" is literalized, from actual dialogue, into an entire plot option. The trick, in a modest enough return to psychological norms, is that he is not dead but only still dying on a hospital bed back in Nam. This is the traumatic uncanny, then, not some supernatural vision of an afterlife. Or at least not yet.

The mechanics of fantasy thereby return us into the real of death, its transcendence ambiguous at best. The genuine ontological trouble comes, and with it the film's generic cross-purposes, when we realize that even the disclosed conspiracy, like everything else, is a deathbed projection of his own imagination. Yet this negative epiphany is then relieved, if only in his mind perhaps, by his being lifted (first by helicoptered gurney, then by the ascent of his domestic stairs) into a refulgent afterlife in reunion with his dead son. This is the son whose pictures had been thrown in the incinerator by the hero's unfeeling mistress after the film's opening nightmarish conversation. This is a moment of preternatural cruelty which we may recognize in retrospect as providing the narrative's first model shot in its unseen matrix of deathbed delusions. For floating rather than falling toward the flames, these pictures are burned up in slow motion. This is the familiar laboratory trope—as with the retarded coming to light of the naked bulb in *The Sixth Sense*—for distended temporality and its artificially deferred succession. In this case, form implicates function at a medial level. The photos may seem submitted to the fire, one after the other, at a pace cognate with the slowed (differential) photograms on the strip that at once project and italicize their descent. Technique is not just flourish in this case but clue. It models the withheld matrix. Time is what you make of it, even if it is all contained within the swollen moment of annihilation. Such is the temporal stylistics of slow motion in this fraught case. Such, in response, is a prefigurative narratography of this one film's dilated plot trick.

When closure does come, long postponed, it is an appeasing relief in *Jacob's Ladder*. And not just at the plot level: not just for the hero. The steadily built-up political pressure is released all of a sudden as well, drained away. The return to equilibrium in the death moment removes all urgency from the suspicion of military conspiracy, all chance of being able to act on it by the hero, all confidence, for the spectator, at least on further consideration, in its being more than a paranoid projection. The movie doesn't want this gutted political urgency, I think. It wants to have it both ways at once: paranoia justified and delusions of hell rectified. But the trick ending outsmarts itself. It gives undue comfort even where it hopes to sustain a cultural tension, lulling critique by narrative ingenuity. This is exactly the danger that was raised only in order to be quelled by the mention of digital fantasy in the new *Manchurian Candidate*. As such, it is an endemic problem of the ontological gothic in the decade and a half since *Jacob's Ladder*. The mode's anodyne effect can accompany even the most gruesome and politically motivated bloodshed.

And the trend continues, usually without even the lingering epistemological ambiguity of *Jacob's Ladder* (as will be noted in this book's last and more recent example, *The Jacket* [2005]). All we need to be assured of, but only after the fact, is that—as if it were all a movie, which of course it is—the bloodied victims were not actually there. In the new ontological dystopias of Hollywood narrative, screen lives are over without having really been lived. The dysfunctional no-place is also achronic, all duration illusory. Perhaps, like Samuel Butler's anagrammatic

Erewhon (1871), the utopic no-where is also meant to be obliquely "topical" as well, to reflect the now/here—the very here and now of contemporary cultural formation. This could be the case in recent screen instances, even beyond any local efforts at political irony or social dissent, because of their very abdication from the real. But even when time present, for the characters themselves, is mediated, reframed, and revisable, rather than irreversibly felt, such fables tend to carry little satiric edge in reception. Instead, in its fantastic excess, this cycle of films serves mostly to represent everything that couldn't happen here, at least not now, not yet.

Measuring the further cultural inferences of these and other films of a more explicitly digital fantastic (inferences for screen mediation as well as human epistemology) is the work of the next two chapters—and then again of the last. Screen evidence shows no sign of abating. Ghostly scapegoats, ectoplasmic psychopaths, robotic simulations, entirely electronic heroes, the suspended animation of every desire: the list lengthens, even while its remaining options might seem to be thinning out. A recent entry on the roster of Hollywood's ontological fantasy, parallel to the digital gothic of virtual agency, is a 2003 film by James Mangold, called *Identity*, that goes so far as to reveal all its characters as unambiguous (if long-masked) doubles of a psychotic central figure years ago abandoned by his mother. The psychopathography that oscillates, at opposite poles of the fantastic, between self and Other, life and its undead surrogates, is built directly into the medial subtext of this one narrative as well, complicating its genre resolution almost beyond recognition. The trouble isn't mainly that this film would be more successful as a black comedy than as a straight thriller. The real trouble is that the story is already a parody before it is even filmed. By who knows what circuit of influence, rumor, or in-joke, the film brings to the screen, remarkably enough, the very idea hatched by the hack twin sibling of the blocked screenwriter in *Adaptation* (2002), made just the year before *Identity* by Spike Jonze from Charlie Kauffman's metafilmic script. There Donald, the easy-going Kauffman brother, successfully peddles a screenplay in which the cop and the girl he's trying to protect from a stalking psychopath are only fractured aspects of the killer's multiple-personality disorder.

As script ideas pass through the innards of the Hollywood machine, let's say at best that *Identity* is a selective adaptation of *Adaptation*'s best joke. And certainly the earlier film is one whose ambivalent evocations of the fantastic hold their own interest in relation to time reframed by screen narrative. What Christian Metz would call the continuous imperceptible *trucage* of dual roles for the same actor (Nicolas Cage) begins to suggest—before any other character has seen the suddenly manifested brother—a vague optical allusion to electronic virtuality and its spectral secondings of the real. Thanks to an uncannily perfect upgrade of the venerable split-screen tricks of former cinematographic "twins," so much do we register the film's new-era digital finesse as a kind of tacit magic that we might well expect the second glib screenwriter to be the phantasmic alter ego of the sleepless, blocked hero. Not so, it soon appears. Yet such is the lingering suspicion that returns in closure, this time in a more allegorical register, when the less neurotic brother is no sooner killed off than his emotional stability is absorbed as a new strength in the immediately more self-sufficient survivor—or at least a new hopefulness troped by cinematography alone, as we will see.

In *Identity*, one can rest assured that the fantasy is uncompromised by allegory, with no conceivable drift toward uplift. What you see is what you get, what you are had by: the shameless replay not just of genre violence at a Hitchcockian motel (though it's that too) but of an already elapsed diegetic mayhem, now rematerialized only in psychotic delusion. The title is its own best clue—as well as its own surest trick—including the way its letters fall away from each other in the lexical deconstruction of the credit sequence. Short of this alphabetic dispersion, what we take at first to be a forensic title, calling out the unknown identity of the killer in the whodunit mode, is revealed instead, in the switch finish, to be a psychoanalytic term. The whole film turns out to have been locked into the mind, the split subject—the definitive nonidentity—of its once victimized villain. It is all a projected fantasy scenario staged by the unconscious of a psychopath to replay his earlier killing spree at an apartment building—a rage that is now taken out directly on refracted facets of his own overcrowded ego.

As each incarnation of his multiple-personality disorder is "folded in" (the script's psychiatric lingo)—in other words, brutally murdered at the plot level—one among the eleven misbegotten splinter selves, a supposed police detective, takes forensic photographs with an off-the-rack camera as disposable as the disappearing bodies turn out to be. The images go undeveloped, as it turns out, perhaps because the radical apparitions of psychosis have no use for the evidentiary force of the fixed image in this fantastic plot. Except after the fact. For the first clue to the ontological disorder of the narrative premise comes when all the hotel guests discover that they share a birthday—and do so by comparing photocopied versions of their driver's licenses kept as security by the hotel night clerk, these second-generation duplicates thus giving the lie to the concept of "photo ID." While the earlier murder plot is still unfolding, though, corpse by corpse, there is no time to be struck still in the present, violently or otherwise, since the entire film is merely the delusional replay of former slaughter—as we might have guessed, if only we had remembered what no one could or would be expected to: that the film began with a photographic model of yet another maternal matrix (as in *The Double Life of Véronique*, Ruiz's version of Proust, *Vanilla Sky*, *A.I.*, on and on). Solving the mystery long in advance by a precredit montage we have no way at the time to interpret, or any reason to recall, the film has opened—after the dissociative letters

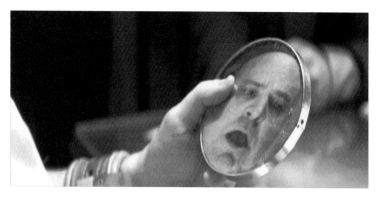

3.21

of the title—with the review of a dossier of headline photos of a boy abandoned by his mother at, yes, a primal motel. These are images cut on lightning bursts in the psychiatrist's office as if they were being traumatically reshot at the moment of original capture. They are followed in turn by a stack of police photos of the victims being riffled through by the doctor—as if they were a protocinematic flip book reanimating the past in anticipatory behalf of the plot's own ruse of embodied replay.

In a film of the psychotic virtual like *Identity*, as well as with other and subtler narratives of the fantastic, having lost one's mother may mean having to mother one's own vengeful or otherwise libidinal urges—to give birth from within to divergent facets of that misnamed singularity called self or identity. In being forced to confront his riven personality, the psychopath at his parole hearing is brought face to face with those earlier photographs, as material proof of his former bloodlust. At first, he doesn't believe the mediated evidence of his senses, doesn't see himself in it. As a result, in the ontological inversions of this revelation scene, one of his incarnated doubles—in the supposed person of the detective-hero (John Cusack) summoned to the hearing—is forced to stare into a mirror at the bloated features of his crazed original (fig. 3.21). "What have you done to my face?" the movie-star double screams hysterically. Time present stands reframed from within its fantasized temporal alternative. In contrast with preternatural and often elegiac effects in European films of spectral memories and traumatic recurrence, where subjectivity is gently ghosted with otherness, this "interface effect" seems on its own ocular terms entirely deranged. In other (if related) terms, the maddening separation from the mother must be replayed in adult psychosis as a long-postponed and now brutal and reinfantilizing mirror stage.

The extended thematic distinction in Todorov between crises of selfhood and perverse relations to the Other, between psychotic delusion and mere neurotic deviance, is found clarified by *Identity* in so fiercely reductive an example that it may appear at first like a genre exception instead. Yet, well within parameters of the fantastic, the film in fact proves the rule of genre alternatives by its very convolution. How so? In the initial and founding dichotomy broached by Todorov, why isn't the resolution of *Identity* more uncanny than marvelous, hence more psychological than supernatural? Given the nature of its story, it might seem to be. But the story is, of course, given to us *as filmed*. The whole embodied cast of this screen narrative, however, all but one of the characters, is evaporated by the point of closure, dissolved by narrative solution itself. If this were futurist sci-fi, some power button controlling an encompassing alternate reality would at this turning point have been summarily switched off. Instead, a Freudian master plot arrives in explanation. Involuntarily, the spectator has all along been asked to witness as visual event the deaths that the protagonist has only fantasized about. A recurrent genre premise, in the form of a familiar rhetorical question, thus rears its head. What if the whole peopled world were not just like a movie but like the multiple-personality disorder of a slasher film rerun in the mind?

Don't believe your eyes—for this, all this, hasn't been taking place at all. Though the caveat and its explanation are certainly psychological, the experience has been, in filmic effect, far more than uncanny. In genre terms, plot's manifestation has been an ocular marvel: fiction raised to the power of itself in a delusional projection. Given the ontological ground it falsifies, as well as the embodied figments it conjures, the apparatus of screen imaging seems in its own right psychotic. In this sense the scale and grain of analytic method again come into question, even with so extreme a gimmick as this film's narrative reversal. To be sure, any variant of narratology is poised to clock such a phantom cast of characters, in serial elimination, through to a re-equilibrated closure, with the time of fantasy dredged up from the unconscious, smoked out, its cover blown, its specters exterminated. Narratography, however, can more nervously trace the surface evidence of such harbored temporality at the very interface between repression and identity formation, where the duration of remembered violence is itself unmasked as fantastic. And not just that: where the virtual is reduced to the *techne* of the *psyche* itself, in what, early on, we may well recognize as its digitally figured mechanism.

Such an apprehension builds as follows. Four times during the prologue of *Identity*—where inexplicable roadside bloodshed reroutes its first victims to the nightmare motel—normal filming has been caught up short, interrupted by a freeze-frame. According to classic models for the conversion of filmic process to cinematic rhetoric, or, in Lyotard's sense, of figure to discourse, this device would work to *image* normal time as suddenly arrested by death or slashing violence. Here it does more—or, in a sense, less. For what comes forward in these throttled images is the admission of constructedness itself, all told and deep down: a stop-time confession that this harrowing sequence of events is all being replayed or rerun in the mind of the protagonist as "spectator" (the modernist condition par excellence for Deleuze)—and this as a fictional simulation, a screen for other and worse (because real even if past) slaughter. Stopping and starting the action, cutting and resplicing it, reframing it under psychic duress, dilating its instant into fixity—all this only compounds the liminal suggestions of the credit sequence, with its stacked forensic photographs suddenly fingered into mechanical motion. It does so by admitting the whole narrative as a combinatory reshuffling of photomechanical images in a grotesquely therapeutic reenactment of its own past and pretext. The violation of the representational by the narrative order (in Lyotard's sense) is thus entirely contained here within the dementia of a self-narrativizing subjectivity.

But there's a medial catch in such narratographic reading. For this quartet of halted images, each blocking the temporal present and tossing the plot back into an explanatory near-past, are images executed not in the photo lab but on the keyboard deck of the computer console. The freeze-frame of photomechanical device has become instead, since the electronic revolution, the selectively delayed pixel shift of a merely prolonged rectangular field. Again we find in action a deeply equivocated sense of the frame itself in the epochal turn from its multiple function in cinematography, twenty-four per second, to its single electrographic status. Images that exclude their own difference under petrifying repetition in the one case become images that freeze from within by numerical nondifferentiation in the other. Unlike the venerable freeze-frame rhetoric effected by photocellular duplication, the piecemeal iteration of the binary array—in a digitally edited version of the optical "hold"—instead produces a stabilized image no different in grain, sheen, or clarity from the comparable artifice of every other moving-image shot. So that those opening freeze-frames in *Identity*, alluding as they well may do to the bloody photographs that precede them in the credit sequence, operate only by a displaced optical allusion to cinematography itself. In image though not in narrative function, they are more like the effect of a pause button (as might be applied at some other point in the DVD scanning of this film) than like a classic photomechanical freeze, with its inevitable ocular evocation of strip, flicker, and photographic texture, its filmic break with the fidelity of the cinematic. All the more so, perhaps, do these digital variants of the freeze in *Identity* yield to narratographic notice as the image of duration under and in arrest, a discontinuity of stop(ped) time as the very picture of fixation.

Here, too, is another case of what structuralist narratology would call metalepsis. This rhetorical function involves the incorporation within plot of its own conditions of possibility—often its own author, or in this case its own readjusted medial options—as part of its narrative discourse. It's what we've noted more than once via a metafilmic turn into the ontological *what-if*, often clearest in the technological hypotheses of sci-fi. In the present case, the question is what the world of duration would look like if it could be braked on a dime by the whims of schizophrenic oversight, not by the mad doctor (or scientist) of futures past but by the mad patient locked out of coherent time altogether. And this has a marked historical—or media archaeological—dimension as well. Seen in this light, the optical allusion of this narrative effect is to its own destiny in response, but a response in this case technological before affective. Stressing the electronic reception rather than generation of screen images in new modes of distribution, Laura Mulvey suggests, in *Death 24x a Second*, that the video pause button of remediated screening has, in postphotographic formats, irreversibly changed the way we think of and see cinema, the way in fact we think its sequence.[5] In which case another variant of the familiar question arises. What if the world were screened as the brain scan of a mind disposed to yank time up short for inspection at moments of traumatic shock? This is to ask in effect, and with special relevance to the "temportations" of the next chapter, whether it isn't the case that the idea of time framed on-screen finds one of its newest models in the home-video apparatus of narrative transmission, operated at will against the flow of the inevitable.

Such an extra wrinkle—or optic fold—induced by those digital freezes at the start of *Identity* seems just right even in its formulaic slasher context. In Deleuze's terms for the new "numerical" dispensation of the image (and its electronic dispensary), these static shots, like all other manifest digital images, are indeed less like the eye's work than like the brain's. In this, as we know, lies their risk of "spoiling," but also their chance of "relaunching," the time-image. All the more apt do they seem here in their ability to encode the psychic damage of a memory that is, in its dementia, *calling all the shots*. Indeed, the relation of such computer-edited effects to a tacit brain's-eye view, without the intermediation of photography and its associated thematics, may suggest the relative rarity of this effect in recent cinema. As the cliché of the photomechanical freeze has passed from common use in its troping of death or memory trace, its facile technical replacement in the computerized "pause" has grown less marked, less feasibly strategic, as a narrative token. It has gone from a feature of discourse to a function of replay. As deployed with new (if blunt) edge in *Identity*, however, and in oblique relation to the elided frames of photographic capture in the cop's evidentiary snapshots later, these opening freeze-frames do unto time as time would do unto plot's central consciousness: undo it, stalling all impulse in the obsessive trace. Escaping from the familiar emphatic gestures of narrative rhetoric, slipping out from underneath discourse to insist on themselves as pure figure, they may in fact amount to the first strictly subjective freeze-frames in cinema history. To vary and combine Lyotard and Deleuze, this is the *un*writing of movement as a figure of time in this film's hall of psychotic mirrors. Narratology can well read the simulated trajectories of eleven separate, but ultimately indivisible, "actants" as an image of traumatic temporality relived and redied. Once the trick is pulled, however, narratography *re*reads the arrests and doublings, the ocular snags and echoing reflections, that have performed en route, and tacitly exposed, this phantom temporality.

One is still left to speculate about whether a fictional character can bring suit for plagiarism, even if the posthumous victory would be cold comfort. Yet one needn't wait for the overripe, deadpan excess of *Identity* to see essential aspects of Donald Kauffman's script, unsuccessfully quarantined by parody, actually reaching the screen. The lurid tropes of his genre-laden brainchild are in fact realized by counterplot, if only allegorically, in *Adaptation* itself. In his own film's larger story, that is, this glib sibling scriptwriter, blissfully ignorant of his second self's doubt, is himself violently eliminated—and in death assimilated—to a newly spirited zest for life, come what may, on the part of his neurotic twin. To this end, an earlier tongue-in-cheek gimmick returns in closure. Time-lapse acceleration had been used before as an optical joke for an inset black-and-white biopic on Darwin. It is as if the scientist were filmed while drafting the theory of selective "adaptation" at his Victorian desk, with all of evolutionary history sped up before our eyes. Another such undisguised *trucage* sets in right at the end, when the screenwriter-hero, having internalized the optimism of his dead brother and positive emotional pole, drives off into the paved L.A. distance. Unexpectedly, a bed of flowers comes into frame in the foreground, its Technicolor blooms snapping shut and popping open again in rapid series along with the ping-pong of daylight and nighttime traffic: nature and culture harmonized in their daily rounds. The movie's long-stalled reflexive scenario has come to rest in a burst of accelerated motion: "This is the right ending, this feels good," the hero has just said in voice-over. And on the world goes, summed in diurnal continuance.

In the mock-lyrical closure of such cornball optimism, traditional narratology could spot contradictions giddily resolved, dualities consolidated, doubles conveniently internalized—all this in an explicit vector of blind hope. More tightly calibrated, narratography finds something else as well in the frenetic "arrhythmy" of the cinematographic track. It detects a slapstick variant of the Deleuzian time-image emerging once more, though by speed rather than the more familiar arrest, in the sheer departure from movement's normal duration. Structure (varying Bordwell's formula again) is accelerated to a point where *style* itself leaves all story behind, opening only onto possibility—possibility pure and simple, in blithe and loony renewal. As the movement-image speeds past all event into a trope of sheer eventuality, it would require a more extensive account of the virtual, as in the fifth chapter, to take the full racing pulse of such an image—and many others less pronounced or farcical. Here, then, is the decisive obverse of the psychotic gothic in *Identity*, where past bloodshed is reworked into virtuality by a trick ending demented rather than merely delirious: an ending by which a history of violence is seen to repeat itself, first as tragedy, then not as farce but as genre fiction and its cheap catharsis. Instead, *Adaptation*, putting the past behind it, closes upon a comic time-image within a single mutating and cyclic rectangle: a case of framed time in perpetuity.

MA(R)KING TIME

Along with the antithetical freeze seizures of *Identity*, such an italicized departure from the normal time-space ratios of projection can remind us how active the medial base of the projected image can be in film narrative, how often recruited from track to anomalous action, how easily recuperated for theme. In this way does the exceptionalism posited by Manovich for avant-garde work seem far too exclusive. Above all, he groups the seen but unrecognized constituents of film's material base (the self-erasing photograms) with the material grain of photography itself (the subliminal print dots). He wants to distinguish both, for instance, from the morphemes that make up the word units of a language. For these linguistic subunits are visible and pertinent in the reading act, even though quickly assimilated to larger structures of meaning. By contrast, writes Manovich: "Neither film frames nor halftone dots have any relation to how a film or photograph affects the viewer (except in modern art and avant-garde film—think of painting by Roy Lichtenstein and films of Paul Sharits—which often make the 'material' units of media into units of meaning)."[6] This is the reigning view, of course. Yet it is based only on what we think we are watching on-screen, not on what we actually see—not, in other words, on what makes the watching possible.

The true story of reception has always been a little different, even in mainstream narrative film. We know that the very first cinema viewers (Bergson among the most astute if allergic of them) were well aware of the flicker effect, aware of both its immediate cause and its historical origin in instantaneous record. From these first spectators to the latest audiences of a digitally mixed medium, no undue illusions need be assumed in reaction to screen illusionism. In the Lyotardian writing "of" or "in" movement between strip and track, film and screen, frameline and framed narrative linearity, the viewer's attention is not by definition so blinkered as to disallow the screen's photogrammatic linkages when they are either manifested or alluded to. Inscriptive difference, that is, can well "affect" (*pace* Manovich) the viewing it also and everywhere effects. This demurral from Manovich, here or in what follows, is not meant to disqualify the application of his work to the coming chapter on temporal mutations in the recent Hollywood fantastic, however. Far from it. For his most important claims about cinema within an archaeology of media, an archaeology whose latest sedimentation is the digital, are directly relevant to the new modes of temporality on the electronically reconceived screen. This is because he situates cinema within a far longer historical span than that defined by the reconstructed piecemeal temporality of the filmic image itself. Only for the length of a century did cinema as machine art provide an automatic "sampling" of real time rather than its represented duration. This is, of course, what I have been calling filmic cinema, based in and sustained upon the photographic index from frame to frame. On either historical side of this temporary reign of automatized time, in Manovich's archaeology, lies the once and future rendering (rather than recording) of time through "animation" effects, formerly magic lanterns and their like, now CGI.[7]

This historical long view is entirely clarifying on its own terms. But if film is essentially a continuous temporal medium, to which the cut came later, so too does the digital image unfold in time. Distinctions too broadly drawn will confuse the issue. And beyond this, one level up from automatic transcription, where the manual returns to film via editing, the cinematic track is hardly (despite Manovich's claims) more temporal than spatial. Even if we ignore the intermittence of the photogram, as Manovich insists we cannot help but do, it is the cellular basis of the filmic that opens the cinematic to narrative possibility. Films as we know them are spatialized through and through. Their images are cut, pasted, laminated, sometimes even reduplicated, sector by graphic sector. Montage is all but defined as the *disregarding* of recording time and its indexical duration. What unfolds over time is not time itself, but something else—something that narratography is quick to register: not the marking of time but its (always spatial) writing with movement. This is where Manovich has drifted into an overly schematic assumption that film is at base a time me-dium, digital interface a spatial one. That's not just too easy, but in some ways backwards. Classic film editing doesn't enchain time. Instead, it exchanges one moment's spatial visualization for the next in a framed plastic series. Montage in cinema actually *produces* (rather than captures) narrative time out of spatialized event, where the graphic series is read not as continuous duration but as an *articulated* temporal sequence. Montage doesn't in this sense take time; it makes it.

Given his questionable assumption of cinematic process, both recording and editing, as definitively temporal in structure and effect, the real difficulty in accepting Manovich's contras-tive medial emphasis comes when what is now montaged and composited on the internal planar fields of the electronic monitor is taken to be something like "time" itself as cinema once knew it on its own single screen: temporality suddenly plasticized. Is it really so, in some entirely new mode of "spatial montage," that as never before in media technology "time becomes spatialized, distributed over the surface of the screen" (325)? Data, yes, and motion, maybe—but time? Why, for instance, is simultaneity rather than succession an exclusive case of the spatialization of time?[8] If Manovich meant by this simply time's arrest and flattening into a digital archive, that would be one thing. But emphasis continues to fall on the new dynamism of electronic montage, where the parallel editing, so to speak, of computer files operates "just as a person, going through life, accumulates more and more memories" (325). Either that analogic "just as" is meant in the loos-est possible sense, or else we must insist that memory is not simply storage, or time the trace it directly leaves. Memory is itself a sorting, a filing system; and time—in being made available to consciousness—is memory's *construct*, a function of deletion and recombination. So that we are in no position to agree with this summary antithesis in Manovich: "In contrast to the cinema's screen, which primarily functions as a record of perception, here the computer screen functions as a record of memory" (325). Leaving aside the vaguely tautological cast of "record of memory," what the next chapter will serve to expose is exactly the way the computer model for memory (entertained by recent plots of traumatic revisitation in the subgenre of the mnemonic gothic) can only betray—in the sense of overthrowing as well as disclosing—the inherent memory effects of an ephemeral, palimpsestic, and discontinuous medium like film.

But this book's parting of company from Manovich begins on decidedly common ground: "To invoke a term often used in film theory, new media move us from identification to action" (183)—which is also to say as well, in spatial terms, from point of view to interface. Here is a distinction made elsewhere by Thomas Elsaesser as that between an enchained "narration" and the random access of "navigation."[9] It need only be insisted, against the claims for computer filing as memory image, that unimpeded navigation isn't the mind's way with time. In any event, when screen plots in the next chapter treat time the way Manovich says the computer does, spatializing it by the interplay of kinetic reframing ("spatial montage"), they do so by a kind of category disorder projected as narrative event. What takes shape as spectacle is no less than a dialectical collapse of spatiotemporal determinates for the very definition of event. In this way, and looking back from Manovich toward Deleuze, such films enact no better than a caricature of the cinematographic time-image and its retrieved "sheets of the past."

The hyperbolic "interactivity" we are to find in the coming fantasies of temportation merely *wills away* its compartmentalized and reframed temporal episodes by alternate if nonsimultaneous access to others like them. In moving from an apparatus of identification to a navigable field, from an articulated view to an operable program, from the ocular to the fiber-optic, there is a slide from consciousness itself to mere data activation (precisely what the hero of *Johnny Mnemonic* would have killed to avoid, and did, as we are to find). What further consequences follow from this shift—consequences for the represented agency of the subject on the cinema screen, as for the narratology of human lifelines—we are then to see. For now, this at least: how the "anti"-hero of *Identity*, antithetically negated many times over by his own disavowed avatars, may be seen almost to parody, from within the throes of psychosis, this functional shift from identification to action—where all action now results from a series of willed and murderous disidentifications. Identity is itself reprogrammed by the navigations of plot. If so, then there is all the more reason to lay stress on that symbolic mirroring interface around which his plot so violently turns (fig. 3.21 again). Although in a vacuum of all gadgetry or electronics, it would symptomize the film's place, nonetheless, in the same contemporary trend that will precipitate hero after hero of the "instrumental marvelous" into the downloaded screens, or windows, of their own selective and often redirected memories. It is there that the "interface effect," stripped of any immediate identification, may be reduced to the video-game action, or byplay, of a strictly interactive (v)id.

BEYOND BODY TIME

If the psychopath of *Identity* is taken out of his body by his vicarious violence, it is only because he is already out of his mind. With their equivalent images long forgotten in the prologue, the photographs reshot by the police detective en route offer only immaterial evidence of this displaced bodiless agency: traces no more visible than the mind's own wounds. Another and far more interesting film of the strictly virtual memory trace, *One Hour Photo* (Mark Romanek, 2002), takes instead a last-minute—rather than encompassing—turn into the fantastic after preparing for it with a more fully contextualized cultural anxiety. The sociohistorical unease in this case has to do not so much with photography's relation to death (the index as corpse) as with its contrast to that unprecedented mode of nonpresence effected by the new media. Coming before our eyes in the film's last shot is a material image that is in fact a falsified index—a strictly virtual snapshot—fantastically conjured in the protagonist's psychotic break with the real. What the delusional photograph traces in the precincts of the fantastic alone, for no optic but the camera's, is the unreal record of a man out of body in his own world, lost and found again only in a phantom imprint.

As with *Identity*, this is a film centered upon childhood trauma and its disfiguring aftereffects. To keep these memory traces at bay, *One Hour Photo* summons in the end the fantastic of photography itself: its power to simulate one past in order to repress another, to still time and its ghosts at once. And it does so by tacitly capitalizing toward the end on yet another forgettable opening, as we'll recover it in a moment: a trick beginning sprung from within a photographic apparatus itself. Immediately after these launching frames of a snapshot camera's internal mechanism, we see the protagonist from behind another kind of lens, this time an ominous oversized police camera in the act of its digital scan. A plot awaits to explain why he is there. Narratography has already anticipated the sense in which he isn't. Played with eerie emotional constipation by Robin Williams, this scanned figure turns out to be a superstore clerk who has made illegal duplicates of a family's snapshots over the years and covered his walls with them, filling his underfurnished life with surrogate family pictures (fig. 3.22). After the police later dis-

3.22

cover that he has scratched out the face of the father in every one of his hundreds of photos, we are led with them to suspect at the climax that he is about to take snuff pictures of the cheating husband and his mistress after cornering them in bed at knifepoint. By now, we may long ago have forgotten that the film is a flashback from police headquarters, sprung by a question about what has provoked the photo developer's rage.

What we are more likely yet to have forgotten is the instigating transition from credits to the opening shot of digital imaging. As if replaying the history of still-image technology, we have begun with the title sequence laterally scrolling past on the film roll, exposed credit by ar-rowed credit as if from inside a reflex camera (fig. 3.23). From there we have jumped to a bulking digital camera in the act of piecing out a mug shot of the protagonist (fig. 3.24), which is then transferred by printout to his file in the Threat Management Division and transposed in turn to his image on a surveillance monitor. Into such a zone of recessional duplication he eventually disappears altogether when narrative returns us to this framing interrogation scene. Deferred revelation arrives there via the film's activated cinematic intertext, Michael Powell's 1960 *Peeping Tom*, where a father's continual filmed surveillance of his boy's life has turned the adult son into a voyeur and sexual psychopath. In a contemporary update of this ob-scene (off-stage) pretext or backstory, the protagonist of *One Hour Photo*—the punningly dubbed Seymour Parrish—turns out to have been subjected to, and objectified by, his father's explicitly pornographic photog-raphy. It is for this brutalization, we suspect, that all subsequent hygienic images operate as a kind of catharsis—until he is reframed again as a captured subject within the telltale rectangular (that is, photographic) dimensions of the police interrogation tank (fig. 3.25). From there he is released only by the fantasized insertion of himself into yet another (this time entirely unreal) family snapshot held to fade-out. The close-up scene of his legal "arrest" has thus given way by lateral displacement and dissolution to the u-topic site of an imagined domestic memento (figs. 3.26–3.27). In another return of repressed narrative detail, what he has managed to do, in imagination, is somehow meld the photo he took of himself in the opening flashback scene at the photo desk—just so as not to waste the last image on the wife's roll—into a single composite frame of a euphemized life as that of *his* extended family.

3.23

3.24

3.25

Bridging two moments early and late in a kind of structural superimposition of its own, the plot's final turn to fantasy is transacted across the revealed photogrammar of the narrative strip itself. For the transitional dissolve is so slow that the split between psychotic and avuncular personalities (fig. 3.26 again) is manifest on-screen—sheer internal difference—as the secret motor of all shot or scene change. Metz's resonant proposition—all montage a *trucage*—has seldom been more openly evoked. Further, too, in the break from sutured into purely hallucinatory space, the collapse of self and its double could hardly be more sharply focused around the very logic of "projection" in the psychoanalytic sense: the mirage of self projected into an illusory group image (fig. 3.27 again). At the moment the hero snaps under pressure, he is released into a snapshot of his own fantastic reprieve. In this metamorphosis, he enters, in short, an archive not his own. This ontological disappearing act doesn't just offer an escape hatch—through the interstices of an explicitly filmic narrative—back into some static photographic imaginary. From the plot's first ominous shot of the police lab's digital camera, and the lineated array of the protagonist's generated image (fig. 3.24 again), his whole role as laboratory technician is shadowed by the encroachments of this particular "new technological medium."

The movie, in other words, already reveals a certain nostalgic cast even before its prolonged explanatory flashback gets plot under way: nostalgia for the reflex camera and its handheld testament to the coherence of the received image in its predigital trace. Instead, in the final stage of his own metafilmic "development" as character, his imaginary self comes only very slowly

3.26–3.27

into view before our eyes, bled off from the real into a fantasy double in the group portrait. Yet he materializes there, in the distended and once quintessentially photogrammatic transition of a decelerated lap dissolve (no doubt digitally implemented in this case), not so much like a photograph from its emulsion as like a libidinal projection from the quasi-digital retouching of the mind's own Photoshop. Becoming its own lateral phantom, subjectivity is here flanged around its very disappearance. Having been recently forced to confess what photography had once done to him, in its mode of the all too true, its victim is at last released, optically doctored but by no means cured, into the virtual.

It is as if residual versus emergent media converge in this moment to elide the vanishing dominant, the indexically traced body: the body that photography used to record and that the digital can now so readily transplant or fabricate. No gainsaying the way in which, by his apparent psychotic break, the photo clerk has become part of an old-fashioned family photograph. The further reverberation is simply that, as a technician himself on the cusp of the digital turn in mass-market imaging, and hence looking down the road to his own obsolescence in the workplace, he might still—even from within the paralysis of his nostalgia—be capable of joining the spectator in recognizing an alternative and otherwise unprecedented way, an electronic rather than a schizophrenic way, by which this illusion—this conflated photofantasy, if you will—could have been induced, indeed produced. Such an oblique cultural reverberation is endemic to the plots to be surveyed in the next chapter. It is the optical allusion by which even the frame time of photomechanical succession, under sufficient distortion, can encode a broader digital climate of simulation and interactivity.

A confession by way of a structural testament. So tacitly saturated is *One Hour Photo* with a residual commitment—and, as personified by the hero, a rather impotent and clinging one at that—to exposure settings and processing speeds, to the very time of photomechanism and photochemistry, that I had lectured and written about the film more than once on just this score, and reflected on the insinuations of its marginal digital technique, without at all remembering the passing bit of dialogue that knots up these matters. For in the opening encounter between the clerk and the mother of the house, there is a nervous exchange that—picking up on the police lab's electronic camera in the frame story—contributes further to modeling the film's whole digital intertext. I like to think I had registered this dialogue, in all its overexplicitness, and then repressed it. In any case, turning hermeneutic oversight into narratographic confirmation, let this cognitive lapse on the critic's part nonetheless confirm his method.[10] For when Sy Parrish first compliments the female head of his surrogate household on her Leica camera, she mumbles that her husband "has been trying to get me to go digital." The clerk's instant response: "Oh, don't do that. I'd be out of a job." That alone might be enough motivation, in the subtext, for his stalking at knifepoint the architect husband, whom we've already seen doing computerized design work while his son plays video games in the next room. At the time of their first exchange, both clerk and wife pass over his paranoid remark with a smile. But it is with a sigh and a whimper that Sy Parrish's role as agent of mediation is, indeed, soon to perish. The wistfulness here seems not just photography's but filmic cinema's own.

Electronic overtones aside, a structure of repression and its returns can in this way, yet again, be more closely tracked, on the track itself, by an apprehension of the film medium's own repressed substrate. Any interpretive account of the hero's final flight from the real in *One Hour Photo* would, of course, engage a scenario of delusional nostalgia that makes yearning for an unreal home into the literalized *unheimlich* (Freud's un-homelike) of an uncanny family snapshot. Narratographically registered, the final slipping away of the real is marked more specifically by a graphic fade from the reframed interrogation tank to the liquefaction of self-image in its ambiguous "interface" passage (Žižek's predigital sense) across its own symmetrical (but not exactly mirrored) doubling (fig. 3.26 again). From within an ethics of the virtual only dimly tinged with the digital, narratography responds in this way, along the very grain of the uncanny, to the medium's own nostalgia-in-progress for its vanishing filmic base. This is a nostalgia for that laboratory construction of narrative's "virtual real" to which the single photogram—or, say, the separability of photograms, whether in the montage work of cutting or superimposition—has always been the essential contributing element.

Just here, and before one can read the transfixing of desire in artificial imprint as a psychotic time-image, one notes in equally Deleuzian terms what this moment of postrealist fantasy has done to that third avatar of the foundational movement-image, mediating between perception and action in his schema: namely the affection-image, with its premium on faciality and close-up. Something very deep in the formative logic of cinema is probed at the close of *One Hour Photo* even under the shadow of its technological outmoding—and even if not deeply probed. Maybe just tapped. The slow overlapping fade from a photophiliac self to his own spectral image—a repetition never subsumed to the same, to self-identity—is digitally implemented in a way that nonetheless calls out, as noted, the constituent but not self-constitutive photo cell of an earlier filmic dispensation. Again, as in the early freeze-frames of *Identity*, the residual technology has turned, by way of medial allusion, to narrative trope. Narratography is the reading of such a rhetoric.

Its recognitions are medium deep. Every image in classic cinema, including the close-up of affection itself inscribed as sign, gives way to another (almost identical) photograph of itself in an infinitesimally differentiated series: the photogram in action. Composed of tens of thousands of the latter, *One Hour Photo* closes by seeming to stall their normal transition—and does so across a dilatory and discrepant chasm, always present but here tactically exaggerated, between the image and its virtual double. Neither in the world, and thus vulnerable to its losses, nor prey any longer to being recorded by that world, virtualization into a solely imagined trace is the protagonist's only escape. And once again, a late-coming plot turn gives the cultural game away. The very predictability of child molestation as the hidden secret, almost the plot matrix, of so many Hollywood films over the last two decades puts the entire audience through a process of unnerving disavowal. He's terribly troubled, yes, but there's no reason to think it's *that* trouble *this* time. Is there? One result is that the actual revelation scene comes with the force of spectatorial repression in a displaced return. Oh, yes, we half knew it all along. That explains everything. In this curious displacement, a narratology of plot confirms the symptomatology of neurotic disavowal. In closer response, a narratography of this same plot traces the scars it leaves, inscribing as they do a temporal discrepancy pervasively at work: the rift in this case between forgetting and regression, denial and revival. The photo clerk's days having been spent sampling other people's snapped time, other people's mechanized memories, the impasse of the mnemonic gothic has this scapegoat just where it wants him. Out of body and out of time, the protagonist slips away to a history not his own and a body never there. To recognize more broadly the now sci-fi, now fantastic subgenre of virtual transport to which such a downbeat psychodrama claims kin, if only by optical allusion at the close, is the business of the next chapter.

FOUR

TEMPORTATION

Means *versus* ends? Scarcely—even from a film's first move. In the cinema of temporal irony, with all its switch finishes, the liminal onset of plot defines its aim in both senses, its goal and its angle of flight. This may be all the more apparent when the aim turns out to be defined as the arc of a boomerang. This chapter thus extends the last two on the score of trick beginnings: a category that continues to describe each of the exhibits at its point of departure. Carrying forward this assumption, we need to track further such narrative devices as they are recruited to new purpose in an altered dispensation of the screen image, where the pixel is slowly claiming invisible precedence, and hence subliminal priority, over the photogram.

The topic of temporal escapism finds among its more potent tropes a formal circularity that replays at the structural level the existential loops of story—and does so by often indirect recourse to a contemporary culture of electronic interface. This is a social imaginary whose explicit evocation is often curtailed so as not to compromise, one speculates, the cinematic manifestation of a given plot. When the digital context, thus hedged, still leaves its visual traces on the screen image, however, narratography may be engaged to situate the optical allusion. Otherwise, plot alone can offer a more efficient—though still symptomatic—cover story, where cultural checkpoints in electronic mediation can go without saying. There may be no denying the widespread digital incursion into film, but there is no reason for a given film to incur its full thematic backlash unless its storyline requires it. As *One Hour Photo* makes clear, the least hints go a long way.

There is certainly no denying this much, at least: that even when the screen image still looks the same, or almost, it can no longer be counted on to bear the same relation to the world it once recorded and rearranged but now partly "generates" from scratch, bit by digital bit. When they don't spew it forth altogether, untold megabytes lately chew away at the recorded image from within its own frame, simulating all or part of its sectored "view"—indeed putting those denaturalizing quotation marks around just that aspect of its illusionism. In this regard the previous chapter arrayed the preliminary duality from which a further clarity might emerge: European psychodramas of the uncanny versus Hollywood ontological gothics. The question once more for Hollywood cinema emerges by direct generic contrast with the spectral "interface effect" (Žižek) of European fantasy: what further transatlantic connection can be brought out between electronically intermixed screen technique and the new fantastic of screen plots, or in other words between the postfilmic and the postrealist texture of cinematic narrative?

The first assumption of such an inquiry, as should be clear from previous discussion, is that the plots toward which computer-assisted narratives tend to gravitate cannot finally answer the question for themselves, and certainly not on Hollywood's own terms. Triangulation is advised. And from its interplay emerges a consistent stress on new modes of temporal imagining, both biographical and historical. In the international drift toward a postrealist cinema, there is European psychodrama on the one side, as so far discussed, its techniques not yet given over in the main to the postfilmic. On the other side lie Asian film epics driven by special effects. And in between the two oceans, the linked but separate preoccupations of Hollywood thrillers. At one foot of such an imaginary triangle—where its (material) base is defined as a spectrum from electronic to filmic inscription—would fall the unlikely marriage of history and fantasy in the digital lyricism of films like *Crouching Tiger, Hidden Dragon* (Ang Lee, 2000) or *Hero* (Yimou Zhang, 2004), where the very format of epic chronicle—of national and cultural history—is rendered virtual in its retelling. At the opposite foot lie those films of the elegiac uncanny in the European cinema since Krzysztof Kieślowski, where the traumas of European history, or at least its present socioeconomic dislocations, often seem internalized in the memory effects of existential psychodrama or their transnational telepathies. And at the split apex of this rough triangulation, we find the recent division in Hollywood cinema. This is the divergence between, on the one hand, digital sci-fi in the futurist mode (borrowing from Asian graphics), in which the postrealist is indeed growing synonymous with the postfilmic, and, on the other, those posthumous gothics whose emphasis on psychic as well as ontological rupture allies them to some extent with the European uncanny.

At the level of plot in both Hollywood subtypes, the very possibility of human time is under siege by extreme forms of disembodiment, whether digital or mystical. Here a continued borrowing by Hollywood industrial production from the electrochoreography of Asian special effects puts a premium, for instance, on the suspensions of "bullet time" rather than the chronological suspense of experienced duration, as formerly organized by narrative. That is a large topic in itself, of course: the Pacific Rim of international comparison. But the other side of the triangle is yet more instructive in regard to cinema's Deleuzian capacity to think time—or at least to conceptualize it by "high-concept" ingenuities. The results may seem more a caricature or travesty of Deleuze's time-image than the thing itself. Yet this wouldn't rule out the paradigm's chance of shedding light on their perverse—and often overliteralized—schematics. In any case, it is the contrast of fantastic storytelling in Hollywood narrative with European plots of amnesia, déjà vu, telepathic transference, apparitional nostalgia, and the like that brings into sharpest outline the temporal loops and delusions that characterize those now typical Hollywood plots (weighted to whatever degree toward the uncanny or the marvelous) of virtual embodiment, posthumous self-recognition, mental time travel, and so on. Along either slope of the Hollywood divide, in fantastic cinemas of psychic aberration and scientific anomaly alike, the archive of human memory is submitted to inordinate and, in media terms, uniquely revealing strain, whether by the vagaries of subjective temporality or by the machineries of its simulation in film, video, laser holography, computer graphics, even photography still: still photography, that is.

TIME SPACED OUT: CINEMA AND THE "NEW TECHNOLOGICAL MEDIA"

Time effects in the cinema, when experienced directly rather than just as an inference from motion, are always, according to their greatest theorist, touched by the virtual. Memory effects included. Deleuze sees this as the overarching aesthetic of postwar cinema: that it revisualizes (and thus rethinks) this imaging of time. In an extreme ontological convolution, what Mangold's film *Identity* does is give us a simulated killing spree in the form of a delusional screen memory credibly present to us in the shape of an ongoing plot. Yet it is a plot that in psychic fact buries—even while it submits to fantasized repetition—an earlier homicidal rampage. The film is therefore quite explicitly *about* the construction of time by and in the unconscious. This is the very unconscious whose still open wounds the artifice of temporality is always mounted to defend by conscious reconfiguration. And for which the filmstrip, even digitally upgraded, offers a handy emblem in its own raw seriality.

More recently than Deleuze, another notable theorist of time in and on film has come to similar conclusions about the figuring rather than the transcription of elapsed temporality in screen narrative. So let me quote from the latest study of photography's relationship to celluloid cinema as a generator of temporal form (or, in other words, of narrative): "I do not think it is too far-fetched to suggest that in the cinema, as in psychoanalysis, time is produced as an effect, at least in part to protect the subject from the anxieties of total representation generated by the new technological media." Such new media, along with contemporaneous and revolutionary developments in neuroscience, would inevitably put the temporality of the screen image at a distinct premium and under special stress. In an epoch of proliferating data storage, time as process must be rescued from sheer accumulation. At the medial as well as psychic level, or at the level of mentality *as* medium, time becomes, as in psychoanalysis, a "symptom" of all the repressed indiscriminate traces that can't be sequenced by mental process—and a protective symptom at that. This is the gist of the recent study I'm still paraphrasing. "Narrative," it claims, "is film's natural way of dealing with its own fear of, and failure at, total temporal continuity"—especially at a cultural turning point when "new . . . media" are contributing to a shift in our whole cognitive paradigm of human perception. Cinema's reaction formation to this burden of the continuous is form per se, imposed rather than disclosed from the raw matter of replayed visual duration. This is the cinematic form achieved by an intermittent deletion or reordering of photomechanical storage: the elisions and leaps of narrative itself.

Imagine a film plot that took something like this narrative function as its subject. Not hard to do, as we know, since there's a new one announced almost every couple of months from Hollywood, and has been for several years: suspense thrillers (of sci-fi and supernatural varieties alike, or even in psycho/slasher formats like *Identity*) where a spatialized sense of the human lifeline makes for an "unconscious map" of memory (to use the lingo of *Eternal Sunshine of the Spotless Mind* from 2004). This is a map of once-lived and now-virtual time whose pathways of recall, in this film and others, are imaged before us by retracing and erasure at once. Where postwar cinema was for Deleuze no longer a cinema of the movement-image but rather of the time-image, we are concerned instead with something since and other—even if it will come to seem a genuine version, rather than a willful perversion, of Deleuze's modernist touchstone. Within a broader social ethos as well as an emergent aesthetics of computer-generated images and digital editing, our topic lies in the way hybrid mediation often seems to shift screen representation toward a cinema of the timespace-image, an image irrationally reified and portable. This is a field of fanta-sized presence both arbitrary and reversible, its duration neither a thrust or vector of motion nor a layered "sheet" of time but rather the glimpsed relativity of each to the other in the immanent technology of the digital array. Desire itself has issued in the hallucinatory retrieval mechanisms of the unconscious, but newly instrumentalized by supernatural exertions of the will.

Time's pace gives way to timespace, no longer carried by analogy on the palpable materiality of celluloid succession in the photogram's automated advance. Once again: a quantum leap from such mechanical frame time of the track to its digital succession in the shifting binary field. It is this latter mode that often takes narrative shape in the framed time of desire's fantastic commerce with a full-screen and reinhabited past. To recognize in such plots how the ethos of electronic similitude plays hide-and-seek with its own medial aesthetics will require several examples of this new Hollywood subgenre and its recurrent narrative cliché: the psychic break of temporal transmigration. To identify thereby the subgenre's leading feature as the trope of temportation is mostly to emphasize the separable timespace-images suddenly rendered commutable by plot.

But in anticipating this outcome of the remaining investigation, let me quickly add that time is out of joint here in another way as well. For I've gotten decidedly ahead of the chronology intended by the source I quoted three paragraphs back. Way ahead. In engaging with Mary Ann Doane's arguments in *The Emergence of Cinema Time*, I've laid stress on her most tantalizing claim: that narrative structure is a protective function bearing out a Freudian sense of time as a repressive mechanism of selection and deletion. This is a temporality that operates by a "flickering" intermit-tence not unlike film's.[1] But I've lifted this claim quite out of its intended historical context and catapulted it a century forward, where it also may seem strangely to belong. Doane herself isn't speaking about the "new technological media" in our sense, a sense grown lately into a curricular designation and a publisher's niche. Or about current neuropsychology. The media in question, and the psychophysics they both test and model, have nothing to do with turn-of-the-millennium electronics. She is talking instead about "new technological media" at the turn of the last century, at the origin of cinema rather than its escalating digital eclipse. This is where the recording capacity of analytic serial imaging, as in the work of Marey and Muybridge, combined with commercial developments of snapshot photography, once they had been animated as cinema—and before cinema regrouped its forces as a narrative rather than a strictly mimetic medium—led to the cognitive defect of a technological virtue: too much automatic storage, too little story.

If cinema, at its advent, was only the continuous "death mask" of the world in time—the faithful, slavish trace of event as attraction, as monstration—then its vitality as an art, if not its viability as a medium, was certainly at risk. I'm thinking again of André Bazin's influential later metaphors for photography: first, as "time embalmed," and then for cinema as that greater astonishment yet, "change mummified."[2] The downside of this alchemical (and photomechanical) magic was all the clearer under pressure of the "new media" (in Doane's sense)—the other new media, that is, besides film: phonography, typography, and especially chronophotography, primarily that of Étienne-Jules Marey. Here Doane has in mind those aspects of Friedrich Kittler's 1900 discourse network that had decoupled memory (as stored temporal events), along with other functions of human affect and expression, from the work of subjective investment and made its inscription automatic.[3]

Doane's interest isn't drawn to specific films that might narrativize such temporal conditions, but their number is legion. Given her context in the innovative technical mediations at the beginning rather than the end of the last century, it would perhaps be a film like René Clair's 1923 *The Crazy Ray* (*Paris Qui Dort*), from as early as cinema's third fledgling decade, that would best serve to enact, within the science-fiction genre, the particular difference film made to the image of time. Clair's plot embodies this difference in a mad scientist who can, with the yank of a lever, do to the world in fantasy what only a film editor or projectionist can do in fact: change the speed of human action, accelerate or retard it or, more often in this case, freeze such action dead in its tracks. He can therefore achieve, so to say, a Lyotardian transfer from the narrative order into the representational order—or, in other terms, a Metzian reversion of *trucage* from syntax to action (in this case halted action). Either way, this is an updated French version of Marey—chronophotographer supreme—at the seized controls of the real. Such a technological irony seems, as so often in later sci-fi, reduplicated at the level of ethical theme. In this case, once the world machine defaults to the stop-time position, Clair's film moves from sci-fi fantasy to a satire of an otiose utopia of infinite consumption, where all the world's material pleasure lies open to the few moving bodies on the scene, those (airborne when the technological spell was cast) who remain exempt from the magic of arrest. The result, even in its comedy, comes close to humanist parable. Time may be money after all, and is spent in this case without pleasure or mental reward. No value accumulates, nothing of worth accrues. What grows clear is that time is only what you make of it. Meaningful duration must, that is, like cinematic narrative itself, be not so much steadily consumed as disruptively produced. A capacity for total storage (or hoarding) is itself incapacitating, an inert satiety. Time in cinema—that is, narrative form—must break from, by breaking up, the repository of sheer duration and its inoperable continuities and instead activate sequence as discourse. Only then does film become cinema.

The many routes by which this happens, between 1895 and 1995, comprise the very history of cinema as filmic medium. Because it is the decade since 1995 that concerns this book, however, we are now looking at the ways—and means—by which the constructedness of human temporality, as a rescue from sheer duration in the psychic archive, has so thoroughly infiltrated narrative structure on both sides of the Atlantic. Why, we've been led to wonder, has it done so—in Hollywood production especially—at exactly the point when the very basis of cinema's own optical archive is no longer dependably or exclusively filmic, no longer rooted in the indexical nature of serial photography, but permeable to computer simulation—or supplementation—at every level? This is where a cross-check with European modes of the temporal fantastic is repeatedly invited. For the uncanny doublings and hypnotic intersubjectivities of European productions during these same years are put into play with Hollywood melodramas of ghostly protagonists and disclosed virtual worlds. Either way, lived time has been bent and attenuated, if not altogether evacuated.

At this point we can look back over a decade to a founding film of what one might be tempted to call the postmodern time-image, one of the earlier instances of the digital-implant fantasy, in *Johnny Mnemonic* (1995), scripted by *Neuromancer* author William Gibson from his early cyberpunk story by the same name, and directed by the avant-garde painter and filmmaker Robert Longo. Gibson introduced only one major new element into the screen adaptation of the source story—a filmic element at that—but one that would seem all but requisite in the intermedial ironies of its screen version. Or would have at the time—on the cusp of cinema's rampant digitization. The innovation in a nutshell: that single framed images are still the key to it all. This takes some review of plot to make clear. In a future rife with electronic data theft, the hero is a secret human courier with a brain implant that crowds out his own human memories. In an updating of the '70s paranoid thriller of corporate intrigue (a transnational conspiracy here resisted by the heroic efforts of data piracy), Johnny unwittingly carries the secret cure for an epidemic nerve disorder, which has been entrusted to him, without details, by renegade scientists hoping to get the word out. The debilitating spasms of this ailment are thought to be induced by, of all things, overindulgence in electronic data and image streams.

We see it coming: the need for a pyrotechnical climax in homeopathic overkill. So far, though, all we know is that this is a sickness whose treatment is more profitable to the global pharmaceutical network than its alleviation would ever be. It is for this reason that the courier is being hunted down by the Japanese mafia for the encoded formula he unknowingly transports. The whole malign scheme to suppress the cure seems implied in the corporate signature (blend of logo and unpunctuated Web-site domain) when finally broadcast in download—Gibson's sly cross between the Derridean and the digital supplement, poison and cure at once: namely, Pharmakom. For the hero with his contraband brain chip, the mafia threat is decapitation and forced retrieval—with its tacit pun on the "hacking" off of the entire cerebral storage system. His only hope is to download his burden to the right people first, once the secret optical password has been transmitted.

Here lies the most striking metafilmic addition to the story as screen parable. For what will ultimately save the day are a trio of artificially stilled photographic images printed off in a row as the opaque equivalent of a second-generation frame strip. The movie plot contrives this photomechanical reflex as follows. When first uploading the secret experimental information, the hero tells his contacts to photograph three random images unseen by him—he actually calls them "frames"—from an oversized video projection running, quite accidentally, in the background of the exposition scene. Any random images will do, and the video screen is simply the most obvious image source at hand. These "frame captures," we learn, will automatically enter the uploaded program at arbitrary intervals, unrecognized by the implanted subject, and their printout will then be faxed to the American downloader before being destroyed. Disjunct serial images: the very definition of a filmic unconscious. And one of them, grabbed from an *anime* cartoon at that, recalls the very basis of cinema in animated frame advance.

Buried without signifying function amid the encrypted data files, this extra sequence of optical imprints rehearses a whole history of remediation. Before electronic data processing, before digitization, before microchip implants, even before televisual scan lines, the individual frames of a film—in their links back, even earlier yet, to the storage mechanics of discrete photography—were always there, genre deep, as familiar metaphors not only for the human mind's indexically validated rather than cybernetically projected relation to the world, but for a founding discontinuity in the psychic memory bank. And it is quite in keeping with other and subsequent films in this latest cycle of eroded human ontology that, in *Johnny Mnemonic*, when all the foreign, unassimilated data have been scoured away from the hero's maxed-out brain, he reverts to a symbolic celebration of biological origin itself. His individual—and individuating—memories return, that is, to video-cam footage, awkwardly jump cut, of one of his childhood birthday parties and the cake legibly decorated with his name: images that had seemed as if they were trying to break through in flash insets along the way. The analog tape again anchors him in the index.

This videograph of the formative self—placed into narratographic relation with other image systems—arrives in a climactic scene that takes as intertext, no doubt, the comparable last scene

4.1–4.3

of Terry Gilliam's *Brazil* five years before. There, too, the hero, similarly encased in bulking digital headgear, escapes from electronic tampering into a world elsewhere. In *Brazil*, the humanizing flight to freedom from the onslaughts of a totalitarian future is rendered as a kind of psychotic break. Narratology would note the escape from one plot into a fantasy alternative. But its local markers are delivered up entirely to narratographic response. No sooner has the hero's pastoral reprieve in a countryside haven been coded as virtual rather than actual—by a match fade from a long shot of the happy valley to a painted version of it (figs. 4.1–4.2)—than the image, qua image, is interrupted by a jump cut to the same planagraphic representation in the hero's glazed line of sight. Yet this image is spread out only for a few last seconds, as plangent fantastic vestige, behind the looming faces of the brutal experimenters back in the urban laboratory (fig. 4.3), whose real background space soon takes over completely. Snapped back from his escapist daydream to the operating theater of involuntary mental experiment, Gilliam's hero exists momentarily (in a full-blown "interface effect") against the impossible backdrop of his own fantasy, hovering between worlds until the scientists leaning over his electronic headset admit, "We've lost him."

By contrast, Johnny Mnemonic, similarly helmeted by electronic apparatus, retreats to the real rather than a fantasy. The time-access functions of memory are confirmed as the very

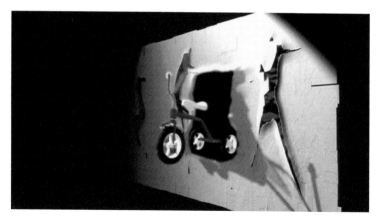

4.4

fundament of the human subject, and this in optical allusion to a benchmark instance of the Deleuzian time-image in the Rosebud thematic of *Citizen Kane*. Instead of the sled as memory vehicle in Welles's film, Longo shows Johnny willing a boyhood tricycle to plunge through the tabula rasa of his cleansed mind (fig. 4.4). A past real can thus penetrate the present, carrying Johnny forward (hence back) to his birthday tapes as internalized home movies. Retrieved again at last, they are life defining. "Happy Birthday Johnny," the remembered childhood cake spells out in close-up, with a real biological mother looking down in human validation at the implied video lens.

The hero thus recovers the moment when his own desire could anchor consciousness in temporal moments that were actually once his to record—rather than foreign data implants to which his mind (rather than mere vehicular brain) has no access whatever. Plot has come full circle, both in narratological and in narratographic terms. In Manovich's duality between cinematic *identification* and computer *activity* (or interaction), here the implanted body that could only activate its own installed interfaces has broken back to self-identification. This is the hero who, in the opening scene with a prostitute—his story's own postcoital "trick beginning"—couldn't tell her, in his lassitude, where "home" is. By the end, he has recovered an image of that point of origin by shaky tape replays in the last scene. Operating against the drift of a digital marvelous in full CGI barrage, this is the recovered past brought to mind's-eye view again in that remediated *heimlich* of handheld video. Either way, though, the old sci-fi question, often turned premise, still persists as formulaic matrix. What indeed if human life were like a movie—or like a digital implant? In this respect it may be useful to think of *Johnny Mnemonic*'s plot, its storyline per se, as marking a transition between the conspiracy thriller and the new ontological gothics of digital futurity. For in the newer dystopian plots, the hero ends up being in doubt about his own original identity as human, rather than just about what he's sold off to digital expediency. Johnny knows he wants to reclaim his full human memory, to clear space in his head for his own past. In contrast with the fate of the same actor (Keanu Reeves) four years later in the role of Neo in *The Matrix* (genre typecasting), there are in *Johnny Mnemonic* no buried surprises concerning the reality of that past, just the suspense involved in the quest for its restitution.

Films at the other end of this first postfilmic decade, however, with human identity no longer imperiled by technology but already subsumed to it, offer narratives that often seem no longer in need of the single-frame trace of presence as either fetish or evidentiary benchmark. The photographic insert is often foresworn as touchstone or secret clue, let alone as indexical validation. When photography appears at all, its very existence is likely to be exposed as fake or phantasmal: the contrived index of a cybernetic (*Dark City*) or a psychotic (*The Forgotten*) past. At the sci-fi pole, many such films tend to press the idea of an entirely virtual or simulated (rather than represented) humanity, trapped in the coils of ontological irony. In these plots the subject is often discovered in the end to have been tricked into thinking of himself as human—or, for that matter, as there at all—whether the trick is technological or a supernatural twist of fate (and fatality).

FREQUENCY MODULATION: WAVES OF PASTNESS

Plot options often alternate in recent Hollywood sci-fi between the digital implant whose passive data elbow out the unconscious basis of human temporality and the digital implant that can upload it wholesale. At the same time an aesthetic of the virtual has lately overtaken Hollywood filmmaking at the level of discourse even when not of plot. The work of computer consoles becomes the tacit and facile model for the escapist—and consoling—reconfiguration of one's own timeline. Before sampling these effects, again a comparison with European practice is instructive—and this again in terms of the "new media" as the nineteenth century might have understood the term: in precisely their capture of duration as inscription.

While inventing the trace methods that would lead to chronophotography and on to cinema, Marey developed that more directly somatic device known as the electrocardiogram. In a recent fantasy film of the erotic uncanny rather than the technological marvelous, the French psychodrama *He Loves Me, He Loves Me Not* (Laetitia Colombani, 2003), an electrocardiograph appears out of nowhere at the film's turning point, detached from any hospital setting as an abstract *mise-en-abyme* of the rectangular film screen, a video trace rather than a celluloid sequencing. This screen-within-the-screen functions in context as a metonymy for death converted to a metaphor of restoration, of revised narrative story. For at midpoint in the film the heroine has turned on the gas to kill herself. Only after we see her videograph lapse to flatline, then her pulse rate return, on this alternate and inset electronic screen—only then does reverse-action footage lift her back to life in a cinematic register. But with a narrative backlash as well as a visual reversal. We had thought we were watching the story of an adulterous cad mistreating our jilted heroine. Now, as the film visibly rewinds to its opening shot of a flower stand ("he loves me, he loves me not," petal by petal), we revisit the stages of their relationship both from new camera angles and from a clarifying perspective. This time we realize the whole sexual relation between the heroine and her neighbor is entirely a projection of her "erotomania," which drives her into the role of a violent stalker in the second half of this exposed *folie d'amour*.

It bears repeating that the technological turning point is not a full invasion of plot by discourse. The heroine doesn't come back to life from a finished death in her kitchen. Replay has been only a discursive function, not a biological anomaly. Nothing is really relived by the characters, as it might be in Hollywood cinema. But the effect is nonetheless an extreme wrench to medium if not mortality. The cardiogram screen has triggered a double remediation, first of electric sensorimotor tracing by photogrammatic cinema, then of cinema by its own succession in the electronics of videotape storage, in this case scanning backward faster than any film spool would permit—and effecting in this way a break from the ocular protocols of the modular celluloid strip.

That's about as techno as the European uncanny tends to get. By contrast, even the hackneyed device of fast-forward footage in the time-loop narrative of the 2001 American cult film *Donnie Darko*, directed by Richard Kelly, marks something strange happening in or to the character, not just to his story. Time-lapse cinematography (unlike at the end of *Adaptation*) is not just a discursive estrangement of temporal progression in *Donnie Darko* but a weird turn of event: the prescient mind racing forward into its own destiny. The pattern is familiar, and not just when made apparent in this film through cinematic devices like time-lapse acceleration. Other strategies of what I would call mnemonic commuting actualize the metaphor of the relived past in modes of more particularly *digital* rather than celluloid distortion. Temportation has a way of rendering literal certain "sheets" of time (Deleuze's famous recurrent metaphor) as layered planes of either past or future, anticipated and returned to by schizoid will. As such, they go beyond the topologies of time in Deleuze to become whole portable topographies. These planes of elapsed or pending time, of past or possibility, are repeatedly brought to full-frame interactivity by the characters themselves through special effects of digital editing: most often a precipitous optic plunge through space itself as if it were time's own inroad. The digital effect known as the "cosmic zoom," so rampant of late in sci-fi editing (as used in reverse-zoom form in *The Matrix*, for example, to mark a quantum shift in scale from microchip to metropolis), is a device of sheer rhetoric that—in shifting from spatial to temporal parameters—becomes metaphysical even while no less clichéd.[4]

The only ingenuity associated with such cosmic zooms comes from the wild variety of diegetic motivations they seek out for themselves. The end of *Donnie Darko* provides a ready example. In the film's opening sequence, the medicated bipolar hero is presumed dead by freak

4.5–4.6

accident in his bedroom until he is found coming awake on a nearby golf course, as if it were all a nightmare. The nightmare comes true when, in the final twist, he is willing to accept this previous death in the name of the girl he loves, to spare her the accidental consequences of his living on. This time we see his deathbed replay from the all but cosmic point of view of his technological nemesis. In (let's say at least) a stratospheric zoom, we track the path of a detached jet engine plummeting through the sky. It will arrive just a second after the camera through a match cut between a hole in the cloud cover and the death-reflecting eye of the Escher *vanitas* print next to Donnie's bed (figs. 4.5–4.6). This is the famous etching that may well have offered the graphic intertext for Julio Medem's corneal mirror in the fatal last shot of *Lovers of the Arctic Circle* as well (fig. 2.13 again). Marking in the American film the final moment when Donnie himself must look death in the face, the print has thus been materialized, with its explicit visual play on the eye's own lens, by a graphic match that evokes, at the end of a digital plummet, nothing less than the old-fangled closural device of the so-called iris shot. That's in a sense the film's own metafilmic nostalgia on exit.

Classic film editing is more openly overridden by the digital in other and more recent modes of spatiotemporal tunneling. In many films of this time-warp subgenre, intercutting (which once allowed the invention of cinematic time out of filmic duration) is replaced, under the overt pressure of "new media," by special effects of funneling rather than shuttling, so that time present is repeatedly vacuumed up by past or future. The spatial disjuncture of the splice becomes instead the temporal vanishing point of the digital chute in a whiplash-like, postfilmic zoom. The smallest unit of ocular change has become not the cellular trace of the photogram but the pinpoint pixel, through whose desperate escape value whole time frames can be "minimized" and remade. This is to say that the instantaneous zeroing out of the movement-image, in Deleuze's original dyad, becomes on the shrinking or ballooning spot what I have called a timespace-image. And need one add that the computer-driven technology thus evoked and refigured is evident if only by allusion? Formal innovation is once again crossbred with sociology. Digital editing as discursive intrusion has, regardless of plot, its recognized intertext in the commercial culture of electronic fantasy. The "travel" function of interactive video, a kind of optic teleportation, combined at times with the "first-person" option of role-playing games, gets translated to the screen in recent plots of the mnemonic gothic, sci-fi or not—and with it many of the more pedestrian frame shifts of Windows graphics. The result, again in Elsaesser's distinction: "navigation" as a breach in the normative succession of "narration."[5]

The watershed moment of the last decade in Hollywood, we might speculate, comes not only with the first of the *Matrix* trilogy but with its immediate predecessor in *The Thirteenth Floor* (Josef Rusnak, 1999), where the hero's "real" is only someone else's imaginary made symbolic, digitized for transference. Given its famous prototype in the ambiguous photographic index of *Blade Runner*, that earlier cyborg anxiety seems by now decades old and worlds away.[6] As it is. In *The Thirteenth Floor*, the cyborgian specter of the freestanding artificial being has been reduced to sheer cyberchip. When the hero stares at a photograph taken of him alongside his partner, mentor, and father figure, and does so at exactly the turning point when he realizes that they are both unreal, the moment marks a final impasse in the ontological thematics of photography in dystopian sci-fi. Yet such a dead end is only a kind of new beginning in the subgenre of temportation.

What seems more surprising in twenty-first-century films since *The Thirteenth Floor* is the detachment of time-travel narratives from the digital thematic altogether, so that all electronic imaging, no longer downloaded within plot, is repeatedly offloaded onto technique—as we will see in our remaining examples. More surprising—but only at first glance. On reflection, it seems like a sustained conversionary strategy. Hollywood cinema thrives now, even at the plot level, on a parasitic relation to the widespread fact of interface culture, a fact it reinscribes in the form of a fantasy. Rather than attempting to satirize or otherwise contain the superficial optic fields of computer gaming, for instance, that have in fact outrun Hollywood in annual profits, the plots of these films effect a repression of the digital competitor while borrowing its effects at the textual or discursive level. Despite the frequent teenage context of their estranged time-travel stories, from *Donnie Darko* forward, the hero's agency is stripped of its natural ambience in the culture of computer games, including the identity exchanges of their interface performances. *Donnie Darko* achieves this by casting its plot back to the '80s, as if for no other apparent reason than to foreground cinema's native capacities for time travel through *mise-en-scène* itself. The deeper motivation may be that here is a hero who can play sophisticated mind

games well in advance of the mass commercial technology that would facilitate, instrumentalize, and ultimately domesticate them. And deeper yet: that here, too, as half a century earlier
in cinema's scramble to best incipient commercial television with increasingly wide-screen and
immersive formats, is something cinema can do even more spectacularly than any home screen.
And for which plots need therefore to be found.[7]

But backdating the cultural setting to defuse the electronic aura of the plot, as *Donnie Darko*
may have intended, is one way to mute the digital intertext—and so float the premises of a less
technological fantastic. A more direct approach is to backdate the technology itself—or, in other
words, to unplug the interface more explicitly. This happens in the next year's intergenerational
fantasia, *Frequency* (Gregory Hoblit, 2000). A plot whose emotional crescendo is based, most
recognizably, on the new frenetic immediacy of electronic "texting" is cast back three decades
to a ham-radio fixation and then given some dubious astral overtones to lend supernatural aura.
In a kind of "trick beginning" of its own, hints of extraterrestrial (or at least stratospheric) intervention are ultimately extraneous to the plot, even if a clue to its electronic associations. It
is only when the aurora borealis is particularly luminous in its electrostatic discharge over the
New York City area that a father dead for years is able to make radio contact with his present-day
policeman son in their Queens living room. But this strained evidentiary gambit arranges that
another kind of luminous airwave conveyance coincides with this mystery, indeed remediates
and confirms it. For we first register the time lag between dead father and still-grieving son
in the difference between the televised image of Dick Cavett in the background as the father
bends over his radio transmission (Cavett interviewing a physicist about the northern-lights
phenomenon) and the staged broadcast of a much older Cavett (with the same now-aged scientist)
seen behind the flabbergasted son's radio reception from his dead father. This is an economical telegraphing in its own right of the plot's fantastic disequilibrium. Yet it is also a thematic
redundancy. Even without this simulation of a later broadcast, the earlier TV image makes a
point all its own in this context. Media comprise the only place where you never die, where
your image is preserved and reproducible again in real time. That's a kind of metaphoric fact.
All the rest of the plot that embeds it is fantasy. And as the dead communicate with the living
not figuratively but literally, the electronic prototype for this 1970s fantasy becomes more and
more palpable even though never pronounced. In their wireless instant messaging, though on
radio still, all it takes is to click and send.[8]

By the labyrinthine plot of *Frequency*, the image of the prematurely lost father will be mi-
raculously restored to its rightful place in a family photo—a trick of photographically reframed
time deployed in *Back to the Future* as well. But this photographic litmus test of past presence

soon takes another turn. In subsequent photos, suddenly it's the mother's indexical trace that is no longer there. In an irony more violent than usual, but quite familiar in the time-loop plots of a tampered past, the son realizes gradually that his rectification of a previous firefighter's fatality and trauma has not paved a direct route to a better future. In the vexed economies of intersecting time lines, a bend in direction has led instead—by the unwitting actions of the surviving father—to his mother's brutal murder by a serial killer in a homicide case that happens, improbably enough, to have been just lately reopened years later at police headquarters.

Answering to and reversing the first half of the plot, then, it is the past that must now intervene instead in the present: a fantastic event here too, certainly, but moving closer to the norms of memory and internalization. Only by getting his father to hunt down the killer in the past they both inhabit will the detective son and hero be able to see his mother's bloody photos disappear from the forensic file at the station—and to find her image materialized again in the family pictures back home. Enlisting the father's aid involves their transferential voice contact across the generations: a Deleuzian *sonsign* of time itself. The effect, achieved by the *opsign* of a marked visual emblem as well, comes across as an unmistakable time-image—on the way to a new self-image for the son.

Given the schematics of this film's convoluted plotline, narratology would immediately note the return to origin as a reauthorization of self. Narratography sees further, mediately, how this return is graphically enacted. Given the technological anachronism of the ham-radio apparatus, the resolution of yet another Hollywood plot transpires as a displacement of the computer console into the consolations of the supernatural. But how has the digital intertext shown its true if veiled colors on the narrative screen? The preparation is deliberate. With the film seeking its own metafilmic equivalent for the radio séance, its communion with the dead begins in standard cinematic terms: an impossible suture across the decades. This treatment via shot/reverse shot then mutates into something even more preternaturally remediated by optical *trucage* (and digital compositing): two discrepant images of both father and son looped over each other (fig. 4.7) as if they were facing off for conversation in a continuous space, rather than "projecting" their voices over the abyss of years.

In this spectral abutment, technique once again inherits the computerized opportunities forbidden to plot (by the rule against anachronism), even while maintaining them safely within an allusion to old-fashioned cinematic trickery (sheer superimposition) circa the 1970s setting of the narrative itself. Visualized within the frame is an intersubjective circuit in which temporal lag (even across the mortal divide) is swallowed up in the graphics of bridged gap—or, in other words, of an interval vanished into interface. No less effective as a trope of temporality

4.7

4.8

in this same film, however, is an instance of momentary optical mystification that proceeds without any trace of *trucage*. The device is far subtler than the laminated faces—and an equal candidate, in its eventual "interface effect" (Žižek again: the self in frame with its redoubled image)—for a Deleuzian time-image. This is the way it sneaks up on us. We may well assume that a spectral crossing of time is at play in an abrupt scene change from a glass pane broken by the father in the past to its unrepaired destiny years later: a planar frame from whose adjacent rectangle the son seems to emerge by time-lapse superimposition (fig. 4.8). In fact, this crystal-image or *hyalosign* (one of Deleuze's key terms for the symbolic refraction of temporality within the screen frame) is only a half-translucent mirror image in one sector of the damaged space divider, not in the frameline itself. We are already back in the present, that is, before the hero materializes there by the normal optics of reflection. By way of a seamless cut forward rather than a lap dissolve, this visual ligature confirms—as well as optically figures from within its image—plot's two-way transit not just in space but in time.

What, in short, does this whole plot sound like a parable of? A disempowered young male, unanchored in history, detached in his personal relationships, his commitment to maturity unfathered by his past, finds alleviation only in conversation with a virtual interlocutor through a desktop technology that is at every chance graphically figured for us, if not exactly for him, as a visual interface with an impossible past—and all this made inexplicably possible only by the mysteries of a scintillating electromagnetic ionosphere well out of sight and mind during the throes of manual contact. In this living room turned haunted chat room, the real allegorical frequency is not shortwave but broadband—and then some.

Nonetheless, beneath the fantasy premise of the plot, there is no final mystification about the image's relation to time. Whether electronically generated or indexically recorded, picturing

4.9–4.12

excludes its referent. Having recognized earlier, in connection with the archived TV footage, that mediation is where you never die, we are asked later in the film to entertain the reverse axiom. This, too, is familiar. Mediation is where you are always and ever dead. In *Frequency*, then, it is only when *trucage* returns from ambivalent mirage (the hero reflected in frame) to overlapping montage (superimposed father and son) that the unsentimental truth of mediation can be laid bare by the film. This happens, out of all stylistic keeping or narrative motivation, in a rapid bravura dissolve (a fourfold visual chiasmus) from a backyard snapshot session in the lost past—across a freeze-frame—to a decomposed corpse in that long dormant serial-murder case in whose solution the dead father must now be recruited, the latter negative image fading at once to positive for the resumption of plot (figs. 4.9–4.12).

With its visual simile of mortal remains, the match cut on the generative negativity of the photographic process itself is a kind of parable on the run: about the heuristic function of mimesis in regard to the actual world whose perception it suspends. Belief in the world doesn't founder on an awareness of the world's death and evacuation in photomechanical representation. That absence is what throws us back, reconfirmed, on the endured real. Deleuze didn't live to see the frequency of films like *Frequency*, or their equivalents in technofuturist virtuality, but he has nonetheless helped us read their *lectosigns* well in advance. One name for that reading is, again, narratography. With *Frequency*, the case is unusually clear.

A narratology of the biographic loop is tactilized, made almost palpable, by devices of the strip that call out for a corresponding narratography of temporal overlap, whether in the doubling or the subtraction of presence, whether in a past recaptured or a moment encorpsed.

Even when not set back into the past like this film from 2000, or like *Donnie Darko* just after it, the pattern can be generalized. In the latest films of temportation, the role-playing apparatus of digital gaming, with its collapsed distinction between subject position and the otherness of image, to say nothing of texting's real-time digital telepathy, is excluded within the ethos of scene and setting so that it can emerge as the pathos, and the agon, of narrative at large. Certainly, big-screen heroes on a role-morphing quest to reprogram their own logged past time can't afford to have this ambition trivialized by mere electronic pastimes on the small screen. Trivialized—or prefigured. As we have seen, the twist ending regularly depends on its not interpolating the interactive or video game model from which the seductions of the digital, both their logic and often their marginal look, have in fact been extrapolated. Only in this way do these plots stay anchored in the realm of the fantastic: hovering undecidably between schizophrenic dissociation and the actual command of paranormal powers, between bipolar disorder and such counterhypotheses as apocalyptic premonitions and telekinesis. And they remain there only because trapped with their heroes, bracketed, artificially sealed off from their full cultural surround. This is to say, yet again, that under pressure of the "new media," cinema a century after its inception still produces narratives whose ingenuities of temporal construction come from defending themselves against the capacities of total storage, now digital rather than photomechanical.

REMODELING NARRATOLOGY

At the century's turn, there is more, too, that doesn't meet the eye: a yet deeper relation between narrative structure and digital templates. In certain branches of cognitive narratology, the protocols of cybernetic programs and subprograms, the latter called up by the former, have been enlisted as a further modeling device for nested or embedded narratives in even the classic mode, as for example *Wuthering Heights* or *Heart of Darkness*. So-called frame narrative suddenly seems like a misnomer. Instead, we find the paradigm of vertical "stacking," where the framing material no longer exists on the same level as the boundary that has been crossed to access it. Rather, the secondary narrative or scene is "pushed" upward—hence forward—to claim more fully the cognitive or semantic field, disappearing only when it is "popped" from view—from the stack—to reveal the founding parameters beneath it.[9] On this model, one thinks of the so-called embedded narrative almost in reverse: as layered over and temporarily eclipsing the host text, which reclaims attention only when the subprogram has been exited. Stacking and popping find their approximate geometric equivalents in the graphic layout of Windows operating systems, of course, where the subfile is layered across the field of view until minimized or closed out. And this in turn has its further graphic equivalent in the cinematic disappearing acts of time present, coded as they are by digital transfigurations of the screen rectangle in recent twist endings, where the ontological "pop" can be lethal.

It seems no surprise that this brand of narratology is most cogently promoted by a scholar whose previous work concerned prose narrative as "virtual reality." This rubric does not apply in the overarching sense familiar from Todorov. Marie-Laure Ryan's point isn't that fantastic virtuality (by being both fictive and interpretively unfixed) is the quintessence—and exaggeration—of the literary, with its defining effects being only dubiously real for the characters themselves. Rather, under the more recent sign of cybernetics, the offered parallel between narrative and virtuality concerns mostly the reader as a kind of interface operator.[10] In neither phase of Ryan's work, however, is the proposed scientific paradigm—whether computer platforms of late, or interactive video earlier—meant to be technologically specific. Nor, therefore, is either model necessarily anachronistic when applied to traditional narrative forms. This is at least her assumption. But there are decided correlations, nonetheless, in recent screen plots: moments where the fit between a formal model derived from high technology and the actual contour of a narrative event grows unmistakable. In recent sci-fi, Ryan's alternate digital master tropes for action in narrative can be converted, that is, from structural latency to dramatic agency. Once again, as with subtextual repression and return in Riffaterre's schema, both sci-fi and fantasy plots find ways for narrativizing (or thematizing) their own organizing logic, turning it from principle to incident. Thus, for instance, does the virtuality of all fictional narration come bearing down on the postfilmic screen narrative of virtuality.

Take the very title of *The Thirteenth Floor* (1999), the unlucky number that many skyscrapers superstitiously forego. In this film, it locates a secret VR lab by *figuring* an interspace that doesn't in fact exist: the hallucinatory zone of its own digitally generated time travel. Edited narration in cinema was, of course, born as a mechanical device to rescue plot from raw duration. Now that the medium can imagine its own eclipse by more advanced time machines whose strata of simulated reality do not depend on indexical transcription at all, but merely on the electronic generation of fantasized lives, the pasts of these lives are not archived as imprints but contrived by pixels. *The Thirteenth Floor* is one narrative approach to this topic. Its trick beginning is in this regard slyly metafilmic—if only in immediate retrospect. Special effects, briefly normalized by expectation, are soon realienated by plot. After ten minutes or so immersed in the digitally enhanced (no surprise

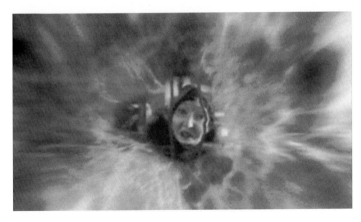

4.13

yet in current screen spectacle) sepia glitz of a reconstituted 1937 L.A., an underlying cognitive plateau bursts into recognition when the fantasy past is "popped." What now confronts us is the zone of an instrumental present rather than a virtual past, of simulating rather than simulated. This zone is inhabited, we learn, by our contemporary scientist-hero in turn-of-the-millennium L.A. In the film's lingo, he has merely been "jacking in" to his own virtualized fantasy of a more glamorous past.

The way up from the confected space, and then back in again, has been, as befits the VR subject matter, technologically figured. Yet the visual image of this transit between levels—and the temporal false bottom through which popping then drops the hero—hovers, as special effect, between metonymy and metaphor. Though linked to computer electronics as digital cause, of course, the visual effect of actual temporal metamorphosis, or temportation, is conveyed more as a metaphor than as a credible picture of such digital migration. The magic access of the hero's "consciousness download" appears by digital effects as a kind of lightning-storm vortex or wormhole penetrating the eye of vision itself (fig. 4.13), whose time tunnel, by a rapid exchange of shots and reverse shots, is then sutured, more fully than with any video game, into a virtual past. It is a familiar case of the cosmic zoom mapped onto time rather than space, or onto the virtual space of a false time.

But wait—as plot will have you do. According to a computer-modeled narratology of the stack, one of the first rules is that pushing and popping can operate only from the bottom up, a shift in levels built upon the stable parameters of its own base. In the kind of recursive formats discussed by Ryan, the radical trick comes when a deeper diegetic ground is inserted below the level we took for the baseline of reality, converting that former real to a phantasmal projection of a more fundamental plane of action. Those unreal photographs of the hero's supposed past, discussed earlier, are the first sign of this in *The Thirteenth Floor*. Without referent of any kind, they are pixilated from the dot-matrix (rather than the halftone-dot) ground up. Yet even as he realizes this simulated trace of his simulated past, the hero is by no means quite prepared to have his own present reality logged off as someone else's elective and contingent interface.

In a narratology of the stack, we are concerned at this point with the violent ratcheting down to a new and negating layer of manipulation, mitigated here only by the self-exonerating sleight of hand in the film's surprise ending. It so happens that the urban present, as we've been seeing it, as in fact we know it in the year of the film's release (1999), is only a "retro" video simulation operated by a sadomasochistic porno gamer in the future. He is none other than a VR headset role-player whose first-person surrogate is, or has been, our hero. In this technological upgrade of the doppelgänger plot, it is a case of the user used, the player played. No problem, though—either for the reception of violent sci-fi or for its video tie-ins. This is what I meant by self-exonerating. As we learn from the *L.A. Times* headline that confirms, with its 2024 date, the hero's glazed recognition of himself as his own future double and prototype, "Crime Rates Are at an All-Time Low." Vicariousness seems to be working its pacifying wonders yet again—and has been all along. This is the case even while our solicited viewing has been stacked and popped.

So it is that sex and violence, always interactive to some extent in the "virtuality" of screen fiction, no matter how much they may overload the viewer's identification, still stay safely within bounds. Here, by parable, this means that such transgressive energies remain inside the navigable parameters of the data system, limited in this case to a gamer's electronic scores rather than any real-life encounters or victims. Still, even at the strictly optical level, suturing an electronic proxy into one's own line of electrographic sight is a short-out in the system of otherness and recognition. Especially from the other and bodiless side of the transaction. Narratology logs the temporal slipknot of storyline. Narratography charts its transtemporal suture as a nugatory match cut: the suddenly disclosed subject brought not face to face but into a strictly virtual alignment—and only across the void of mediation—with his interactive surrogate.

In the realm of the virtual, of course, plugged or unplugged, anyone's present is always someone else's perceptual past in the making. That's time's catch. Todorov, as we know, is quick to acknowledge that an antecedent model for his own differential paradigm of the fantastic—as a vanishingly thin line between the uncanny and the marvelous—comes from philosophical definitions of time present (among which he would no doubt count Bergson's). This is the present when conceived as merely the sliding difference between an accumulated past and a looming future. By converting this philosophical premise to an ontological loophole, numerous recent films position their subjects, heroes and viewers alike, in this mutable limbo of strictly differential time. It is there that temporal dividing lines erode, centers don't hold, and things repressed collide with things deferred on the sliding and elided ground of a relativistic and layered present. These are narratives, then, where both cognitive paradigms of the stack and genre paradigms of the undecidable sliding spectrum are each made to swallow their own tails in the trick loops of closure.

Whatever their ethics of desire, and however far from electronic gadgetry they may contrive to keep their heroes, if not their technicians, these are also narratives whose at least tacit aesthetics of the virtual cannot be prevented from toppling over into a politics of the unreal. Even with the best of intentions—even, that is, when committed at the level of character to embracing the real at all costs, as happens in the closing conversionary experiences, say, of *Vanilla Sky* or, as we'll see, of *Eternal Sunshine of the Spotless Mind*. Plot, or at least technique, has usually done other work, pulling the rug out from under its own ontology to that point. With the earlier mode of conspiracy thriller from the 1970s, Fredric Jameson taught us to see how the paranoid logic of deception's totalizing network made for a formal as well as a narrative category: a kind of double complot. Now, with the virtual thriller, whether in electronic or supernatural form, whether populated by ghosts in or out of the machine, the conspiracy has caved in upon plot as the exposed evacuation of its own lived content. That's the real paranoia now: that there's never been anything really there. As often in these matters, the paranoia seems justified. It is as if Johnny Mnemonic were to discover that the lost interval of his youth, once disinterred from the digital wreckage of his brain, was not an interval at all, vouched for by video duration, but just another implant-generated bionic interface.

Oppositely, in one climactic and epitomizing instance of the virtual already examined, the perversity exerted over the structures of memory itself, and its mechanisms, is complete—and exemplary. This is the moment of fantasized photographic memoir at the end of *One Hour Photo*. In that film's final break from sutured into purely hallucinatory space, the collapse of self into its double could hardly be more sharply focused around the very logic of "projection" in the psychoanalytic sense: the empty mnemonic of a self artificially welcomed into an illusory group image (fig. 3.27 again). *One Hour Photo* ends in a wish-fulfillment fantasy materialized by a trope of the medium itself. The man for whom the inertial momentum of time bears always the pressure of intolerable memory is assuaged by absolute stasis, however unreal.

More often of late, at least in Hollywood's ontological gothic, the unreality is pervasive and entirely unwilled, imposed upon a protagonist in the form of his or her very negation. A recent essay, wishing to recuperate a wide array of these films in the modes of hallucinatory and digital illusion alike, to rescue them for ethical weight, identifies their shared premise as that of a "slipstream reality."[11] But besides the content of such monitory fables, my point is that both the structure of fantasy as genre and the manipulation of screen image as medium can mark ethical parameters as well—or their forfeit. The continuous equivocation of the real, even when it keeps viewers on edge, can blunt their sense of consequence. In the grips of the immaterial, nothing finally matters. If the character we took to be a hero is only an imaginary figment or a dream double or a ghost, we leave the theater with a certain indifference, absolved of identification, of credence. We accede to the unreal, which is in its own way a highly political act. In evaluating the cultural valence of the fantastic, then, one must often look past cinematic content to a filmic form that may reconfigure the premise of a given plot as an axiom of cinematic mediation itself. And that mediation is still filmic enough at this point in cinema history, however much electronically doctored, for reflexive ironies to take root in the "slipstream" of the filmstrip itself, its enchained photographic traces. It is at that level where screen narratives often replot their own normative condition as a fantastic deviance, the medium reasserting itself as ontological nemesis. This is because photography vanishes into its own filmic ghost as cinema: montage as apparition.

Yet this is in turn because photography, so we have seen, has a dimension of the fantastic in its own right. In picturing that woman there, from way back then, photography is, as noted, the moment of her transcribed past becoming my perceptual future across the indexical interstice of time present, time presented for capture. Duration is cancelled within the image and displaced onto its aftermath as object. Film then redoubles this function by giving us duration itself as visual object. It offers, and is defined by, that continuous reanimation of the photogram that takes the form of its relentless erasure: the mutual vanishing and eventuation of image in the mode of motion. Stilled time is recirculated as time still—the time of spectacle rather than its sequential registration. All cinema knows this. But certain films find a use for showing it, rather than just depending on it. In the present context, what comes through is a second-order recognition that is often routed back into plot as its final undoing. Over and above the inherent uncanniness of photography, that is, we find in cinema the fantastic of the still image in spectral animation.

TIME(D) OUT

Examples of this recognition mobilized by plot keep coming, one after another, in Hollywood's current retoolings of the virtual. Even turning the paradigm of virtuality on end does not necessarily dislodge its stranglehold on the filmic stratum of cinematic effect. This is to say that even reversing the whole ethical problem—showing the real as under assault rather than under suspicion—may well fall within the same logic of virtuality. In *Eternal Sunshine of the Spotless Mind* (Michel Gondry, 2004), once again a trick beginning colludes with a twist ending (the two here filmically identical) even while corkscrewing round to undermine the plot's own premise. The narrative leads off, misleadingly enough, with the hero waking to a chance encounter with the woman who will become his lover, only for both of them, once the affair has gone bad, to contract with Lacuna Corporation for selective "brain damage" to remove the unwanted traces of pain. This passive and effacing temportation—an exile from one's own constitutive prehistory—finds first a photofilmic, then a digital, touchstone—and then, at the end, a quasi-photogrammatic one.

Once all photographs of the former lovers have been confiscated, according to contract, the brain can then be scoured by machine. Only when the film's first waking sequence is replayed in the penultimate scene, through the device of exactly repeated film footage, do we realize that, the first time around, it had in fact been something like a proleptic flash-forward to the couple's accidental second rather than first meeting, taking place now on the morning after the hero's electronic surgery. But this procedure is portrayed as being far more traumatic than the psychic scars it has sought to remove. Digital erasure is, we are shown, only the dystopian obverse of digital fabrication in other recent plots. In the ethics of the virtual, the real must be valorized, embraced, even if only after the fact. Which is why the unconscious of the couple resists their deliberate intention to forget.

En route, the plot thickens as its images begin disintegrating in a race against time: a race into memory against the violations of time present. In no more than metonymic association with electronic surgery, the ripples of digital editing become independent metaphors for an eroded reality. In the bookstore where the heroine works, the placard Fiction and Literature on the shelf behind her is rapidly effaced as the first sign of the mental erasure undergone by the hero, converting the very story of his time with her to fleeting figment. Behind the couple, book covers and titles are blanked out in corruscating waves, almost subliminal, of kinetic Photoshopping. All those books gone blank: all those neurological traces dismissed now as the overwritten or whited-out text of subjectivity.

Later, in a different, and indeed more medium-nostalgic mode, technique surfaces in a way that compromises the film's own gesture of resolution. On view in the very last shot is perhaps, along with the final lap dissolve of *One Hour Photo*, the clearest divulgence of a filmic photogrammar in recent—and by now digitally hybrid—cinema. This is the case even if one recognizes in this effect a transfer from the computerized experiments of the director's earlier work in digital video and its sinuous editing devices. Stylistic innovation puts the old under nostalgic erasure at the level of technique itself. For with the hero and heroine fleeing into the distance of a snowy landscape, a twofold loop begins, takes its slipping hold on the image plane (figs. 4.14–4.15). Along with the freeze-frame, such reprinted footage is perhaps the quintessential disclosure, from within cinema, of the edited cellular chain: the photogram as a signifying increment. What its overt manipulation of discrete imprint serves to image in this case, even with its overlap being digitally facilitated, and whether as hallucination or metaphor or both, is the couple's urge to start out—and up—all over again, and then again, in their willed escape from self-inflicted erasure. Yet this figuring of revived desire appears in a filmic manifestation so dubious that a potential figure for renewal, at the level of cinematic rhetoric, gets thrown back into plot in a form more

4.14–4.15

likely to evoke another psychic recursion, a stalled vector of will. As if violating Lacuna Corporation's first contractual stipulation to surrender all photographs, this last staggered thrust of desire seems unwilling to surrender the photographic index of suspended time, even in the familiar form of its serial forgetting as filmic photogram. Instead, it holds motion itself to sheer repetition—at least until it fades to the pure white field not of snow but of projected light, unimpeded by imprint.

This last brazen *trucage* allows the film's closing artifice of editing to resist, or at least postpone, the normal mode of filmic erasure and its traces. The effect is snagged somewhere, in this instance, between the photograph that cannot finally still time and the speeding track that cannot really mummify its change in passing. In *Vanilla Sky*, too, reprieve from digital intervention (in that case, electronic implant) took, in closing, the form of a filmic montage ceding to a blank white screen. In this later fable of the digital unconscious, the vicious (or mitigating) circle of the narrative's closing double loop—and its fade to white—offers a spliced succession running in place to nowhere. The Lyotardian "arrhythmy" of this scenic stutter once again operates, by optical allusion, to evoke the celluloid order from within the order of representation, but shadowed now by the digital. This filmic confession persists, as noted, until the looped grain of the snow-bleached figures fades further into the tabula rasa of the narrative's ultimate title shot: the spotless blind frame of sheer projected light. But limited repetition has in itself pried open a sense of possibility. As originally designed, the closing lap of this recursion was going to run continuously behind the credits, over and over again.[12] A last-minute decision rescued this narrative impasse for at least a viable ambiguity of plot. Without this endless ironic treadmill, the viewer is at least allowed to speculate. Will they make it this time or not? Isn't desire always in need of revving up again in some imagined elsewhere? In medial terms, nonetheless, we have watched in passing only a last instance of frame(d) time: the photogram *over*exposed in every sense—even if its manipulation is itself a computer simulation of filmic lab work. For here is a metafilmic succession entirely bleached out in its pyrrhic victory over the digital.

More sophisticated than most recent films about how to image time as more than duration plus action, Gondry's film is almost by definition more Deleuzian. In turn, Deleuze's concept of the time-image is itself most Deleuzian when it connects with the governing ideas of his other work beyond the two film books. Associations readily converge. The time-image is space "deterritorialized," become available for the nomadic traverse and trespass of rootless impulses unhinged from action. In the particular case of *Eternal Sunshine*, the hypertrophic visualization of a figurative time loop arrives as a kinetic variant of the Deleuzian "fold" (his term from Liebniz), where space is tucked into itself under the pressure of time rethought. In the process, Gondry's two-ply image figures the Eros of characters trying to burrow beneath mere recollection to a repetition that is never static duplication, but always freed to newness by a difference from itself.[13] In all this, narratology may well be cued to recognize in the open-ended closure an ethics of acceptance. But in the grain of the final and fading image, it is narratography that sees the rub of the unresolved.

In another ontological gothic of memory from this same year, an equally uncommercial ending is ameliorated, its whole closural logic neutralized, in an equally revealing way—and again by a revised finish that invokes in the process a dated filmic function. For *The Butterfly Effect* (Eric Bress and J. Mackye Gruber, 2004) closes in and down on "primitive" film footage from 16 mm home movies projected within the digitally enhanced screening of the plot's last scene. Both films from 2004 thereby inscribe film as the return of the repressed in a multimediated cinema. The hero of *The Butterfly Effect* has a genetic case of psychosis inherited from his father, who thought that he could use a secret album of photographs to hallucinate his way back into the past in order to redirect the course of his own repeatedly ill-fated history. The son learns of his father's photographic fetish only late in the film, having used his own handwritten diaries instead of images for a similar retroactivation of the past. He is now told by his psychiatrist, obviously just to pacify him, that those notebooks, just like his father's photo albums, never existed. He knows better. He believes what we have seen: how the convulsed handwriting of his private pages always gives way, by digitized blur, to a full-screen cinematic image of the past event.

Beyond this, and before the benign home movies at the end, film as medium has also been explicitly on this film's mind. It barely needs mention, here or in the narrative itself (such is the ubiquitous cliché of Hollywood trauma lately), that the hero's redemption of the past through metafilmic access is set up in contrast with a malign use of imaging—namely, a child pornography subplot in which his girlfriend's pedophile father (with yet again, as in *One Hour Photo*, the

4.16–4.19

implicitly absent or negligent mother) had once taken naked videos of them with his new camera. That, too, can be fixed, if only the scene could be inhabited again and refused. This defiance of origin, this overthrow of precedent, indeed this confrontation with the obscene father figure once removed, is one more variant of the Oedipal twists that are the common thread of such earlier time-travel narratives as *The Terminator* (1984) and *Back to the Future* (1985).[14] Such ironies make their appearance as well when the hero of *Eternal Sunshine* finds that one of the few safe places of escape is, at miniature boyhood scale, beneath the busy 1960s miniskirt of his mother in her kitchen.

In *The Butterfly Effect*, it does come to pass that the girl's father is thwarted in his abusive visual intent by the hero's temportation. But other things in the future imperfect of plot's narrative grammar continue to go amiss. The institutionalized hero's last chance at rectification is to get his mother to circumvent the psychiatrist and sneak old home movies into the hospital for him. Here, as with *Eternal Sunshine*, we circle round to an exact replay of the film's opening scene, which at the time we couldn't make sense of and which in the meantime we have inevitably forgotten. Escaping from security guards, the hero has barricaded himself in an office, scribbling what seems like a kind of suicide note. The rest, we now realize on replay, has been a plot-long flashback. What the identical repetition of this episode leads to now, at last, is the projection, on an opposite hospital-room wall, of film footage taken at the original backyard picnic where he met the neighbor girl who was gradually to enthrall him (fig. 4.16). By the digital rustle and whoosh of a tunneling image plane (figs. 4.17–4.18), he is suddenly vacuumed into the time of record, emerg-

4.20–4.21

ing by reverse shot, and a similar precipitous reversion of the frame, into his own englobing past. The present temporal layer of defended subjectivity is thus vertiginously minimized or "popped." The screened childhood image once before him, scratched and flickering, has thereby been "pushed" (again the computer argot) into a new and full-frame potentiality, coded by 35-mm resolution and a more richly saturated palette (fig. 4.19). The past is, in short, "remaximized." In this newly afforded interface, the hero is able—once (upon a time) and for all—to scare the girl away, threatening the life of both her and her family. Heroically, he thereby foregoes his own history, his very life with her, for her own eventual good. What follows this self-sacrificing restraint is a coda where he sets fire not only to the home-movie reel he's just screened (counterevidence of the virtual "rewrite" he has accomplished by force of visionary will) but also to some expendable, because no longer "real," snapshots of the life he might have led, and once did lead—and from whose now nonexistent film they can no longer be imagined extracted.

But that wasn't the original plan. Before studio executives talked the directors out of it, the film was to end with the same revisionary and digitally figured plunge into the filmic past—so far so good—but this time into a further mediated recess. Here, too, we may realize that the opening chase and barricade footage (the same in both versions) is the discursive equivalent of a repressed memory: a strictly narrational point of departure that, replayed in closure, pulls the trigger of an ensuing trick ending. The originally intended ending was in fact to have been triggered by the same digital effect as the one seen in the theatrical release: sending the oscillating image of the hero funneling at warp speed into the past, evacuating the plane of the present by the field of the relived. But this time he is thrust back to a yet more primal scene: the actual moment of his birth. The interface between fled present and embedding past is made clear by the full-screen image of his mother smiling on the gurney in the delivery room. But no sooner does our own perspective worm inward for its streaked focus on the pregnant belly (figs. 4.20–21)—call it the microcosmic

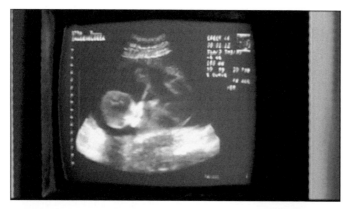

4.22–4.23

zoom—than a reverse shot also reveals the father-to-be as the cinematographer of the very film we are eventually watching.

It is at this point that the whole optic field of recovered immanence—and fetal imminence—is invaded and violated on the spot by retroactive desire. Here is the film's real child abuse, prenatally administered even while filmed by yet another father. A shift from the faded grain of the home movie, across its resuscitated past in wide-screen clarity to the splotchy frame-within-the-frame of the ultrasound monitor (fig. 4.22), yields to a now fully "cinematized" image of the once-recorded but now-relived past. It is there that intercut shots of the mother's horror and the panic of the hospital staff accompany the open-eyed action of the prenatal body, on that inset monitor, as it elects to strangle itself with its own umbilical cord (fig. 4.23). Not just kissing off the girl next door but denying her and your own mother your very existence, this eradicating gesture passes beyond all figurative or nonfantastic versions of the Oedipal in a voluntary and quite literal return to the womb. The Deleuzian moment where time is experienced, in prospect, as its own virtualization by memory is here knotted off at the source: a "genetic" temporality under negation.

With *The Butterfly Effect*, then, it may seem all too clear why the bipolar hero isn't given a computer or video game addiction, whether to vent his dissociated sensibility or as training ground for his temporal and psychic trespasses. The contemporary prosthesis of subjectivity in computer graphics could not afford to be ransacked for local color, let alone irony, because it could so easily contaminate the overarching fantasy plot. Instead, the film's protagonist brings the virtual and bracketed role reversals of electronic performativity dragging and kicking, or at least digitally lurching, into the real itself, whose very ontology is at the same time nipped in the bud. For this to carry the impact it is reaching for, the impact of the unprecedented, the film surely doesn't want us to remember how much on-screen dying commercial culture's video gamers, including fans of first-person shooters, have become inured to in the ludic mortalities of interface death.

Remediating ultrasound tracing by 16 and then 35 mm celluloid film, and the latter by the torqued timespace of electronic distortion, this ending is a shock to the optical as well as the narrative system. What we see here is the human body rescuing time from its remorseless storage and returns only by robbing that same body, from within its very image, of its own life. Sound familiar? What can keep us from taking this image, extricated from its fantastic plot, as the précis of a yet wider assault on the lived time of the body under the reign of electronics? At the least, this too disturbing and thus reshot last scene in a commercial studio film represents an extreme pole on that spectrum running from the temporal uncanny of European cinema, where mental life is oddly intersected by other psyches and other times, to the mnemonic marvelous of Hollywood production, where the mode of ontological gothic works in the end, the trick end, to neutralize anxiety. It does so—whatever immediate jolt is conveyed through such reversed expectations—precisely by rendering entire lives unreal in closure.

Discussion so far has meant to chart the various means by which proliferating themes of virtuality, both ghostly and digital, have in a sense, over the last decade, been playing catch-up with medial transformations themselves. And there is a further wrinkle to this fabric of transformation. We learn about the original ending of *The Butterfly Effect*, it should be noted, only from the DVD remediation of this digitalized cinematic artifact. This is repeatedly the case with the special features and alternate scenes of digital reissue. Is it far-fetched to see in this commercial phenomenon a self-mirroring case of "alternate worlds" domesticated within the very paratext (as Gérard Genette would have it) of cinema's new distribution channels? Narrative is always recognized to shape expectations for the imagined arc of our lives. The hero of *The Butterfly Effect* might well have cut his eyeteeth on just such a digital model of roads not taken, parallel universes, virtualities unactualized, as has brought us to understand the sudden alternative to his own plot's closure. Take it as another case of Deleuzian temporality gone wrong, twisted out of shape. Properly understood in relation to an open whole and its mutable horizons, lived time is narrowed and contorted by such narratives to the triviality of a remote "menu" choice.

Back, though, to the more general point about the way technological advance precedes, but not always by much, its thematization within plot. Hints of this, as we've noted, are leaked whenever the special effects of discourse trickle back, often under disguise, into story, or in other words when photocellular and digital technique—or, more often, the equivocation between them—is manifested in preternatural plot turns. What happens is that theme is to be found catching up with medium only at a mutual vanishing point of the bodiless and unreal—or, in short, the timeless. And this takes us round again to the indexical overkill at the origins of the motion-picture medium. Faced with continuous record, technologies of narrative soon imposed the system of cinema as we know it upon the unchecked mechanized unconscious of filmic storage. The process amounted to the turning of image into signifier, attraction into narrative action, optical field into plot, the *jouissance* of image into the business of suture. But if certain recent plots point us back to origins in this way, they also push on through ends: cinema's end, at least, as an essentially filmic operation. Then, too, original dichotomies associated with the discovered options of a new medium return as it wanes in the form not only of ironic vestiges but of revisionary disclosures. As we began this chapter by seeing, similar terms apply at far poles of the historical arc, if only now under the most drastic of technological and cognitive reconception. Unlike early narrative cinema, where the suppressed relentless seriality of automatic record was rendered susceptible to discrete recombination; and unlike Freud's sense of consciousness as a discriminating and defensive perception, with its self-interested processes of selection and deletion (Doane again so far)—unlike either machinery of time and its traverse, digitization arises from no protean, shapeless, totalized, and presignifying unconscious. It is one.

To plug into that digital unconscious for narratives of temporal transit and psychic recursion in the hybrid medium of postcentenary cinema is all but inevitable. For the digital morphing of the image opens a gap in the finite and consecutive into which the infinitely mutable can be inserted. When this happens in *The Butterfly Effect*, a final suggestion dawns. At the zero degree of optic mediation, aren't the continuous trace methods of fetal imaging perhaps now the last place where the image of the human body, before emerging into—let alone retroactively invaded by—agency, can come before us without suspicion of editing, simulation, or enhancement? In biological rather than technological terms: the gestated rather than generated image, tangibility in the making, morphology before morphing. In all this the intertext is so unavoidable as to seem intrusive, almost unwelcome. In terms of medial rather than species evolution, we have come just a few short years beyond the epochal sci-fi date of 2001. Yet the last shot on the hospital monitor of *The Butterfly Effect* (fig. 4.23 again)—conveying the open-eyed suicidal intent of the unborn—marks a far greater conceptual distance from both the technique and the temporal thematic of *2001: A Space Odyssey*. The wide-eyed fetal panoptic of Kubrick's legendary last shot in his 70 mm film of films from 1968 (fig. 4.24)—the Nietzschean world made new again, on the far side of human time, only when transcending the Cyclopean eye of the digital computer—is an apocalyptic finish recalled only by reversal in the digitally sprung regress (and timespace odyssey) of *The Butterfly Effect*: as the heroic move to induce a stillbirth of the real. Narratology might have every interest in marking the long shadow of this Kubrickian intertext as it falls over the visionary stages and blockages of the new plot. But, in that final shot of the monitor, another degree of attention—and this to the technical remediations of the image (the image qua image)

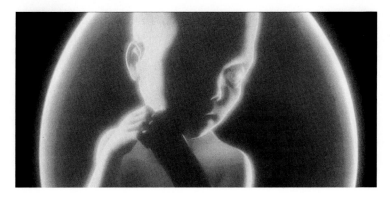

4.24

across three modes of screen materialization (electrosonic, filmic, and digito-
cinematic)—is called up. This is where
narratography would measure the inset
dissonance of media and their cultural
static at once, all within a layered *textual*
system open to reading.

The abortive trauma of self-infanticide at the end of *The Butterfly Effect* marks
an extreme telos not only in the fantasy
plot of temportation but in the narratively unmotivated digital inroad into film: here in the form
of its quite literal inrush—but only as trope of revision—into the dated celluloid flicker of a
low-tech home movie. The electronic discourse of metamorphosis can't remain safely stylistic or
merely figurative, however. It buckles the celluloid image as cultural symptom as well as metaphor.
In just the sense that cinema began as a time machine, it is now under siege from the material base
of its new digital constituent, from pure binarity with neither succession nor direction. But what
else is new about cinema versus the "new media"? Narrative was always a time travel against the
grain of the strip, symbolic pattern overwriting index, traced motion overtaking first stillness,
then brute continuum, in order to shape a duration into story, to make time itself signify. The
only difference—only, or total?—is that digital incursions are now everywhere available, if
mostly on the monitor or theater screen so far, for the imperceptible transformation, when not
absolute simulation, and hence for the ontological erasure, of human time itself.

IMAGE AS NARRATOGRAM

If narratology sweeps the image field, mapping more than it actually navigates, narratography
closes in to chart the increment. In its own way, narratography is a kind of sonogram of narrative
pulsions and dissonant tonalities. *The Butterfly Effect* offers this metaphor, of course, in ultrasound
form—but offers as well, and more immediately, a ferociously clear example. By holding close
for as long as it takes, looking hard, narratography would find in the grain—quite literally, the
graininess—of that film's climactic and twice remediated image (ultrasound monitor by 16 mm
and then by digitally enhanced 35 mm film) a further thickening of significance. The whole idea
of structural scale is reconfigured by this mode of tighter calibration—yet with no surrender of
cultural scope. Quite the opposite.

Able to discriminate among the broad structural patterns of genre cycles across national cinemas, for instance, the familiar maneuvers of narrative analysis brought us to the present considerations. Once narratology has sorted, however, narrratography moves to sift. Yet what is it, more precisely, about a narratographic grasp of plotted image that renders its response a more sensitized cultural register than an otherwise perceptive and historically informed approach to narrative systems cutting across separate media? It is this: that in the actual surface features of a medium lie not only the texture but the abiding narrative tropes of its own sociological context. The closer analysis comes to this, the farther back it finds itself able to stand. Narratology takes bearings, certainly. But narratography bears down—and in the process can occasionally push through, or hope to, to the further cultural horizons of a formalist account. These horizons may rise to delineation in medial analysis, in other words, even when they are occulted by plot.

Hold the present example of *The Butterfly Effect* open to such narratographic inspection for a moment longer, then. Exactly where the digital inflects the metafilmic in the transfigured home-movie rerun of a life under revision, a whole contemporaneous history of electronic mediation is retold in disguised or fabulated form. Everyman manqué in the age of interface, the hero of *The Butterfly Effect* has found a quasi-digital and subelectronic way, by the neural routes of inherited psychosis, to make biological time autobiographical, to render duration itself user friendly even if suicidal. Narratology might locate the "forking paths" structure of the alternate-universe plot (David Bordwell's term) as it is undercut further by a psychopoetics of foreclosure in the death drive (Peter Brooks).[15] Narratography, however, is always medium specific. As it did with the videographic retrieval of maternal presence in *Johnny Mnemonic*, it would therefore chart the optical micromanagement of this finale. It is there at the end that the enframed palimpsest of rescreenings would be found to reflect a far more pervasive and diffuse condition of mediation—one for whose outer limits of electronic interface our hero is both unspoken pioneer and scapegoat.

It is a working assumption of this study that the culture of interface, as well as of digital simulation, can pervade a narrative idiom even when not spelled out by it. One film above all, Spike Jonze's *Being John Malkovich* (1999), makes this tendency inescapable. It does so by cordoning off its actual use of the digital in a lackluster special effect while unfolding the rest of its plot as if it were what it never actually becomes, a monitory narrative of electronic brain implants and digital cloning. Such contemporary anxieties about the integrity of self and its mnemonic constitution, as brought to explicitness in films from *Dark City* to *The Final Cut* (the latter film's textbook summation of these phobias to be taken up in the next chapter), remain in this earlier narrative part of the troubled Zeitgeist—rather than the actual plot crises—of an entirely techno-free fantasy that bears immediate comparison to the comparable avoidance of sci-fi technology in *The Butterfly Effect*.

That 2004 film images time at the quite literal *disposal* of the subject. This is the imputed heroism of its lead character: the willingness to forego himself. In *Eternal Sunshine of the Spotless Mind*, from the same year, the fantasy of such erasure and disposal is instead a nightmare. A wittier and more surprising film than either, and a far darker one, by the screenwriter of the latter (Charlie Kauffman), turns both of these routine fantasies on end. *Being John Malkovich* isn't interested in what it would be like either to erase or to reclaim your own past. Instead, it entertains the more fantastic possibility of erasing the future life of another to make room for your own rejuvenation. In this respect, it offers one more instance of the trick beginning: the full-frame puppet theater that quickly grows irrelevant to the manifold complications of plot, at least for most of its length—except as it is the metaphoric key to them all. The uncanny feeling that one is not oneself today, that some other force seems to be pulling the strings—that uneasy human mood—is literalized in Jonze's film not only by teleportation but by a bizarre reincarnation scheme that bypasses natural death altogether. And among the further surprises of this film is that, in the optical allusion of its twofold trick ending, it recruits the canonical photomontage of life's elapsed moments as a decisive visual emblem for mortality overcome rather than mourned.

Like the rest of Kaufman's cartoonish, tongue-in-cheek narrative, the ending is at least as extreme, though less solemn, than *The Butterfly Effect*'s umbilical suicide. It all depends, a good while back, on the premise that a magic tunnel has been found leading from a cramped office building directly into John Malkovich's body. This is a psychosomatic exit path from the real discovered by accident, proved for us by the iris masking of a POV shot within the star's apartment (fig. 4.25), and quickly marketed to malcontents. Self-image is easier when borrowed, it would seem. As no viewer could conceivably remember at this point, the identifying shot of Malkovich preening before the mirror as host to an invasive POV echoes the opening puppet figure reaching out unaccountably to touch its own alienated image in the mirror (fig. 4.26). For like a puppet, Malkovich is now seeing not only through his own but also through the camera-like eye of his operator. Although the route to his POV is imaged as a plummet down the rabbit hole of a new psychosomatic wonderland, more a magic well or chute than a window, the film's word for its hokey digital effect is *portal*, not *porthole*, despite the rear-entry ironies of the initial male-to-male incubus plot: *portal*, with its overtones of a computer port—including the weird capacities in this case for wholesale jacking in. The POV implementation of this fantasy is beyond surveillance—even as it evokes electronic invasiveness of all sorts. More than bugging, it is a kind of buggering from within.

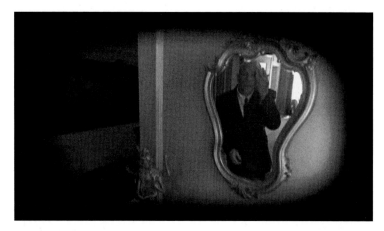

4.25

4.26

Anything like normal cinematic "identification" with Malkovich has been perversely replaced—to recall Manovich's dichotomy for the shift from cinema to newer media—by sensorimotor "action" (or interactivity) instead. All of this fantasy transpires, of course, under the abiding protocinematic model of puppetry's artificial animation. As with film viewing, although the vacation from the actual to the virtual self may seem total, it is only temporary. So far. But another, noncommercial scheme is slowly disclosed. What we discover is that the spatial displacement from one body into another is designed to be a full-scale temportation of the mortal subject into a new embodied duration. The astounding "portal" is not, as previously assumed, unknown to the lecherous company president, one Mr. Lester, in whose offices it is hidden (on floor 7½ rather than 13). He and his aging cronies are only biding their—and

Malkovich's—time. Not interested in *being* John Malkovich by vicarious access, they have instead selected his body as vessel for their own collective metempsychosis, which must happen no later than midnight on the star's forty-fourth birthday, before his neural circuits grow unreceptive. Like all great actors, he seems to contain multitudes—but his days in the limelight are numbered. True to genre form in the "hyperbolic marvelous" (Todorov), the fantastic plot simply literalizes these twin idiomatic clichés.

And it does so by photomechanical parable. Such a group invasion of the body snatchers is figured here, in respect to its violent truncation of a single human lifeline, by the artificial animation of a photographic collage into a feverish montage that seems to be bringing snapshots of the past at least halfway back to presence. This happens when one of the three main characters, the puppeteer's wife, gets her first clue about Lester's involvement from a room in his mansion, full

4.27

of photographs of Malkovich—images from childhood on through high school yearbook pictures to playbills and publicity shots from the various stages of his theatrical and film careers—that she mistakes for a fan's obsessive "memorial" (fig. 4.27). More than shrine, for instance, her word for this archive unwittingly suggests the middle-aged terminus ad quem that rapidly approaches. After a panning shot takes in the array of photos at a first sweeping glance, cinematography then takes them up—at a different pace, and in a more kinetic register—by quick cuts across their intermittent chronology. Gradually bringing the more contemporary snapshots into full-scale alignment with the screen frame until they become for a moment all there is of our own movie, editing seems to animate them into a decisive biographic trajectory—yet one whose end is near.

The effect repeats itself at, and as, the film's second major twist. In its penultimate scene, we flash forward to an aging and renamed Malkovich, now inhabited invisibly by Lester and the rest. We find him explaining to another friend the latent possibility for a subsequent fresh rebirth, illustrating the waiting vessel, once again, in the form of photography's mechanical rather than magical arrest of time. Malkovich's photos have been replaced by the new candidate for collective reincarnation in Lester's gallery: the girl fathered by Malkovich while (unbeknownst to him) his body was inhabited briefly by his female lover's own lesbian lover—it's all just as funny as it sounds—rather than by the male protagonist who originally had access to it. A second rapid montage of such ensconced photos now brings the daughter to the brink of her teens—decades before she will be expected to yield up her preassigned body to the cadre of self-survivors. The last of her snapshots, fawned over in particular by the new Malkovich, shows her behind a chain-link fence with her mother: literally *screened* for us in her potentiality as full-frame surrogate in this life-is-like-a-movie scenario of vicarious presence. We then cut at once to a live-action coda with the girl full frame at poolside in the company of her doting same-sex parents. Before Spike Jonze's screen realization of it, Kauffman's original script made mention of such a photographic archive only in the initial scene in Lester's mansion, where there was only a single iconic picture. No wall-spanning collage in the screenplay, no panning animation of a photo chain, no metafilmic spacing out of photographic time frames—and no return to such an image repository for the penultimate scene. As befits the narratography of a given medial realization, only when actually filmed does the movie actor of this metacinematic plot become an explicitly filmic figment, his biographic arc engendered by photomechanical jump cuts—and after him, the developmental arc of his successor in waiting as well.

Narrative yields now to this offshoot in the new female vessel of longevity. Natural generation and regeneration—the father hoping to "live on" through his daughter—have found here a mad travesty in this trick finish, even as it has yet one more turn to go. Once the photo has given way—under Malkovich's doting, projective gaze—to that final filmed shot of the poolside girl, we cut to the kind of masked reverse shot (fig. 4.28) that has served repeatedly as this movie's POV code for the psychic inhabitation of a foreign body. As cued further by voice-over (or is it voice-under this time?), we realize suddenly that it is the puppeteer himself who is presently inhabiting the girl's body, that of the daughter he wishes were his, and trying to speak forth his desire to the lesbian mother, who has rejected him in her affair with his own wife. The typical generational ironies and loops of all other time-travel plots converge upon this morbid double twist. In its lampoon of paternal yearning, the grotesque incest allegory has the father figure actually entering the body of the daughter in order to get closer to the mother. Figuring further the co-optation of the girl's human form by temporal manipulation, the closing credits begin as she dives into the water and floats weightless for the last several minutes, suspended in a kind of sensorimotor limbo behind the roster of her image's own industrial production as moving picture. Seeking her natural element in groundlessness itself, she has become visibly what she in fact is: time's puppet (fig. 4.29), hovering before us in a paradoxical moment of both slowed and proleptic duration. In this closing optical allusion, this final narratogram, her being is thus held in abeyance somewhere between still image and time's normally paced screen montage.

Clearly, more methodologies have been at play to this point than my own suggestions about narratography—and at stake, too, even when not directly at odds with each other. Narrative anatomies both structuralist and semiotic, from Todorov to Riffaterre, along with their postmodern cognitive descendants in the work of virtuality theory and computer modeling (Ryan and others), are only part of the contest of explanatory frameworks provoked by the cinematic evidence brought forward. The psychoanalytic neoformalism of Žižek on the "interface effect" (quite apart from its more technological home in the electronic interactivity of Hollywood sci-fi) crosses paths, or wires, with the claims of suture theory—whose grounding of the look in the "absent one" is submitted, for instance, to diegetic parody in the ghostly appropriations of POV in *Being John Malkovich*. Further, in this confluence of methodologies, the early-cinema historicism of Mary Ann Doane has been implicitly confronted in these pages by the philosophical metahistory of De-

4.28

4.29

leuzian periodization, with its emphasis on the postwar emergence of the time-image: temporality as sign function rather than sheer measured duration. Along the way, whole theoretical paradigms have been found rebuilt from within by plot, explanatory schemata engineered from the track up by the disposition of a given film scene. Deleuze would say that such is only film's appointed aesthetic role in doing an aspect of our thinking for us. Filmic narratography is one way of thinking about just this. And for a not surprising reason. Explicitly or not, any of the theorists convened and interviewed here, at least the visual theorists, do a good deal of their definitive work at the point of intersection between image and narrative. That's why the temporality of a given image can be so directly caught up in the articulation of story time.

This is hardly less obvious when the story time is itself multiple and contradictory—as in the favored test cases of what we might call exceptionalist narratology. Bordwell is following Borges, for instance, in designating as "forking-path" narratives films like *Run Lola Run* or *Sliding Doors* (1998). These are films whose parallel time streams are exceptions that prove the rule of normal narrative cognition, in and beyond cinema, as Edward Branigan has further helped show.[16] They bespeak in eccentric form narrative's normative repressions and elisions. In responding to Bordwell, Branigan cites narratologist Gerald Prince on the actively "disnarrated" in standard practice (rather than merely nonnarrated).[17] This involves the choices and moves *not* elected by the characters, with all the structural possibilities thereby foreclosed.[18] Such is the phantom narrativity that gives negative or "overwritten" texture to traditional story time: we half know in going forward what might have been, even as we want to know what in fact will be. These, then, are two modes of the virtual that are tightly coordinated in normal plotting.

But we are concerned in these chapters with abnormal—and particularly under the excuse of paranormal—plotting. With its not just revisionary but also reversionary flips, temportation is one name for the maintenance, against all odds and logic, of time's alternative force. Such plotting often manages to retain the road not taken as a still present possibility. But when the option ordinarily foregone by linear form is self-consciously trodden by the hero himself in corrective retrospect, we have verged instead on that alternate narratology previously discussed: the "garden-path" framework of the false lead and the trick ending, from, let us say, *An Occurrence at Owl Creek Bridge* to *Jacob's Ladder*, *Vanilla Sky*, and *Minority Report*.[19] Then, too, in the recent plots of an implied "interface" virtuality ascribed instead to the psychotic powers of mental "defection" or to the technologically unaided magic of metempsychosis, time is not just traveled through but vehiculated in its own right, slipped out from under itself, transposed and inverted. Or, in *Being John Malkovich*, begun anew in the exchanged body as husk of the colonizing ego. Call it temporal desire as exportation—through the "portal" of a cult fantasy.

Details and instances aside, the principle should be clear. It is not only that films playing tricks with time in order to flee the exigencies of mortal fate take on a whole new valence in the digital epoch. Beyond this, the on/off binarism of electrographic representation has risen to the level of narrative toggle itself. Its either/or logic of alternate realities often derives in turn, however indirectly, from the programmatic options of interactive video. This seems the case even with the predigital imagery that closes *Being John Malkovich*, an imagery that, pressured from behind by plot, works at least to whisper the unspoken: a deep sociological unease over biomedical technology in an age of the simulacrum. How artificial will insemination eventually become? Can one donate one's own DNA in another's sperm—and just by *willing* it? And the metafilmic dimension of this plot brings such black-comic anxieties home to another medium of reproduction, another mode of longevity: filmic record itself, filmic reanimation. In more explicitly optical and "projective" terms—by the concatenation of the film's closing shots in the move from photographic inspection to a specialized cinematic POV—it is as if spectatorship itself has been shoved down into the film's own unconscious.

Readable there, across the disjunctive progress of these last narratograms, may well be cinema's own institutional fears, in 1999, of its pending eclipse as an indexical cinema: a cinema of star performers within a humanist culture (and a cult) of valorized personality, even if the stars are indeed mostly recycled (like Malkovich here) in their midforties. It isn't just a case of cinema encoding the death of the human in cloning's perpetual-motion technology. It is cinema imaging the end of the psychic mechanism on which its own narrative power has been founded. By extension three years later, it wouldn't just be Sy Parrish in *One Hour Photo*, then, but Robin Williams himself who, faced with the thought of his industry going entirely digital, might well say, "Oh, don't do that." Such anxiety is made legible in *Malkovich* only through the manifestation of itself as fantasy. Forget genetic science, then. There is a more unsettling and specifically institutional edge to the farce. Whenever first-person interactive video and the erotic chat rooms of vicarious encounter render the movie theater's more passive identification no longer sufficiently intense, the commercial regime of the gaze is threatened. The empire of suture yields to the assembly line of desktop surrogacy.

Yet again: questions of screen technique enter the realm of social diagnosis. Along the course of narrative construction itself, as we've seen, narratography reads certain theoretical positions, as well as numerous cultural symptoms, back to us in a sometimes clearer translation. It is either a circular truism or a hermeneutic dare, this: that films often encapsulate the logic of the theories that best address them. Narratography goes on to assume that this embedding of principles is most likely to be found, for cinema, in the articulations of structure at the plane of the image—camera angles, framing, editing, shifts in film stock, special effects, masking, even perturbations of the movement-image surfaced from the flux of the strip into the flow of image: to be found there, that is, rather than in an encompassing plot paradigm. Narratology is preoccupied with story forms, narratography with the particular forms of their visual story*telling*.

How the latter approach, as we've been attempting it, might enable a comparative graphing of effects across the virtualities of the fantastic genre in both European and Hollywood cinema: that's been the question, and the test, so far. It leads to another question that I hope will find at least an acceptably tentative answer in what follows. For in the next chapter we look to see how the most far-reaching theory of cinematic virtuality might, precisely because of its own *predigital* terms, bring out the deepest common denominators of these paired, even though divergent, national trends in an ongoing screen practice of postrealist cinema. For it is in the films to come, as well as those we've just surveyed, that the equivocations of frame(d) time continue apace. In them, again and again, the biographical past is recalled, recovered, and somehow redeemed—or, in certain cases, put altogether beyond such salvation. In this way the forked meanings of the verb *repair* continue to make their mark on fantastic plots of temportation. For repairing is a returning-to as well as a renovation, in either sense a kind of recovery. Such, then, is the narratology of recuperation in the mode of the virtual, even when ruled out by a countervailing ethics of the irreversible. A more closely gauged narratography can bear down on the optical allusions of an increasingly postfilmic cinema in just this regard: the fantastic distortions of a time trans/fixed.

FIVE

A. Hakmann : 51 years : 186 days : 7 hours

In tracking the temporal warpage and buckling in recent fantasy plots, attention gravitates to a narratography of the virtual. This is true for surface effects and deep structure alike, and more importantly for the relation between them. The implications of this connection deserve a further airing. This takes us, once again, to the imaging of time on the narrative screen. And then, in the final chapter, to the protocols of its analysis in a posthermeneutic critical climate. Virtuality is the genuine crux; its archaeology versus its interpretation, the strained and artificial divide. In each chapter, a final test of method is offered—or provoked is more like it—by films released right down to the wire of publication. These are screen narratives, variously mediated and hypermediated, that suggest an ongoing distinction between the transnational uncanny of European cinema (Michael Haneke's *Caché*) and the supernatural ontology of Hollywood production, even when the latter is crossbred with the Anglo-European avant-garde (John Maybury's *The Jacket*).

On the hope that there was a perceived heuristic gain, once before, in calling in anonymous testimony on the temporality of screen narrative and the evolution of its editing—and allowing the claims of this testimony to reverberate across the medium's whole history before associating them quite specifically with the waxing rather than waning days of celluloid projection, or before naming their author as Mary Ann Doane—I'll take the liberty of a similar experiment a second time. So I call to the witness stand, anonymously again, another study that will be subpoenaed, for the length of a few paragraphs, out of all identifying historical context. The hope, once again, is that a historically embedded theoretical argument, introduced at a skewed angle to its own intended context, may eventually take us, even if somewhat round the bush, eventually to the deeper roots of the issue as a transhistorical problematic.

To begin with, the work in question is an account, like mine, that divides film's treatment of temporal limit cases and their ambivalent metaphysical horizons between "European humanism" and "American science fiction." Here is a cultural binary that can be seen to prevail even when the humanist end of the spectrum—under a mounting antimetaphysical strain—tends to contemplate the multiplicities and splits of the unconscious as definitive features of the ego in action. This is an ego open to the uncanny, therefore, at every discontinuous turn—as well as to cybernetic incursions in a climate of the posthuman. What follows from this dichotomy between humanist temporality on the one side, the technologizing of time on the other, is an increasing polarity in the consciousness of lived duration. Published, as it happens, during the sudden global upsurge in video and digital mediation, the study in question is concerned, like Doane's in her look at an earlier and precedent moment, with the "emergence" of new temporal models across all modes of screen narration. One result, for both European and American cinema of this later period, in their equivalent break from normative realism, is a frequent detachment of agency from its spatiotemporal surround. A new distance is installed between perception and somatic engagement, seeing and doing.

Actors within a spectacle have been transformed into spectators of their own programmed actions, with a new premium thus placed on virtuality. In the aftermath of a character's autonomous "I," subjectivity is more overseen than internalized. Collapsing the old paradigm of an image medium with which only the spectator, not the film's narrated participants, had to *will* an identification, "it is now that the identification is actually inverted: the character has become a kind of viewer," and this of a reality in which "the imaginary and the real have become indiscernible"—or, in other words, indistinguishable from each other. A further result is that plot is no longer stabilized by temporal progression. According to the broad-ranging analysis prosecuted in these terms, the "new cinema"—as innovative as it is involuntarily caught up in historical change—has arrived at a point of temporal crisis where "chronos is sickness itself." What this amounts to is an undermining, one might say, of the chronological by the chronic.

A directly imaged temporality of this sort "is the phantom which has always haunted the cinema," the study is quick to admit; but "now"—and the point is *suddenly*—cinema has found a way "to give a body to this phantom" in a whole new mode in which, "if virtual is opposed to actual, it is not opposed to real, far from it." In all of this, whether in European cinema's new fantasy-imbued "humanism" or in "American science fiction," a massive decentering of the psyche has been less effected overnight than belatedly recognized. For it is now, and manifestly, "we who are internal to time, not the other way round," with subjectivity relativized and dispersed across a duration not its own. Though this recognition, along with the new modes of perception it brings in tow, is inevitably alienating at first, it is in the long run the only way of "restoring our belief in" a world otherwise removed from us by mediation.

The argument I'm paraphrasing so sketchily, as one has no doubt guessed by now, is that of Gilles Deleuze.[1] But he concludes his second volume, as we know, with an explicit demurral. He is not really speaking of the "new images" at all, "tele" and "numerical." His filmic "new," his "now," is backdated to a modernist moment that may—or may not—extend to include them. For he has been talking all along, even with his emphasis on virtuality and spectatorial mediation, not about CGI and its VR plots but about a cinema from Rossellini through Resnais, Hitchcock through Kubrick, in which the indirect image of time derived from action cinema, or in other words from the movement-image, has been displaced by the direct time-image. Temporality is entirely reconfigured in the editing room. Succession yields place to a more radical and irrational seriality, the motivated shift to the arbitrary splice, the montage juncture to the irrational cut. His main evidence thus precedes mine often by half a century. Yet in his remarks on films from *Last Year at Marienbad* to *2001*, from elegiac virtuality to an apocalypse of temporal cognition in the "world brain," Deleuze anticipates everything the current spectator must contend with. This is especially clear when his analysis brings home, *avant la lettre*, the many interlocking ways in which the divide between European "art" cinema, in all of its humanist (or posthumanist) reflection, and American action film is with us even yet. Though strained almost beyond recognition in some cases, the operation of a shared temporal problematic across both cinemas remains still within the conceptual precincts of a subjectivity that, while bereft "now" of constitutive action, has become the site of a pure spectation suspended in a time it can no longer call its own.

Deleuze on the overthrow of logical succession in screen editing by the irrational series might resemble the latest *Screen* article on Christopher Nolan's *Memento*—except that he is in fact speaking of Straub-Huillet and Godard. In remarking on the discontinuities of virtuality's "new" editing style, with its retroactive divergences in the time frames of action, memory, and conjecture, Deleuze suggests that "the forking points are very often so imperceptible that they cannot be revealed until after their occurrence, to an attentive memory" (*Cinema 2*, 50). In this, though, his evidence is not what would come first to the contemporary spectator's mind. He isn't referencing that spate of alternate-world plots and "forking-path" narratives so familiar lately to screen viewers. He isn't thinking of *Sliding Doors* or *Run Lola Run* or *Swimming Pool* or the sci-fi premonitions of *Minority Report*. He is thinking of Welles and Fellini. At the same time, the "attentive memory" that he finds requisite for the spectator is often displaced onto characterization as a deciphering function of lived time—so that plot agents struggle heroically to retain and interpret the very shapes of time. Characters themselves become not just viewers of their own existence but sleuths of their lives' elusive crossroads in the rearview mirror of virtual replay. It is in this sense that Deleuze's grasp of the high-modernist moment of postwar cinema quite strikingly anticipates the trick solutions and retroactive adequations of much more recent films. These are plot devices, as well as visual figurations, common not only to the ontological gothic of American thrillers but to the mnemonic uncanny of humanist fantasy, with its erosion of mental borders and its preternatural relays of consciousness.

Deleuze emphasizes the way the "mental world of a character" becomes "so filled up with other proliferating characters" that psychic autonomy breaks down—"hence the importance of the telepath" (8). To say so doesn't require, of course, any reference to the oracle in *The Matrix*, or the clairvoyant child in *The Sixth Sense*, or the blind medium at the séance in *The Others*, or the child's eponymous telepathy in *The Shining*, or the reanimated mummy and his human prosthesis, the telekinetic androgyne Ishmael, in Bergman's *Fanny and Alexander* (1982), or even the uncanny transference across the parallel-universe divide of *The Double Life of Véronique*. His example is the performance-artist mind reader in Fellini's *8 1/2*. Writing of a temporality that floats anchorless in "any space whatever" (5), or even when mapping the cognitive infrastructure of the "brain-city couple" (267), Deleuze isn't led to the arbitrary digital simulations of sci-fi VR but to the anonymity and anomie of postwar urban settings.

Moreover, in drawing on Bergson's interest in déjà vu and "paramnesia" (79) to typify the modern perturbations of this dysfunctional brainspace, Deleuze is not describing, with a two-decade clairvoyance of his own, the plot twist of *The Forgotten* (Joseph Ruben, 2004). But you can't tell this from just reading him. For in that recent supernatural gothic, a mother whose son has been abducted by aliens, with even his photographic and video traces erased by extraterrestrial electronics, is half tricked (along with us) into believing, for the first hour or so, that an extreme case of the uncanny, rather than a tabloid case of the technological marvelous, is at play, and that she has simply fantasized the boy. This is a not uncommon symptom, says a psychiatrist in league with the aliens, of none other than "paramnesia" (Bergson's uncited term as well). If one were to credit this charge of alleged fantasy, time is warped by forgetting the merely virtual status of what never was.

Concerned instead, as Deleuze is, with magically animated photographs rather than their magical disappearance, he writes of the inseparable fold between the virtual and the actual in that form of crystalline apparition called the "mirror-image," commenting that it is "as if" a "photo or a postcard came to life" (68). With Zanussi's *The Structure of Crystals* (70) in view, he makes his point without needing to mention the animated wall photo of Medem's *The Red Squirrel* or the 4-D stereo card, involving the eerie dimension of time as well as depth, in Raul Ruiz's version of Proust. Nor does he need us to recall, on the other hand, the transfigured photographs as registers of death and its eerie reversals (images of family members disappearing or remanifested within the frame according to the vagaries of time travel or supernatural intervention) in fantasy plots from *Back to the Future* through *Frequency* to *The Forgotten*.

When Deleuze insists that time manifests itself most directly in cinema by layered, tectonic, stratigraphic space, he is not speaking of the planar downloads of electronic virtuality in current Hollywood, or of the "sheets" of the past in the metaphysical double loops of the latest European uncanny. But he *could* be. In demonstrating as a signal cinematic moment a character's paraphrased confidence that "I am not dead because I have not seen my life pass before me" (68), he doesn't need reference to films employing this trope from *Jacob's Ladder* through *Vanilla Sky* to 2005's *The Jacket*. He has only, as we've seen, to cite the 1980 *Slow Motion* by Godard. Writing of a cinema in which "there are no more flashbacks, but rather feedbacks and failed feedbacks," he is not speaking of what he has called "American science fiction" at all, then or now, since for him the playback in question needs "no special machinery" (266). Nor does he have in mind the fateful recurrences and psychic overlaps in the latest European films of a doubled or split subject. He points instead to the "newness" of a cinema like the New Wave. When he therefore sees modern cinema as having "killed the flashback" (278) through the more direct presencing of memory as virtual, its "birth" a "function of the future" (59), he is not adducing the time-loop plots of an alternate past in embryo. He is commenting, rather, on prophetic flashbacks in Mankiewicz's Roman historical films.

The legacy of the time-image runs deep—and wide—if sometimes thin. The hallucinatory breakdowns of humanist melodrama fall under its rubric, as well as the speculative plots of dystopian cybernetics. Everything we're dealing with now as postmodern (or better postrealist, even postfilmic) circles within the orbit of, and tests to their limit, the temporal axioms of high modernism as Deleuze conceives them. In his forward-looking conclusion, commenting on the way cinema, defined as a mechanical automatism in its own right, "confronts automata, not accidentally, but fundamentally" (263), he means this as a gloss on the idea of cinema as "spiritual automaton" (263). True, he doesn't instance in this light the electronic ghosts in the machine of the *Matrix* trilogy or the "precog" zombies in *Minority Report*, or *A.I.*, or *I, Robot*, but rather Dreyer's *Gertrud* and the sleepwalkers of Antonioni. But the tracks of interpretation are firmly laid. Or say that he has read the *lectosigns* (his term) of these more recent films in a precognizant vocabulary all his own. Again and again, cinema since the modernist apogee of the time-image has found ways to literalize, parody, or further instrumentalize the conceits of temporality as the operable deceits of the actual in a new virtual real. Such is the cinema of either the elegiac uncanny, where virtuality is often at one with mourning, or the instrumental marvelous (Todorov's category for equipment-heavy fantasy)—as divided most often, as we've seen, between European *psyche* and Hollywood *techne*.

Of course Deleuze, writing twenty years before the decade that concerns this book, had no way to consider the films of uncanny or magic temportation under discussion here. These are films that send embodied characters back and forth across the arcs of their desire, by either digital cause or merely digital effect in American cinema, or by the bridging of irrational interstices in its European counterpart. But Deleuze's writing on the medium can in fact be read as one long effort to come to grips with the philosophical coordinates of such films well in advance of their actual production. Recent subversions of the somatic by the digital, of lived movement by pixilated trace, may well fall within the regime of the time-image. This, at least, is tentatively foreseen in the cryptic last suggestions of Deleuze's second volume, where the postcinematographic image, as we've seen him suggest, may be recognized as much to "relaunch" as to "spoil" the modernist time-image. But with the actual films before us two decades later, a proviso is needed: one having to do with the perverse spatializations of time that this new imagery may smuggle into the screen frame. It is thus, in the new grammar of the fantastic, that "time travels" becomes a clause rather than a phrase.

DREAMING TIME

In plots concerned with the travels or spatial trajectories (and deformations) of time itself, or in other words the mobility (and plasticity) of past and future with respect to the present, we are closer to a travesty of the Deleuzian time-image than to its revisionary narrative ratification. Whether technologically empowered or psychotic, characters seem to *territorialize* the very idea of the temporal in order to transport its timespace as a mobile sector of willed retreat from the continuous. Time's own traveling—the hallucinatory transport of whole time frames—is thus a misbegotten image of time traversed. Hence the proviso. If what I am calling the fantasies of temportation are to be subordinated to the time-image, they must be used to rethink the category, in its truest lineaments, or else they will sabotage it from within. Temportation, in its emphasis on a mutable timespace, or in Bakhtinian terms on a "chronotope" gone entirely virtual, carries cinema toward a new dialectical crisis. For in its mix of action and virtuality, of both inhabited space and (in a profound double sense) *occupied* time, it grows clear that the timespace-image of digital interface does not so much synthesize as elide the ground of *durée* and of consciousness itself. It locates not the emergence of a subsuming third term—a habitable zone beyond place and time—but an undermining of previous coordinates in received notions of extension itself, both temporal and spatial.

To spatialize or *frame* time in this fashion is to pit against each other, in categorical sedition, such essential Kantian priorities—rephrased most recently by Bernard Stiegler—as calendarity and cardinality, or in other words the bearings of clock and compass.[2] This chapter will return to the Hollywood implosion of time-space coordinates in a single recent example of science fabulation, the reflexively titled *The Final Cut*, after addressing two further European instances, *City of Lost Children* and *Bad Education*, in themselves polarized, that set the most important distinctions into firmer place. For all three films, though, the category (con)fusion suggested by the very term *temportation* is a sustaining premise. In *Bad Education* and *The Final Cut*, time is topographically mapped and retraced, now in libidinal circuits, now in cybernetic ones, and then bent—out of a spatial shape it never properly had—to subsequent intent.

National tendencies certainly inhere and persist. Movies continue to be made, if mostly in Europe, that seem to be self-consciously operating within the orbit, as epicycles, of cinema's mnemonic uncanny, where the very existence of the unconscious is not so immediately problematized as in VR sci-fi. And where, across the traumas processed by that unconscious, the old vocabulary of "humanism" may remain sequestered within an otherwise paranoid scenario. That is certainly one of the orientations hidden, in plain view as it were, by the hidden-camera thematic of Michael Haneke's *Caché*, and one of the best reasons for this chapter's closing in at the end on the tortures of temporality (both real-time and virtual) in that harrowing film. Then, too, concerning the specific separation of powers between transatlantic screen tendencies proposed by Deleuze, it must be acknowledged that the twain do sometimes meet—the temporal horizons of humanism versus technofuturism—and on theoretical ground. In bringing the countercurrents of recent Anglo-European filmmaking to terms with this earlier Deleuzian

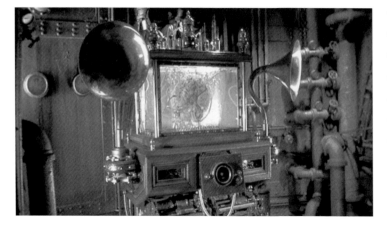

5.1

terminology, it seems almost inevitable that his postwar "cinema of the brain" would find its explicit futurist parable in a deliberate cross between what he otherwise sees, in regard to temporal limit cases, as European and American tendencies toward the psychological versus the technological.

Eternal Sunshine of the Spotless Mind might be an American example of this crossbred mode, this intragenre conversation: a film where electronic brain technology is mostly an allegorical prosthesis of the psychic drive to forget. On the other side of the Atlantic, there is a more bizarre equivalent. Before French director Jean-Pierre Jeunet is taken up by Hollywood for the sequel machine of *Alien Resurrection* (1997), what is imperiled by science in his dystopian fantasy, *City of Lost Children* (1996), is precisely the "uncanny" in its own right. Virtuality itself is dramatized as the explicit victim of bad science. The case is gargoyle clear. The central character is a disembodied cerebral mass in a fish tank, receiving the world not only through antique ear trumpets but through an old-fashioned, wood-mounted camera lens (fig. 5.1), whose shutter action we see (and see through) in repeated POV shots. This "migraine-ridden brain" is also the narrator of the opening plot exposition. Eavesdropped upon by the film audience, he gloatingly tells a self-conscious "fairy tale" to the five cloned male offshoots of a mad scientist who also happens to have created the detached brain itself (himself?) as "confidante." What especially needs explanation in this "tale" is that the clones, by the nature of genetic multiplication, have no access to an individuating "inner" life, to dreams, to imagination.

In this, the familiar dyad tacitly reasserts itself: movement versus time—and its failed amalgam under such perverse genetic circumstances. The omniscient brain is altogether beyond motion because beyond body. He is all apparatus: disembodied cognition plus optic prosthesis. By contrast, the cloned doubles of his maker operate primarily in a regime of motion, with all but the sensorimotor brain functions atrophied or deleted by replication. They operate, that is, only at the level of conscious perception, with no unbidden drives: as if all the world were for them no more than someone else's film. These multiple doubles of one another cannot, in short, reach the shifting basis of self-identity in the unconscious. They cannot dream, cannot reach back to that alembic of temporality itself—of memory and repression—known as the imaginary. As a result, they must steal the dreams of birth-born children to occupy this void. With no unconscious, time does not live in them, or they in it. They have no nostalgia, no premonition, no irrational impulses of any sort. So that even within the shared hyperboles of dystopian science fiction, that other familiar contrast comes to the fore across national tendencies. In American sci-fi the typical plot turn threatens human identity with a world of deceptive virtuality, a subterfuge of images detached from the real of action. By contrast, in this French counterpart the tables have been quite drastically turned. It is now intentional motion without virtuality, unadulterated perceptual response in the Deleuzian fields of action and reaction, that is the ultimate fear realized: the living nightmare come true for those whose consciousness is unalleviated, even as untroubled, by dream time.

ROUGH CUTS: MEMORY UNDER THE KNIFE

For a high-contrast pairing of examples that should permit summation of much preceding them in these chapters, and help orient in turn some final remarks on methodology, we can take up two films from the same year (2004) whose only obvious common denominator is that they have muted almost beyond notice the fantastic extremes of the uncanny versus the supernatural that have shaped previous discussion. Pedro Almodóvar's *Bad Education* (*La Mala Educación*) and Omar Naim's *The Final Cut* appear to have settled comfortably into other if still opposite genre niches, pure erotic psychodrama versus pure science fiction. What brings them into comparison, and at this late point in the run of various material we've examined in the genre of the fantastic, is not their reflexive commentary on cinema, almost ubiquitous across both plots. It is, instead, their overt concern with the virtual as a psychological rather than a technological issue. Certainly *Bad Education*, more than any other film, has found a way within European humanism to thematize the digital as the elegiac virtual—not only the virtual past but the virtual future.

In the cimnemonics of a neo-Felliniesque film like *Bad Education*, there might be good cause to expect that both technique and theme would eschew the logic of electrographic effects—or their tacit cultural intertext. Reason enough could be found for a less contemporaneously embedded celebration of cinema as a memory machine. But global crosscurrents in an age of optical electronics operate otherwise—and have made their dramatic mark on Almodóvar's style. Though far from anything associated with digital VR in the Hollywood mode, the generally low-tech, quasi-autobiographical melodrama in *Bad Education*, negotiating en route a strange hybrid of realisms both magic and psychological, does break stride at the end. It suspends its fervid metafilmic visualization, after a final freeze-frame, for an electronically doctored last shot whose computerization is impossible to miss. And what we find there is a rehearsal of the happily-ever-after (or not) of future-tense plot summaries typical of certain biographically minded narratives. This time, however, these prophetic flash-forwards (entirely verbal rather than visual, and hence confessing the entire film as text) are given in explicit optical allusion to the layered devices of Windows interface. This despite numerous previous allusions, in quite a different mood, to the more psychologically invested ocular mediations of photography and film. Only an immediate feel for the film's baroque plot can make sense of this shift in technical orientation and historical bearings alike—and send us back as well to its technical prefigurations earlier on, both in tricks of digital editing and in explicit morphing technology. In the unfolding of its convoluted plot, Almodóvar's film becomes a metacinematic education in its own right, both filmic and postfilmic.

Ignatio, hero(ine) of the long second-half flashback, is a drag queen whose death we first hear about when we see a framed photograph of his stage persona in her mother's house. It is noticed there by a film director, Enrique Goded, who had been Ignatio's boyhood lover in Catholic boarding school—and who goes on to direct a film, partly our film we come to realize, based posthumously on the memoir written by Ignatio about his abuse by a priest. This evasively concurrent and dovetailed film-within-the-film is a work in progress whose typescript often fills the frame. Even so, its realized scenes come intermittently before us as either envisioned or produced (we sometimes can't tell which). Often its episodes appear in fantasized transit between the director's invested reading in Ignatio's typescript and an executed *mise en scène* on the film's eventual soundstage. The plot of Almodóvar's narrative is, therefore, its own imagery in the making. This is an imagery that is virtual (in the mind's eye of the reader) even before it becomes materialized as *screened* illusion. Thus, its true predecessor is, in Deleuze's terms, that benchmark of "genetic" temporality, Fellini's *8 1/2*: a film that sustains a plot-long "seed-image, in process of being produced, which feeds on its own setbacks" (76). So does *Bad Education*. Unfolding as the cumulative facets of its own potentiality in rough-cut versions, the narrative works to actualize itself en route in a plot-long gestation. Like Fellini's film, the psychic trajectory of Almodóvar's narrative, which also begins with an escape from priests to a cult of voyeuristic sex, suggests that autobiographical temporality becomes fantastic, its realism magic, when reimagined through the memory of desire. Not only that, but Almodóvar's film helps us recognize more clearly yet what Deleuze sees in Fellini's, why it would illustrate for him a prototypical manifestation of the time-image. Like Almodóvar's directorial self-anatomy, mov-

5.2

ies about the deferred coming into being of their maker's own past as present narrative, about elapsed history as image, become fables, no less, of time's elusive submission to consciousness.

The opening scene of *Bad Education* has the film director clipping a newspaper story that he thinks might make a great movie episode one day: a motorcyclist chased by the police even after he has frozen to death in a blizzard—a tale of sheer rotary motion divorced from life. Cinema itself in parable, no doubt. More specifically, though, this brief and dispensable episode installs the model for the whole film's hidden matrix: namely, the death drive. It is this libidinal generator of theme that inflects the eroticism of the narrative from the start, anticipating the impersonated survival of death that constitutes its major plot twist. For the portrayed director, Enrique, is first approached with the child-molestation story by a young man pretending to be its author and his former lover, Ignatio—only for the director to find out in due course, along with us, that the young man is actually the dead writer's brother, a would-be actor who wants the lead part and who has in fact murdered the drag queen with deadly pure heroin to escape from the burden of the latter's addiction.

By this disclosure, in the language of computer-modeled narratology from the last chapter, one extended plotline about a lost and found erotic object is "popped" by another. The director's former lover exists now, after all, only on his typed pages, peddled under false pretenses by his brother. The underlying "data platform" thus surfaced is one in which text per se (script, then film) must substitute for Enrique's short-lived dream of recovered passion with the impersonated former boyfriend. All that is left of Ignatio, and of Enrique's past with him, awaits not behind, in time, but forward, on the eventual film screen. Art only comes at the cost of life, memory at the expense of desire, the time-image as the death of the real. To say the least, of course, the substitute affair that ensues is no less sexual for being steeped in fantasy. And it serves well the framing metacinematic purposes of this film-in-the-making plot that the director's voice-over about Ignatio's fraternal surrogate now starring in the movie, and thus annihilating the original by image as well as by homicide, should appear just where it does. The monologue is recorded over a shameless close-up, lasting several minutes, of the film camera's inner mechanism, its pistons and thrusting rotors caught pumping away (fig. 5.2). "He allowed me to penetrate him repeatedly," goes the voice-over, "but only physically . . . his mystery was still intact." The

5.3–5.4

5.5

"enigma" of Enrique's replacing Ignatio with his brother and murderer will not come clear from mechanical transcription onto celluloid, but only from the corrective retooling of the film's larger plot.

What we gradually discover is that Ignatio's brother and his secret lover, the defrocked priest who was once Ignatio's abuser and had recently been blackmailed by him for sex-change money, deliberately leave Ignatio the time to overdose by hiding out in a movie theater. The feature is a rerun screening in, bluntly enough, a film-noir festival, an unnamed film whose every line of criminal dialogue "seemed to be about us." Yet this is an almost deliberately flat false lead. For life has imitated art, as we've already seen, in far more invasive ways until this late point in the narrative. Rather than facilitating an escape into filmic illusion in order to "kill time," as the brother here proposes, the illusionist mechanisms of Almodóvar's cinema have previously resuscitated various sheets of reminiscence and layered them over the present. These appear in scrimlike planes of virtuality and nostalgic time travel, as when a decayed wall of film posters fades backwards into freshness (figs. 5.3–5.4), or when a reading of the film script dissipates into the past visit to the already decaying movie theater it details (fig. 5.5). In all of these transitions, the libidinal *reframing* of the past in the director's conjuring eye is literalized by a fluid reversion from contemporary wide-screen ratios to earlier projection formats, with lateral black masking used to mark the frame's tunnel vision into the past. Then, too, in the schoolyard retrospect to which the script eventually leads, various "arrhythmies" (Lyotard) of the track—slow-motion eye contact between the infatuated boys and

5.6

at one point a freeze-frame on a blocked soccer kick—suggest the imposition of memory on lived time.

No instance, however, is more marked and notable—no instance of motion sectored as space, frozen as time, and segued back to duration—than a bravura image from this same formative childhood sequence. Its graphic—its electrographic—extremity acts as the true forking (indeed splaying) image for the time machine of the whole retrospective plot. In this visual conceit, pastiche invades the body itself. It does so when, at the moment of Ignatio's lost sexual innocence, the whole screen frame is transformed from montage to a torn collage in motion. The sequence begins with the priest staring at an actual photo of the boy (fig. 5.6). It is this backlit portable framing of the desired body that sets us up for the violent tearing in half of that same imaged body in real space. The device, if not the idea, is in my experience unprecedented. It amounts to the pop graphics of the title sequence being internalized by plot—yet not by photomechanical (that is, cinematic) realism in the mode of the ocular fantastic. Operating instead is a digital uncanny as hard to decipher technically as it is to miss rhetorically.

It happens this way. After Ignatio's first sexual assault by the priest, the boy falls and cuts himself in trying to run away, so that a thin rill of blood bisects his forehead in close-up. Articulating for us the director's parallel reading of the typescript that recovers and records this incident, Ignatio's voice-over remarks that, like his bloodied face, so was his life to be "always divided in two" from that moment forward. At which point the whole image plane—resembling something more like a photographic still than a freeze-frame—is optically torn in half as a psychosexual trope (figs. 5.7–5.9). Revealed behind it is the uncanny reverse shot (across a gaping hole in time itself) of the older priest reading of his own former lust and cruelty (fig. 5.9 again). In quick succession we then cut forward, less roughly, more conventionally, to the later site of the director's own reading of this whole confrontation scene (fig. 5.10), the lover's desire aligned, by inevitable identification, with that of the abuser. As implicitly illustrated both by Lyotard in "Acinema" and by Deleuze in *The Time-Image*, the ongoing temporal violations of montage, its breaks with duration and continuity, can sometimes find manifestation as a violence within the represented scene. In this case it is the *cut* itself as a slicing in half of the hereafter divided self. And this is where narratography shows its hand as a method of both Lyotardian and

5.7–5.10

Deleuzian syncopations, with the *interval* projected across the visual field as a signifying graphic *interference*. Medial apprehension of this sort stands vigil at those pressure points where, in outplaying both motion and duration at once, in exceeding together both the taking of action and the taking of time, the image instead, as Deleuze would have it, takes thought.

Thus has the film's more frequent syntax of temporal progression, the lap dissolve, been almost grotesquely over-ridden in this case by a visual pun on the scissoring moment of the editor's normal laboratory cut. *Trucage* as symbolic grammar: releasing the future from within the ruptures of time present. All this is actualized, of course, only in the reading mind's eye of the director's hypervisual imagination—Enrique's, that is (quite apart from Almodóvar's). It is as if self-division were innovatively filmed by the former, the director as character—filmed on the spot as a dead metaphor turned to visual trope. A digital trope at that. For in echo of the priest's archived snapshot of his lost love, photochemistry is evoked only by the rending of the two-dimensional image, not by any freeze-frame that permits it. This is the most remarkable, ultimately uncanny, aspect of this irruptive magic within a realist retrospect. For motion isn't arrested before the self is sexually bisected. The boy's eyes are still moving, pained and searching. They are looking up (as if, at one point, directly through the camera) at the source of his violation, even as

5.11

they separate across the fissure in his brain. Only digital imaging could produce this razor-clean irony. What happens in this spectacular two-ply *trucage* of a sacrificial subject and his surviving reader is not just the split psyche in formation ("my life . . . divided in two"), as it might be meta-phorically captured in textual retrospect. Pictured by digital emblem—and across the double interface of its twice-figured reader—is a gash in time itself.

Other schematic bifurcations accompany this (de)formative moment. Escaping from the sexual coercions of the pedophile schoolmaster into a movie theater, yet finding in the screen's sensuous version of religious iconography an alternative route to sexual obsession, the young Ignacio and Enrique masturbate each other for the first time while watching an actress (Sara Monteil) they both find desperately beautiful in the role of an ex-nun, "Mother Soledad." With the boys silhouetted from behind against the huge female image—and a lone adult male visible between them and the screen, and more completely dwarfed by it (fig. 5.11)—what emerges is a complex distillation of cinema's eroticized apparatus of psychic projection, even as the scene becomes in its own right a "time-image" of sexuality's forking future paths. Beyond the star's feminine perfection, nothing is mentioned—at either pole of the boys' eventual desires—about this enthrallment by the screen image. In yet another version of that absent, disembodied, but somehow incorporated maternal image from the cycles of fantasy we've been tracing, the boys are outed nonetheless in their creative (or self-creating) impulses. The one's libido is so inflamed by sexuality's link to female performance that he will want to *become* the woman; the other's is so cathected around the display itself that he ends up being a director of such enacted screen spectacles. That faceless third man, the general audience writ small, is left to his own—presumably more normative—devices of fetishistic pleasure.

It is not up to the inferences of framing and rear projection in this scene alone, of course, to mark the divided fates of the boys—or even their initial separation. Let alone their internal division from themselves. We have noted the show-stopping transection, right down the middle, which emblemizes the split impulses of the eventual transsexual. Digital ingenuity is also called upon shortly after, and then once again in closure, to figure time's forked courses: first in a multiple distortion of the image plane, later in its surrender to sheer text. More explicit than in the implied digital editing of the split skull, these two moments, even if not linked in the

5.12–5.14, 5.15–5.17

spectator's mind, together foreground the uncanny power of the new electronic injection—and in the first case, a true infusion—into filmic cinema. But they do so in a way as far removed as possible from the "instrumental marvelous" of comparable Hollywood effects. It is, for instance, a subtle, almost imperceptible mix of digital morphing and traditional cinematic superimposition—almost imperceptible even in a frame-by-frame scan—that captures the sundering past as restitched into the textual present. The priest has arranged for the expulsion of Ignatio's new boyfriend, and the moment of separation has come. As Ignatio looks longingly at our eventual director in shot/reverse shot across the barred schoolyard gate, he is transmogrified into his future actorly self by a "time-image" leap of desire. This is conveyed by an indiscernible blend of pixel and photogram in transition—with, by metafilmic logic again, the floating-open of the frame from earlier to present ratio (figs. 5.12–5.14). Once his boyhood image rapidly mutates into the adult actor looking back across the years—and across a fantastic suture—on the still-young friend, it is the latter who then morphs in turn into the adult director for a meeting in the shared present space of his studio office (figs. 5.15–5.17).

5.18

Think of it this way. The digital transform does not just substitute for the photomechanical, but refigures its work. Photography stops time. Cinema puts fixities back into flux as imaged motion. Only digitization can then seem to make *time itself happen*, rather than just exist as duration. Only the new screen uncanny can appear to picture time's own motion within framed space. From the POV of reception in the director's lenslike mind's eye, a piece of duration textualized by the ongoing script allows a lost past to arrive fulfilled—across that bridging morph—into its own destined present. Such is the digital technology more often used in Hollywood thrillers to reverse this transit. Usually it is found buckling or dissolving the whole image plane in order to provide the blocked hero with an escape into his own mutable past. In Almodóvar's case, instead, digitization achieves a limited and strictly uncanny (rather than supernatural) deployment for the time-lapse maturation of bodies in the reciprocal eyes of desire.

Yet morphing, in its supplement to photogrammatic superimposition, is in fact deceptive here, since Ignatio hasn't grown up into the handsome actor come courting. The young self has only been, as the digital trick may end up suggesting in retrospect, usurped by the later surrogate. *Trucage* as montage as subterfuge—and precisely in this evasive blend of cimnemonics and digitime. The mirage of maturation from yearning boy to sexually available man has in fact elided, that is, the as-yet undisclosed murder of the actual morphed and transsexed body. This we are given to see in the film's second use of slow motion, long after the suspended athletic choreography on the soccer field: another Lyotardian "arrhythmy," figuring this time the violent foreclosure of mortal duration itself. The buildup to this slow-motion sequence has an unusual narratographic density. Ignatio is blasted by an overdose just after we see his face miniaturized in a heroin spoon (fig. 5.18). Here is an "interface effect" (Žižek) caught this time in a more than metaphoric "anamorphosis" (Lacan via Žižek) of the real—as noted in less flamboyant connection with the distorting train window in *The Double Life of Véronique*. Having completed the story that will become our film, and already in the process of typing out a note to its director, the dead writer in profile stares wide-eyed and visionless into the void until he falls forward onto his keyboard. Symbolically passing out of range of any concave reflection in the spoon, the addicted narcissist is lost at last to all self-image—even as his fall sends into useless activation a knot of slammed and meaningless keys. Equally symbolic, and answering in the end to the rhythmic aberrations of the recovered past in the schoolyard sequence, this descent into death is wrenched from the normal movement-image. It is clocked, beyond lived time, by a slight slowing (without in any way breaking) of the fall. In the vanished meantime, that bent refraction of self in the accoutrement of its death becomes the Deleuzian "mirror-image," degree zero. As one mode of time's crystal faceting, it marks the emptying out of all presence by its reduced and disincarnate double.

The plot will seek its point of rest only when that typewritten note is delivered to its intended source. In all the director's obsessive "paging through" of the inscribed past comprised by the film's retroactive plot, with its sinuous trespass from one diaphanous "section" of time to another, it is only in the last shot that the door is finally closed on time past. This happens when Enrique locks the gate of his mansion on his lover and star—once that imposter has been discovered to be not only Ignatio's brother but also his murderer. This closing off of access is not my metaphor but, rather, the film's own. The huge, rectangular, screen-scaled gate now fills the narrative screen as the lone remaining character, Enrique, is fixated by freeze-frame after reading the single truncated sentence of Ignatio's last letter, held up in permanent arrest by the director figure at screen left: "I think this time I have succeeded," meaning kicked his heroin habit, is a testament that we know has been broken off, in effect, by his lethal overdose—and punctuated by the mangled intersection of those arbitrary typewriter keys. Slow motion brought him down. A freeze-frame on the recognition scene now answers. Here are Lyotard's two rhythmic violations in a single link: variable pace and stasis, each an "écrit en mouvements" tied in this metatextual case to an incomplete typescript that serves to cancel out the self. And there are more inscriptive technologies in play here than just those of the image track. The film's last full-screen typescript, that one-line letter, dates from the early 1980s. It was thus written on the very brink not only of word processing but of other electronic systems that together inflect the discourse rather than the plot of Almodóvar's own film text. Film text, yes—and one that is about to become sheer text processing in its own right, electronic rather than mechanical.

But why the freeze-frame on the director as transition? For one thing, any such stop-action image—even when digitally produced—involves, by optical allusion, a fairly direct return to the actualized substrate of traditional cinema. It is there that the reduplicated photogram necessary to prolong the frame into apparent fixity is, as we know, the ultimate limit case of the filmic: lifted from strip into track with an arresting vengeance. To put it again in Lyotard's terms, an effect of motor "writing" is manifest as the jolt to rhythm administered by stasis itself. As few films could make more unavoidable than Almodóvar's, such consideration solicits a return to

5.19–5.20

Deleuze's second volume as well. This is not just because the freeze-frame locates the most obvious overlap between motion and time in the paralysis of the former by the latter. If the freeze-frame inevitably provides, on Deleuze's model, something like a canceled duration-image, then such is also the case with other visual anomalies in Almodóvar's film—as in other instances of the European uncanny over the last decade. The freeze thus makes common cause with any number of optical deviances that retain, retard, or displace motion even while relaying its cross sections along a shifting field of temporal planes and composite frames. And here in Almodóvar common cause is made, in addition, with electronic discourse itself, sampled en route within the mimetic register and now spelled out as an unequivocal effect of writing: a quite literally predictive "writing on the wall."

How so? The effect is so abrupt that we may forget its form in the closural satisfactions of its content. Out of the geometrical panels subdividing that closed gate now erupt, one after another, by digital *trucage*, a sequence of interface "Windows" (almost like rectangular and computerized cartoon bubbles) that inscribe in simple electronic print the fates and future careers of the principals (fig. 5.19). After these proleptic data subfiles are maximized and deciphered, they recede to their place in the segmental grid of the gate's whole finalizing impasse. Nonetheless, the narrative discourse, if not exactly the film, having managed to reach forward into the early '80s from its '60s boyhood setting, has now stretched further ahead into the present. The last thing we hear about the decades that have elapsed since the truncation of our visualized plot is that the director has continued to make movies with an unabated "pasión"—that single word filling the screen in Spanish orthography and digital graphics at the end (fig. 5.20). Yet there are no passionate images left before us now in the plane of foresight without visualization. Barricading the past and anticipating the future, the apparatus of time travel has itself shifted, against the backdrop of this impermeable gate, from the photofilmic sheet to the quasi-digital interface. Screen storytelling is thereby reduced to sheer lexical text, framed view to mere textual window, vista to data grid.

In reflecting on this closural textuality as a response to the revealed death moment just before, narratology is well equipped to study the flashback structure that routes us round to the murder as a kind of detective story. Narratography, by contrast, would find its own further clues in the subterranean interchange of images that link this dead end of writing with the film's own way of writing beyond its end. The emblematic vacuum of typography for the consciousness removed from plot by death is answered and corrected by plot's own self-fulfilling prophecy in digital inscription and its windowed text files. Narratology (in the mode of psychopoetics) might see Thanatos transferred ironically from violent murder to narrative closure, or in other words to the decease of image once Ignatio's story is over. Narratography would dwell within the textual system of this transfer as in fact a discovered common denominator between the imprint media of script and film, surfaced at the last in a strange new triangulation with the information stream of electronic interface. In the end, then, this one film honors its fallen writer by becoming always and only written. It also refuses the virtual future (simulated images rather than transcribed words about human outcomes) in deference to the virtuality of the past as remembered. It thus holds the time-image to the normative field of humanism after all. And in valorizing the Deleuzian *lectosign* in this mode of reductive bluntness, it doesn't simply make us grateful for what we've more richly seen up to this point. We are made aware at the same time that we have not been alone in our access to the past as spectacle. In ocular alignment with the director's own imagining of the story's route from narrative typescript to film script to staged scene, we are made further aware that, in our own way too, we have all along been *reading* the story simply by watching it unfurl.

POSTMORTEM SLICES OF LIFE

If death is the advent of the strictly illegible in the jamming of typewriter keys, it is also, in another register, the moment of the exclusively readable, where life has become only the words of a script. That's what gives the pervasive elegiac cast to Almodóvar's film. Death surfaces as the wholly inscribed within the no longer lived. That's in turn the climactic idea of the film whose Hollywood-thriller devices await comparison now on just the score of time, memory, and virtuality. If the postcentenary decade of the temporal fantastic—in its high-concept Hollywood incarnation—were to be seen, in historical hindsight, to have come suddenly to a finish in 2005, then the DVD reissue of *The Final Cut* (after an abortive limited release in the fall of 2004 in satellite projection at selected AMC theaters) would make perfect sense as its dead end. More than the out-of-body elegiacs of *The Jacket* (considered in the next chapter), it pushes the sci-fi format inward, to its organic terminus ad quem. In this strangely direct and overly expository plot, the mysteries of the fantastic are entirely explained away by commercial technology, not just as tacit intertext but in explicit plot details. All is now lucidly exposited, both as high-tech extrapolation from present marketing ideas and as broader cultural anthropology. Gone is the waver between uncanny and supernatural explanation, hallucinatory flashbacks versus instrumental marvels. And in being straightened out, the tropes of temportation have gone flat. All those heady superficial themes of divided subjectivity, uncanny interface, voyeurism, disembodiment, and vicarious death that have been jumbled together in frenetic and superficial permutations ever since *Johnny Mnemonic*, and with renewed contrivance in the last half decade, have been reconfigured instead in what is for the most part a straightforward blend of sci-fi and social satire. *The Final Cut* is Evelyn Waugh's *The Loved One* for a cybernetic age.

The main drive of the plot is clear, its unlikeliest turns derived only from subsidiary intersections of separate subplots. The punningly designated corporation EYEtech has the second *E* facing backward in its rebus logo to suggest the reversibility of that autobiographical (rather than autobiological) vision it facilitates. The company has flooded the nation in the near future with its "Zoe implant." This is a device, digital but "wholly organic" and undetectable, that records an entire life from the viewpoint of the subject in whom it has been inserted at birth. Its replay, whether on a funeral monument or at a public screening, is called a "rememory." Yet in that very neologism there is an epistemological catch that turns ontological. For this synthetic record resembles so-called photographic or eidetic memory only from a postmortem vantage—exclusively, in other words, for the survivors, the mourners. It doesn't enhance memories or redouble access to one's own life in progress. It isn't in this sense embodied memory at all, only embedded—available in absolute retrospect and then solely by proxy. Picking up on this irony in the familiar reflexivity of the DVD packaging, the film is brought to us as if distributed by the EYEtech promotion department rather than the video distributor, so that the scene-selection format is identified as "rememories." The irony is airtight. Selective memory is like a digital menu. More than this, supposedly focalized experience has become a movie someone else selectively replays. And that someone is you.

As such, the work of the implant, once edited, is both less and more than a typical funerary image. The kind of iconic record it supersedes as mourning ritual is alluded to in vestigial form when a photographic portrait of the deceased is seen to accompany, at a typical "rememory" service, a book for the signatures of the funeral guests. And there is another kind of super-

5.21

seding that seems inscribed here as well. The Zoe device is no doubt named by explicit verbal allusion (not in this case optical allusion, but etymological) to the original Zoetrope ("life-turning"), where manual rotation of a slotted wheel produces for the outside eye the apparent motion of a band of images attached to its inner curved surface. With the new Zoe device, the manual has become the wholly incarnate: the body as its own ocular record of change over time. Indeed, all the patent wars among the early American movie companies seem to cash in on certain narratological assumptions on the way to their genre distortions. Vitagraph, Mutoscope, Biograph: lifeline, mutation, biography—all on the road toward holography and beyond. Sci-fi narratives have always found ways to upgrade these "time machines" into magic prostheses within their cautionary plots.[3]

The magic is perversely constrained, however, in *The Final Cut*. For the actual projection of the Zoe disk in "rememory," the only mortuary images of lost loved ones available for recognition would be shots of them aging over time in their private mirrors, shots mostly edited out before screening. Reversing the closural logic of the self-aborted fetus in the previous year's *The Butterfly Effect*, the film's first sample screening, for instance, begins with a shot of the mother presented to natal consciousness, rather than the infant to her, at the moment of his birth (fig. 5.21). Here and afterward, we see what they, the now dead, saw, not what they looked like. In other words, we see a superficial parody of what we meant to them, not what they signified to themselves. This is the built-in human alienation of the device that has led to protests against EYEtech by increasingly violent crowds.

But our plot takes us for the most part behind the corporate scene. Faced with thousands of hours of POV frames, expert "cutters" are required to edit the digital record down to a presentable feature-length DVD. And, as we overhear one of the cutters explain, this editing is a willful narratological intervention. Out of all the trivia, distraction, and ugliness of any life, an appealing "storyline" must be developed to make sense of the events, a chain of however artificial and flattering causality. It might so far sound like the ordinary way one tells one's own story to oneself—unless, of course, one has drawn the model of such narrative shaping from the "alternate endings" of DVD releases (as may be suspected with a film like *The Butterfly Effect*). But here it is in hands other than the subject's, in the service of other interests than self-recognition. In any case, no preservation of the life in time takes place. What instead takes its place is the construction of time across the lacunae of a life. To put it in a related pair of seasoned terms, we are so far removed from anything like Walter Benjamin's aura, or self-aura, by this cybernetic upload of a life that Benjamin's almost equally famous sense, elsewhere, of storytelling as sanctioned by deathbed reflection, where meaning can be finalized in retrospect, has been altogether displaced from the site of lived coherence to an audience of self-interested survivors.[4]

Humanist culture rises in backlash on the streets, but not before the master cutter himself gets his own lifeline caught up in the tangles of the text he is editing with the help of his "guillotine" program. And who better to play the anonymous cosmetic surgeon of another's biographic profile than Robin Williams, fresh from the reengineered photochemistry of his own fantasy in *One Hour Photo* and ready to retool as the electronic sculptor of other people's imaged lives. In both roles, he presides over the death of a Lacanian real, in one case escaping into the imaginary, in the other case turning the unconscious of total memory into the symbolic of an encapsulated narrative. Hakman—dismemberer supreme of other lives, as his name makes inescapable with its crude double thud—is a man who calls himself a "sin-eater." He is singled out among his colleagues for having unusually few qualms about his euphemistic deletions of life's recorded "garbage." And this, for a single reason meant to explain everything. Traumatized as a child by secret guilt over the death of a playmate, he forgives everyone anything, more than willing to help delete the unwanted memories of another. His life is so entirely squelched and withered that he has enough repression left over for everyone else's life.

This is a marked departure from the normal thriller plots of recent Hollywood formula. These would allow his childhood trauma to surface in fragmented flashbacks, coming together in a scene that could then, in the extra twist of temportation, be returned to and repaired. Instead of this, here the whole formative scene of past violence is done away with at one stroke by the (at last) correctly remembered, rather than drastically corrected, record. This happens, with bland improbability, when the cutter recognizes his supposedly dead playmate full grown and thriving in another life disk he has been given to edit. As he will later discover by hacking into his own chip, the boy was only dazed by a fall instead of killed; the remembered pool of blood was only an overturned paint can. His vision, so to say, had recorded more than in his panic he had actually seen. Here, then, is a full cultural inversion in the temportation plot. Widely shared social and juridical anxiety over false witness in the act of reconstructed trauma is turned round on itself by the chance to return to and rescind the scarring (but misattributed) guilt. In a Deleuzian vocabulary, sheets of the past are recovered as corrective *lectosigns*.

The real dramatic problem comes with the rest of the seedy material in Hakman's new editing assignment. For he has been hired to finesse the sexual scandal of a top lawyer for the EYEtech corporation itself. Naturally, as recent films in this genre go, the secret de rigueur that must be doctored is the lawyer's molestation of his preteen daughter—and her mother's implied complicity in it. Child abuse, especially sexual, is the traumatic norm in everything from *Identity* and *One Hour Photo* to *The Butterfly Effect*, and even in *The Sixth Sense* (and by more than the daughter-poisoning video subplot, according to one critic).[5] What is unusual in *The Final Cut* is that the protagonist's own repression of familial incest, erased to suit the needs of the bereaved, has already been elided from the film's own images, cut away with remarkable tact. This despite the fact that everything depends on the hero's having seen that molestation—seen and unsuspectingly *recorded* it.

This is the film's first real twist, its deepest irony of hypermediation. At least an hour before, in another version of the trick beginning, the "Cutter's Code" is flashed before us just after the credits—credits which themselves fade away sequentially without fully disappearing, leaving an indiscernible blur of their own graphic traces, layer upon layer. Among the prohibitions outlined by the code is the caveat that no cutter can have a Zoe implant of his own. Yet Hakman discovers that he has been implanted, without his knowledge, by parents who died before they could tell him. So his own supposed guilty secret from the past will be recorded there for all to see: that's his initial horror. He cannot know at this point that the Zoe device will in fact redeem him. At serious risk of lobotomization, as in *Johnny Mnemonic*, he is the one subject able to retrieve, while still alive, a truer encoded memory of the distant events that bedevil him. This is exactly what is not possible for the device in its commercial marketing, where memory is entirely seeded in the unconscious rather than searchable there.

So now the obligatory double twist. If we thought good could ultimately come of the "instrumental marvelous," we settled (back) too soon. By an intervening plot complication, the lawyer's disk has been accidentally destroyed. The only way for the anti-Zoe forces in the counterculture to discredit the corporation now, in the film's revealed conspiracy plot, is to kill Hakman himself—and siphon off his remediated images of the lawyer's past sexual crime. Here, then, is where the backdrop of social unrest and protest is brought to the foreground. As Hakman is hunted down by these reality activists, we watch his flight and interception, ironically enough, through the intermittent flashes of his own Zoe chip in operation. By the oscillation between Tak Fujimoto's amber-saturated cinematography and this full-frame antiseptic video, we thus monitor the turning of present into past from second to second until transcription closes in on the definitive moment of such conversion: the vanishing last instant of life snagged and filed

5.22–5.23

5.24

away as optical record. This happens when Hakman sees his own bloodied hand after the final shooting, rather than the chest wound that has drenched it (previously visible to us, not to him)—while it is only we, through a more deeply embedded POV, who register the subscript record of his digital life clock (fig. 5.22). We do so at the very moment when both the visual image and its text pass from clear resolution to an electronic breakdown into tessellated fragments—and this as a trope of mortal disintegration (fig. 5.23). These are fragments echoed in turn by the bathroom tiles behind the "rememorized" hero as he has once unconsciously recorded himself straightening his tie in the mirror (fig. 5.24). For that is the last image we see of him in our film, with his own Zoe transcript being posthumously combed through before the final credits by the lead saboteur. He is still looking for incriminating clues in the same "life" (the lawyer's) where Hakman, seeing his boyhood playmate grown up, had come upon his own exoneration.

More than irony, there is a kind of allegory at work, slowly consolidated. From the start, the coexistence of middle-class picketers and fiercely tattooed hippies at the "rememory" services, with their updated version of right-to-life banners in the repeated humanist slogan "Live Your Own Life Now," has been a common cause hard to factor. At one level, these existential Luddites seem banded together like closet Deleuzians, resisting the cyborgian (rather than cinematic) virtualization of embodied temporal succession, the continuous retracing of the organic memory trace. For Deleuze, every image of the moment, even from within its duration, virtualizes it. EYEtech perverts that organic process by involuntary instrumental storage. But why the scarification of its detractors, the disfiguring tattoos? Why this violence to rather than within the body? It's not, one comes to realize, that punk energy is closer to nature than bourgeois culture and its rituals of perpetuation, or even that what these street freaks do to the body is less violent than the invisible intrusiveness of the chip.

Plot again shores up the allegory. Taking the sanctity-of-life crusade into the realm of cloned consciousness (rather than the cloned bodies of contemporary stem-cell debate), these organicist neoprimitives make the case for an unmediated relation to one's life. At a crucial turning point, what we discover is that disfiguring lesions on the tattooed among them are the result of surgical strikes against nothing less than the separate audio and visual tracks of the Zoe device itself: postoperative insignias of heroic refusal from former subjects of the implants. What we have previously mistaken for flesh art is found operating instead in an open defiance of mental wounding. It is a way of insisting, and visibly, on the body as the proper site of one's own experience—rather than surrendering that body as a mere vessel for the prosthetic device of a digitized mental supplement. This protest is thus a way of writing on the body that resists its total reinscription from within: a visible protest against the invisible discursification, one might say, of all desire.

In Deleuze, the virtual image is life lived under the sign of its own potential as memory: a process inevitable to consciousness. Whenever we see a cinematographic POV shot replaced by the video-log image in *The Final Cut*, we find again, and instead, the technologizing of the virtual: not the "spiritual automaton" (or thinking machine) of philosophical cinema, but an overt simulacrum of mental process. Under the auspices of the Zoe implant (and even, to some extent, of the initial satellite distribution of the film itself, where it is all there, somewhere, just waiting to be downloaded), we revert to a condition of total storage that requires, as at the beginning of cinema, the incision of the cut—or, in other words, deletion and recombination—in order to make narrative. In this most narratological of premises, where life is contractually remastered as a commercial DVD requiring a streamlined plotting, the real traction of narratography comes with those precipitous overlays of digital upon cinematographic sequencing. These involve sudden shifts from photogram-based capture in the film's own "omniscient perspective" to its instantaneously pixilated reconstruction in the implanted archive. The POV logic of such optical disjunctures reaches its cul-de-sac in the tessellated breakdown of Hakman's death moment, half metonymy for a damaged implant, yes, but at the same time half metaphor for a failing vision of the mortal world.[6]

Narratography also responds with an intertextual tremor to one moment of overt and

5.25

strictly metaphoric *trucage* in *The Final Cut*. This is a digital tricking of the image that installs an unmitigated echo of a straightforwardly realist (and all the more disturbing) scene from Robin Williams's previous scopic vicariousness in *One Hour Photo* (fig. 3.22 again). What is more, this moment also involves a further reversal and travesty of Deleuze (and through him of Bergson). Figuring Hakman's immersive obsession with the cross-referenced images of a life-line for which he is the digital editor, the camera, facing toward him rather than his monitors, pulls back from his console to reveal its subfiles and the body that reads them floating together before a vast field of the discrete "clips" formerly contained and processed, respectively, by those subfiles and that body (fig. 5.25). Images once generated and ingested by the cutter's work now swallow up and relativize him as one more digitally edited screen figment in a hybrid projection system. Like the abused photo clerk dwarfed by a collaged archive of a family life not his own in the previous film, here the guilt-obsessed digital editor is engulfed by "time . . . spatialized" (Manovich's term again, with a new, perverse aptness). This is the elapsed and re-sorted time of another's life, for whose traces our hero will eventually die.

On borrowed time in both senses of the idiom, here is a totally instrumentalized subject neither living in time nor letting it live in him, just going over and over it again as mosaic object. Further, too, this puzzling-out of a life in a gridwork mapping of rectangles is the macro image to which, at a kind of fractal scale, pixilation offers the mostly unseen microcosm—until, that is, the mosaic disintegration of the single POV rectangle at the point of death (fig. 5.23 again). But where is the real crux of the problem? The protestors in the streets object to the crippling "self-consciousness" induced by the Zoe implant, which they claim prevents one from "living your life now." But immediacy isn't the real issue, self-consciousness not the keenest danger. It is really consciousness itself, psychoanalytically understood, that is sacrificed to record in this unholy commercial bargain. For it is consciousness that must fashion any functional picture of itself—in and across time—on precisely the deletions and compressions, the leaps, free associations, and protective erasures, that in this case only the "cutter," and only after the fact, can bring to bear on desire and duration, even his own.

In transatlantic sum, then, what have we seen so far? For both *Bad Education* and *The Final Cut*, as for many films before them in either the mnemonic uncanny or the ontological gothic, death as human cancellation is often the endpoint toward which plot, in its deferrals, is just biding its time. But how? By what different increments and suppressions? There lies one of the major cultural and aesthetic differentiations between two related practices in a postmodern fantastic. Such tendencies are seen to diverge, for instance, when morphing—now digitally implemented, now digitally *themed*—would come before us either as a metaphor or as a metaphysical transformation. In any narratology, of course, much, if not everything, depends on timing. In the unabashed ethical humanism of Almodóvar's *Bad Education*, child abuse is the overriding motor of plot, from which all other dramatic ramifications are paced to follow. That violence, both psychic and physical, is unfolded from the start in the character's own memory, not lodged as a repressed secret of structure itself. Alternately, in either the very slow leak of disclosure or its bottled-up denial until the last minute, Hollywood follows instead what we might call the *Peeping Tom* archetype in its treatment of repressed abuse, from Michael Powell's 1960s British psychodrama down through *Identity* or *One Hour Photo* to *The Final Cut*. Given their deliberate twist finishes in the genre contract of prolonged disavowal, the contrast with a European tradition of humanist intersubjectivity, as brought to explicit sexual display in Almodóvar, couldn't be clearer. Linked to the uncanny in the Spanish film, desire confounds the very epistemology of otherness, fusing actors, sexual agents, and their intimacies across time and space. By contrast, and often linked to the scientific marvelous in Hollywood plotting, remembered desire can be excised by machination, the whole traced ontology of the subject reconfigured for a spectacle both internal and external.

Concerning the "Hakman" himself of *The Final Cut*, we noted how his own hidden guilt is caught up by, and distributed across, the guilty secrets he keeps—by cutting away—for others. Sapped and flattened by the overexplicitness of this one storyline and its bland inversion of the norm, nonetheless the recent Hollywood prototype is to be decisively glimpsed here even in shadow form: namely, the trick ending that reverts to abusive origins as the biographical equivalent of a narratological disequilibrium. We are faced again with a broad cultural contrast, between, on the Continent, the transferential exchanges of desire and identification and, in Hollywood, the garish twists of suddenly "recovered" memory—indeed, in that paradoxical neologism (and pleonasm) of *The Final Cut*, of "rememory." As graphed in progress by the porosity, overlays, and disjunctions of the image plane itself, with all its obtruded secondary mediations, one can therefore resummarize the contrastive narratography of the European and American fantastic on the basis of our latest paired instance. The elegiac eroticism of Almodóvar's plot maps the gendered byways of a same-sexed libido through to the alternate fates of death and temporal renewal: the reflected image of the heroin fatality followed, in a diametrically opposed "interface effect," by the digitally windowed glimpse of the future in prose. In Hollywood plotting, it is instead the repressed matrix of forbidden sexuality per se (displaced in *The Final Cut* onto a tangential character's molestation scenario) that is likely to turn the whole plot into a death drive, clamping down on desire, sublimating it by introversion and negation.

LECTOGRAMS: TIMED FRAMES TO FRAMED TIME

Virtuality, not actuality, has been our theme in this chapter—but as a function no less of reality. Before turning to a stringent instance of real-time ethics in a parable of remediation, certain inferences of the medial "real" bear review, in precisely its divorce from the actual. The virtual memories of *Bad Education* linger by residual filmic technique, even when digitally abetted, in the zone of what I am calling cimnemonics. By contrast, *The Final Cut* goes over entirely, by theme and device, into the precincts of digitime—where life is virtualized by record in the very place, instant by instant, where it might have been lived more directly in the actuality of its transience. Humanism versus science fiction yet again, with time as crux. It is only by delinking virtuality from its strictly electronic sense, therefore, that lines of comparison can come clear between the European cinema of multiple realities, conditional memories, and trespassed limits, of possible worlds orbiting the actual and warping its gravitational field, and their alternate genre pole not only in supernatural gothics but in the VR plots of sci-fi.

In one broad understanding, virtuality for Deleuze is duration's functional transfer from moments elapsing to moments imagined as eventually recaptured—yet recaptured only as reimagined. In this respect memory is a cinema of the mind, with virtuality its native mode. Time is always and already behind us in being brought forth as mental picture. Even off the screen, enacted moments, when falling under the sign of an eventual retrospect, are virtualized on the spot. The here-and-gone of somatic perception is replaced by the heretofore-merely-potential: the evanescent "now" overtaken by the possible on its way out the back door to "then." Given this, we've seen again how a continuously vanishing present takes on an aura of the fantastic, a "now" slung over the abyss between the surrendered and the still only possible. Films that make this temporal process into the stuff of fantasies either elegiac or technological, and thus break with all naturalizations of the time-image in psychological realism, are films that bring the issue to the fore. Just as with the free play of literature for Todorov, which is always virtual (his word is instead *unreal*) and always undecidable, so—all the more so—with a medium whose spectral linkages and montaged associations, regardless of topic or plot, will always outrun the actions and actualities whose images they borrow, elide, and recombine. It is thus for Deleuze that film is one isolated site, as art, where the virtual can be readily accepted as such—and, as it were, *practiced*—well before any return to the actual.

Much depends, for Deleuze, more than is often noted, on the concept of the *lectosign*. This is where movement, no longer self-sufficient, drifts over into meaning, hence into interpretation. Nothing more clearly locates Deleuze within the "linguistic turn" of cinema studies after all, despite the apparently irreconcilable standoff between his phenomenology of immanent motion and Lyotard's insistence on cinema as, at base, a "writing with movement(s)."[7] Deleuze's indifference to filmic cinema's tactile substrate in the photogram is in this sense entirely tactical. For he puts to brilliant use his conceptual—as well as material—blind spot with regard to the underlying discontinuities of filmic motion. And he is relentlessly polemical in setting this out. Movement is *given* in cinema, to it and by it, rather than constructed. It is only when cinema puts some distance between this raw matter of duration and mobility that it finds its will to art in a modernist impulse to resist this automatized medium of sensorimotor images. Only then—though Deleuze never quite puts it in either Lyotard's terms (discourse subsuming figure), or in Metz's (the imaginary signifier)—does cinema begin to write time, to write in and with it. The *lectosign* appears when movement no longer simply is, but means—means something other than itself: where it passes absolutely from the imaginary into the symbolic, turning its picture plane into *figures* rather than instances of duration. When cinema is found asserting its sign function over against its recording function in this respect, it becomes text by any other name. Fictive text—which is (as Todorov stresses in genre terms) synonymous with the virtual. Under this dispensation, time is there on the film screen not just to be experienced in and as action, or inferred from it, but to be read—and read into. Narratography is one way in.

This bracketing of the movement-image in modernism is why the actor, no longer in action, becomes himself a viewer, proxy for the passive audience. Almost by default, the suspension of the movement-image leaves the time-image to fill the vacuum. Almost by necessity in turn, the lack of spectacle in postwar modernism, the flight from the very fascism of the spectacular, leaves the inactivated viewer deciphering rather than just watching. More perhaps than any film before it, *The Final Cut* travesties this interpretive distance from the sensorimotor given. Its implanted subjects inscribe while they watch, but never in order to read time for themselves. More in the Deleuzian spirit, the Eurofantastic from Kieślowlski to Medem and Almodóvar bears out the regime of the *lectosign* by carrying its overtones forward into the second century of an increasingly postfilmic and certainly postrealist cinema—so much so that, in Almodóvar, the director figure is, for a large stretch of the film, reading along as we watch, imagining (for us, in both senses) what we are given to see. The lectogram would be the smallest measure of this medial envisioning. If, in this sense, the pixilated breakdown of POV in *The Final Cut* is the reductio ad absurdum of temporal consciousness, a European film of the next year transfigures the video track of hidden surveillance into an unparalleled time-image of the diagnosed flight from such consciousness.

BROUGHT TO LIGHT: THE ETHICS OF REAL TIME

Such a narratological thematics of time consciousness as an ongoing reading act has never been more pointedly used to refigure a life of self-spectation (the movement-image stalled in the time-image) than in the layered temporality and real-time hypermediation of Michael Haneke's latest film, whose protagonist is a professional book reader and TV star. Such a final European exhibit in this chapter should cement the (still-Deleuzian) contrast between the current Hollywood mode of technophobic fantasy (with its ontological assumptions about organic life) and the new transnational uncanny (with its overlapping spheres of private with historical and communal consciousness). Here, as elsewhere, the liminal irony of a threshold shot—along with its later twists—continues to operate in the mnemonic uncanny of European production as much as it does in the Hollywood ontological gothic. Haneke's 2005 *Caché* (translated as *Hidden*: for hidden cameras, as well as the secrets they goad into the open) is a film produced the year after both *Bad Education* and *The Final Cut* by an Austrian, German, French, and Italian collaboration that tacitly implicates the colonial and genocidal pasts of the New Europe in a violent historical uncanny filtered through the new rootless telepresence of video mediation. Whereas films from *Peeping Tom* to *One Hour Photo* imagine the psychoanalytic trauma of a patriarchal gaze, *Caché* suggests, in its remorselessly ambiguous ending, a more ethical trauma—and a more historiographic panic. This is the gaze of the son rather than the father, so that the sins of the past become a cross-generational primal scene.[8] Here is humanist "temportation" at degree zero, where the past must be "visited" both by and upon the present.

Haneke's riveting (and often visually riveted) film establishes its optical tenor straight off with a mysteriously inert liminal shot, its fixed frame held for the entire length of the credits and beyond. The narrative later closes with a "switch" ending that involves a similar camera setup, across whose fixed field underspecified explanatory events come and go. In the mode of a trick beginning, the prolonged first shot turns out to be a surveillance videotape being watched not at the moment of recording but in subsequent playback on a domestic TV monitor, where we see its image variously degraded—in a baffled search for its purpose—by the horizontal striations of fast-forward and rewind functions (fig. 5.26). This first several-minute shot is therefore focalized,

5.26

even before its temporal manipulation, not from the POV of a hidden camera but from that of the subjects whose house is under surveillance, the so-far unseen couple heard only in voice-over. ("Well?" is the film's first word, interrogating the event-free exterior shot still in front of us.) The trick at the end, with an answering last shot of their son's school steps—in all its comparable fixity—is that we can no longer confidently decide whether we are watching just filmic narration or some continuing, subversive remediation via surveillance. There's every reason to wonder, but no telling. In this sense, the revelation we may have expected from the prolonged detective plotting—who has been videotaping their world?—is emptied out by the final turn. By now, the whole distinction between being and being filmed has been worn thin by the allegory of political denial: the flight from self-inspection. In semiotic terms, that initial model shot—a deceptively present street scene twice remediated in taped replay—has installed an unsettling subtext in which we are never immediately sure, from then on, whether we are watching a narrative film or an inset video. In the context of family shame, what develops is almost a running metafilmic pun on the political ambiguities of first- versus second-"generation" imaging.

The backstory, in short, won't stay back. Georges Laurent's farm-owning parents had wanted to adopt the bereft son of two of their immigrant laborers, who were killed in a political massacre of Algerians in Paris. Jealous, the natural son "lied about" his would-be stepbrother (he admits years later to his wife) until Majid was sent away to an orphanage. The triggering moment was telling Majid that his parents ordered him to kill the family rooster with a hatchet, only to report instead that the adopted boy did it to scare Georges himself. Not accidentally, in the film's contemporary historical context, the childhood transgression takes the form of an empty charge of terrorism as preemptive strike. It is this "lie" concerning the subaltern threat that is perpetuated into adulthood in the attenuated but no less poisonous form of a willfully erased conscience, at which point a cauldron of trans-European uneasiness is stirred up by Haneke's plot.[9]

The native son, pampered in solitude, has grown up to be a famous media "personality," host of a literary roundtable, whose intellectual chat is televised against a wall of simulated, title-free, backlit books. Off camera, Georges is impersonal, nearly mute. On camera, he mostly moderates interpretation—while privately avoiding it. Cued by the blank spines and illegible texts, this film's ongoing time-image unfolds as a true Deleuzian *lectosign*. When the mysterious series of videotapes starts appearing that show Georges' house under exterior surveillance, his wife's first thought is a stalking fan: the video personality cornered as such in his private life. So far, the uneventful time loops of these rewound and uselessly scrutinized videos only force upon him the replay of his immediate, colorless past, the diurnal round of driving to and from the office. But when the tapes begin to display scenes of his boyhood home, and then a lower-class Parisian neighborhood, the frazzled hero tracks down his childhood victim in the latter and accuses him of criminal threats to his family, a "campaign of terror." That the clue to the low-rent neighborhood to which the tape directs him comes from a pause and step-through of the video, until the street name Rue de Lenin can be blurrily made out, is one of the few heavy-handed notes in the film's political irony. Yet this, too, is part of a subtext that recalls our glimpse of the technological pun on Rue des Iris in filming's first break from the fixed frame of the opening surveillance shot—for a pan of the side street down which the camera must have been hidden. It is, however, the time of secondary rescreening that most preoccupies our mediated view from here on—and that secures its theme. What Laura Mulvey calls "delaying cinema" under the regime of the pause button, that new video and digital access to the subunits of the moving image, has become part of the plot itself in this enforced replay of the indelible.[10]

And no matter how much we expect it, this delaying action (or an opposite acceleration, for that matter) can still take us by surprise. In scene after scene of the film's reflex hypermediation, its diegesis is regularly one step ahead of us. In this sense, it taps the founding narratological paradigm of the detective plot, going over again the tracks already laid down. At one point, the image we've been watching as a development of our screen narrative is abruptly fast-forwarded, so that we realize we are privy only to its secondary replay as evidentiary trace, under analysis by other spectators. At a later point we see Georges' television show, full frame, as if we have just cut to the studio during taping. Or we might guess that we are actually seeing the show being broadcast in some as-yet undisclosed diegetic space, perhaps back in his own living room. But suddenly a stop-action image interrupts the dialogue. We find ourselves, instead, in the digital editing room, where Georges is ordering technicians, after a fast reverse, to splice out remarks by a guest that are becoming "too theoretical." Yet were we altogether wrong in associating this image with his real-time existence? Even in the main film we have been watching, his laconic character is always exercising veto power over expression whenever it might cut too deep. Once again, hypermediation folds back into allegory. It is in this context, too, that digital facility puts analog recording at a narratographic distance, so that the dated "home video" quality of the tapes, with their streaked indications of tampered temporality, can be taken to assault the media professional as the clumsy return of a technological repressed. Every past has its residuum.

Yet allegory, as we know from Todorov, is the enemy of the fantastic. All that is genuinely uncanny (certainly not marvelous) about the tapes—namely, their indiscernible point of origin—is matched by an entire failure to search for the hidden camera with anything like normal curiosity, either in the street across from his townhouse or later in Majid's apartment. For the contemporary teleconsciousness of a broadcast star, perhaps, the ubiquitousness of visibility seems less an issue than the mystery of its local intent. Then, too, problems of intent are not easily contained by plot. What does it all mean, this aggressive haunting by his past? The issue isn't so much a matter of guilt and reparation as of recognition, acknowledgment. Hence the emphasis on seeing. The postwar modern hero as "spectator" rather than motor agent of his own life in time: that ultimate Deleuzean prototype has boxed itself further into a hall of video mirrors, where virtual replay is indistinguishable from immanence itself.

Where is the abiding blindness in all this watching? As a defensive six-year-old boy, Georges was too young to take full moral responsibility for the exile of his nemesis. To this extent, he's right in his belligerent defensiveness. The chain of incrimination does not stop with him. It is the parents, surely, who should have been more cautious, asked more questions, looked through the lies, exercised less racial prejudgment. For the young Georges, there was mostly self-protectionism, not maliciousness. But the consequences are no less real and disastrous. Solipsism may be forgiven, but only in remission—only if its effects on the Other are admitted, only if the Other is finally seen in light of the self. No longer accusatory, Georges' lies persist nonetheless. They take the form of denial now, a refusal to feel. And to see. That's the real governing irony. Surveillance is not just an invasion of privacy but a punitive inversion of it. Georges must replay tapes of himself because the whole point of the exercise is that he should recognize his actions from the outside at last—and hence see his way, via historical response if not direct responsibility, into the place of the Other.

Majid, denying all notion of the tapes (which may have been made by his teenage son), is only trying to force acknowledgment. His final act is to repeat the decapitation of the rooster, as scapegoat ritual, by slitting his own throat in front of Georges (recorded in turn on video—as we assume even at the time, given the fixed camera position). His last words: "I just wanted you to be present." For presence is no defense against the past. The time-image makes all duration virtual in the now—and this as an ethical as much as a psychoanalytic imperative. If mediation has been keeping reality at a distance, only real death can cut through it with a single swipe of the penknife. Instead of remaining present in the wake of the suicide, however, and without even looking for

5.27

the camera which he must surely suspect, Georges flees to a movie theater, as do the murderers in *Bad Education*, and again implicitly to "kill time." But from the marquees at the cineplex seen as he exits, we would guess it to be less than a fully escapist venture. For at least two of the film titles bear down on his situation with another irony unspeakably blunt, were it not fortuitous: *Ma Mère* and *Deux Frères* (each in release in 2004, just when Haneke was filming the exterior locations in Paris).

Despite Georges' attempt to escape the gaze of the Other in the form of its fictional dispersion by screen projection, what asserts itself against him at this turn is the uncanny of the real itself.

What there is certainly no escaping is the optic of the unconscious. Even in the closing moments of the film, when after a double dose of sleeping pills Georges goes naked to bed in broad daylight, he dreams his way back into guilt.[11] His nightmare replays the forced exile of Majid—after the boy's last-ditch attempt at running away from his institutional captors, screen right (fig. 5.27). The long-held image is shot as if spied upon by a fixed camera locked into place at the back of the barn, the incriminating hatchet still resting on a stump at screen left, other barnyard fowl awaiting their time. We don't have to assume that the young Georges was in fact hidden (*caché*) there as the scene originally transpired. The camera angle is itself a trope. Narratologically, it is a flashback coded as nightmare. Narratographically, it is a victory for the POV of the Other in the mode of surveillance, a new and suddenly inverted ethics of the gaze. Georges' own recording by an unseen camera has by this point been wholly internalized, dream deep, as an ineradicable time-image. For the screen viewer it is the *lectosign* of repression in its inevitable return: the fixed frame, in short, of an unshakeable fixation. Majid's revenge has been presence itself: presence in the face of disavowal. The ultimate work of the tapes is thus to model, for Deleuze's "spectatorial" modern subject, none other than the haunting recursions of the time-image—and to do so precisely by "projecting a camera" into the unconscious itself.[12]

As cued by narratography, in this case centered on the geometrics of camera placement, genre contrast again throws internal differences into relief. The force of remediation in *Caché*'s denuded humanism operates as it might, for instance, in *The Final Cut* from the year before—if, that is, the cybernetic implant and guilty childhood archive of the American film were not an actual pros-

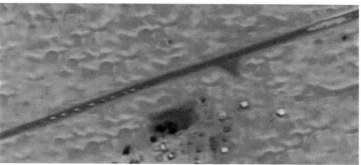

5.28–5.29

thesis of consciousness but a mere metaphor for it, building toward an expected switch finish that turns out to be only an ingrown turn of an allegorical screw. Or, put differently, it is as if Georges Laurent were forced to jack in to the playback of his own digital chip, rather than keeping it protectively switched off.

From the same year, if from an opposite perspective (a stratospheric one this time), there is another Hollywood twist ending concerned with digitime—under the sign, now, of global mediation in the geopolitical sphere. This is a film that takes the ironies of real-time surveillance, familiar in other Hollywood modes as well, to a newly topical level without crossing over into sci-fi.[13] In the hypermediated climax of *Syriana* (Stephen Caghan, 2005), we have been following a sheik's motorcade in the desert, intercut both with stateside corporate intrigue and an escalating terrorist plot on Arab soil. But the cruellest cut is soon to come. A commercial American adviser to the rebel emir, whose puppet brother is preferred by Washington to inherit control of the oil flow, looks out the window of the sheik's car (fig. 5.28) and realizes aloud that "It's really very beautiful out here"—*out here* on the land itself, in its natural rather than its national expanse. But "up here" versus "down there" is the axis that the conspiracy plot now has in mind instead. For at just this point a jump cut—more like a global leap cut—takes us to a heavily digitized satellite image of the automotive caravan (fig. 5.29), from whose full-frame screen we soon pull back to see a cadre of CIA agents zeroing in for a stealth-bomb assassination.

An ironic loop in the film's technological subtext is thus closed by mediation. Early on, the domestic surveillance system of the emirate's pleasure palace was shorted out—during a proud display to Chinese oilmen of its multiframe monitors—by a technological failure that led to the electrocution of the American adviser's son in the swimming pool: a lethal squandering of water and energy in a political circuit that puts a keen premium on both. Now, at the end, remote-control surveillance leads to the death of the emir's own son at the hands of the father's

5.30

manipulative American allies, a death to which we cut after a last shot of the satellite image framed by the faceless CIA operative on the left—with the sheik's car in his computer sight and his hand on the trigger at right (fig. 5.30). The fourfold grid of that early palatial surveillance system, able by digital remote control to fill the frame with any one view of the estate at a time, has been, in fact, a graphic model for the film's whole quadratic structure: a model with, once again, a tacit—and in this case tentacular—matrix. For with the energy consortium's parent firm being named not just by a near anagram of Exxon but by way of a covert pun, Connex lends an all but explicit voicing to the film's unspoken politics—and metanarrative logic—of global "connectivity."

In counterpoint to the final convergence of the Arab-royalty plot with the CIA plot (in that internally reframed assassination image), the remaining two strands of the alternating narrative are immediately knotted off by the irony of temporal rather than spatial convergence. An oil industry banquet has reached its final award speech just as—by another lethal jump cut—terrorist recruits speed to their suicide bombing of the newly merged company's Suez-based tanker. From which we then cut away at the moment of impact—and just a microsecond before detonation. In such nonnegotiable confrontations, conflagration goes without showing. And by such far-flung plot correlations, it is film history—not just industrial modernity—that is replayed under the graphic disposition of framed time. This is because more totalizing economies are at work than just those portrayed: economies narrative as well as geopolitical. As reflexively enhanced now by digital imaging and digital editing, D. W. Griffith's early innovation in the mode of both narrative simultaneity and thematic equivalence—parallel montage—has found a postmodern apotheosis, via networked capitalism, in the collapsed space-time differentials of global technology. And, of course, in the resistance to global capitalism by a deliberately regressive violence. It is by such desperate means that terrorism would, as it does narratographically here, stop the clock.

Just before this final explosive shock cut, what we confront with the climactic shots of digital surveillance in *Syriana* is a technology of (geopolitical) omniscience that operates like the "instrumental marvelous" of time present. In the hypermediations of *Caché*, by contrast, we are made to watch the suicide of the racial Other rather than see his murder from a perpetrator's telescopic line of sight. The only power that electronic spying has over the cornered self in

Haneke's film—and it is a devastating one—is to return the self's image to it in real time, return it not for subterfuge but for reading: the kind of reading for which, in another sense, he is a media star rather than victim, connoisseur rather than object. With no recourse to fantasy, the *unheimlich* of mediation—with its videotapes of a house that is not a home—gives way to the alienation of subjectivity itself: real time continuously reframed in the eye of its own depicted agency in a life caught merely going through the motions.

A more extreme relativism, however, characterizes the contingent protagonists of the new Hollywood plots whose travails and reversals we've been gathering up. These modern actors are not simply voyeurs of their own being or even editorial readers of their own duration. Beyond this, the pages of their lives can often be turned back, even glossed and rewritten, or simply erased. That's no doubt the reason why—although the hero's psychotic father in *The Butterfly Effect* stared at photographs in order to remotivate time lost—the second-generation protagonist fixes instead upon the lines of his diary until he can rewrite their past signifieds in his mind's eye. Or why it is textual inscription in the bookstore that takes the first brunt of electronic erasure in *Eternal Sunshine of the Spotless Mind*. Or why the amnesiac hero of *Memento*, reading his own tattoos in order to piece together a probable chronology, is annoyed in flashback by his now dead wife's willingness to read the same novel over and over again as if it were always fresh and suspenseful. Or why the heroine is reading a book about dreamed premonitions even in the French film *Irreversible*, as we'll see in the next chapter—and this in a narrative whose violent scenes, following each other in reverse sequence, are also punctuated occasionally by nightmarish flash-forwards. Characters of such plots actually inhabit the same stratified, figural space that we, in seeing it, are ordinarily solicited to read.

In a principle that cinema had for decades been luminously good at illustrating, life in time leaves a running mnemonic trace whose comprehension comes only in the wake of its sensorimotor impression, comes only by way of virtual replay. Taken too literally in narratives falling under the new interactive dispensation, this rule becomes the stuff of fantasy. Characters cruising the microcalibrated expanses of illusory digital environments, or enduring the unreal existential counterparts of such spaces by mental displacement, characters from Johnny Mnemonic to Neo to Donnie Darko and beyond, are in effect reading—and sometimes rewriting (rather than re-experiencing)—their own elapsed temporality, planar sheet by sheet, layer by layer, the way the spectator is. Such a postrealist "chronotope" (Bakhtin's term, again, for the localized narrative space of the novel) is a timespace that subtracts the real of action altogether from its image, leaving a temporality in itself relative, aleatory, and more than ordinarily unstable. Extending the high-modernist apogee of the cinematographic time-image, screen(ed) time is more recently steeped—by cognitive association, even when not by technological fact—in the strictly differential nature of computer-generated images no longer indexically coupled with a world they record. If science fiction film began with mad scientists answering our question about what life would be like if it were more like a movie, in the new postfilmic cinema the impossible interface with one's own past answers the question about what life would be like if it were more like an inscribed playback loop: not just an encompassing video apparition but, again, a recursive interactive id.

With *Donnie Darko* as well as *The Butterfly Effect*, each tending toward an ambiguous schizophrenic uncanny within the genre of the fantastic, or with *Thirteenth Floor*, squarely in the digital arena of explicated scientific marvel—as with many other films around and between them—time has been graphically reimagined. Quite apart from anything cinematic except its exhibition before us, temporality has become, as we've seen, not only plastic, malleable, relative, but willfully edited. Edited—or forcibly reread. And not just by definition as a narrative function for the screen viewer, as would always have been the case. But reread by human agents within the plot who can almost literally turn the page on their own unwanted past. Through the figured portholes and vortices of consciousness itself, temportation is a deportation from the subject's native ground in succession. In Deleuze's terms, the irrational cut of the modernist time-image has thus been rationalized away again by plot. As the most extreme case so far of this fantastic jump-cutting as a motivated leap into the past, *The Butterfly Effect* achieves, so we've seen, a "genetic" temporality in reverse. Plot recovers origins in order to cancel them, seeding narrative not with its own potentiality but with its sudden impossibility. The film accomplishes, in fact, an infanticide of lifeline and plotline together. And narratographic attention has, with this film among others, revealed the underlying logic of its tricked narrative reversal. In its temporal variant of the spatialized "cosmic zoom," digital editing in *The Butterfly Effect* serves paradoxically to territorialize time as a portable topography. The instance is prototypical. Rather than having human temporality figured in part by a founding evanescence on the strip, it is lately the work of morphing or digital editing that offers the prevailing metatropes for any and all manifestations of—or tampering with—the plunge of consciousness into time.

Place itself in the logic of many new plots, place as the common denominator of movement (across) and duration (within), has become merely *spatialized*. It is no longer confronted and recorded so as to be transmuted. It is merely generated. Time, too. And then more navigated than really occupied. In the new "action cinema" of the digital thriller, what we think of as time is now merely an effect of pixilated space. No longer the vanishings and eventuations of embodied duration, temporality is now just a shuffle of time frames—hypnotic, operable, but unreal. In what we have isolated as the implosion of timespace bearings in this facile dialectic,

the result amounts to a weird structural inversion of Deleuze's metahistory of screen dynamics. The movement-image, with all its emphasis on the "vehicular" in Deleuze, having ceded
historically to the time-image, comes back as the return of its own repressed. It does so when
time itself has *become mobile*. And the dazzle of the how is supposed to put the why from mind.
Recent plots thereby suppress the cultural currency of the interface through the fantasy, instead,
of its impossible realization as extramachinic marvel. Or once more, by transatlantic comparison: unlike in *The Butterfly Effect*, for instance, in *Bad Education* morphing remains a figure of
time—even a thematically deceptive one (the source of an impersonator's misrecognition). At
the climax of the American film, however, and many others like it, it is timespace itself that is
morphed by a character exiled from the present, who finds a way to mutate and finally excise
his own past. An ethics neither of deception nor of self-deception quite catches the tone of the
fantasy, where it is the audience itself that has been tricked, but only by genre contract.

    And behind this "semioptics," if you will, of narratographic response, special pressure seems
exerted on a single stress point of narratology. The new fantastic, existential or technological,
humanist or cybernetic, its mysteries epistemological or ontological, intersects in this sense
with narrative theory's most suggestive version—or revision—of spatial form: namely, the
pivotal insight of Peter Brooks's *Reading for the Plot*. In his Freudian sense of the bindings and
recursions of plot as a quasi-therapeutic repetition and working-through, he notes the fact that
echoing moments in the plotline of narrative, even when they are figured for characters as the
return of the repressed, are often rendered ambivalent at the structural level. The reader can't
be sure, even sometimes in regard to heroines or heroes, whether the past is catching up with
them or they retreating (or otherwise returning) to it. Nothing could make this spatialization
of time clearer than in *Caché*. Despite the prompt delivery of videotapes, what ensues is the
protagonist's frantic tracking down of his own past, which waits intransigently for him across
Paris in a squalid flat—waits to replay its own violence in a more lethal ritual yet. In Brooks's
definitive formulation, parallelism in plot is undecidably a "return *to* or a return *of*."[14] Definitive in its very ambiguity: because this is exactly the way the unconscious works, for Freud,
in constructing time in the first place, time and the spatial imagination we have of it. In the
psychopoetic understanding of fictional form that Brooks erects upon this Freudian axiom, it is
inevitable that a given symmetrical return across the unfolding extent of text—an extent spatial
and temporal both—marks either, and uncertainly, an unexpected fold of psychic chronology
or a willed transport of the mind, a coming back or a going back. Riffaterre might agree. For
in this light the recognized "model," as generated by a first avoidance of the unconscious matrix, is taken up at various points along the track of the "subtext" in such a way that the text's

underlying semiosis is built up at once by converting and dispersing the model and by reverting to its unspoken fundament in the matrix. What results—in terms compatible with Brooks's ambiguous holding point between repetition and recursion—is a veritable lectography of narrative time. More important yet for the medium of postfilmic cinema, Deleuze might agree here with Brooks as well. For the difference between the return of the past and the return to it is one way of grasping the essence of its virtuality.

Such virtuality is always the double-edged sword of plot itself in these films. The temporal transports (in the weak sense of that noun) that characterize the European fantastic at the far reaches of its humanist innovations submit the epistemology of the subject to uncanny ambiguities. These make for filmic experiments that hover in this zone of the virtual with a frequency that seems to confirm and extend Deleuze's postwar trajectory even while elucidating its terms. In the suspiciously more explicit sense of temportation in recent Hollywood film, the attempt to actualize and hence access the past in narratives of marvelous return—a spatiotemporal displacement realized increasingly by postfilmic devices—is a radically different thing. The whole Deleuzian dichotomy may well seem imperiled—or outdated. Or, again, travestied. In the maturation of the cinematic medium, movement first implied time, then *figured* it as the troped import of the framed image. Now time often defers to movement. Temportation throws over the virtual time-image for that new movement-image I have been identifying as the (com)mutable figure of timespace, where past and future are willed into a motility and plasticity all their own. And where temporality, once having been spatialized, can itself be morphed. The interval has been replaced by the interface.

In contrast with this temportation subgenre of Hollywood fantasy, one still finds the most energetic experiments of "European humanism," even at their most mordant (as in Haneke), struggling to image a temporality that deflects (and rethinks) the real without needing resolution in an abrupt revisionary finish. They work in this way toward a sense of imaged time that keeps virtuality open rather than foreclosing it in simulation or delusion. This is to say that forms of consciousness remain the content of these films. In much "American science fiction" and its curious gothic siblings, narrative pulls the rug out from under its own content. After ontological mystifications of whatever sort, the case is closed in comeuppance and undoing. With assumptions not only escapist but reality denying, such films of temportation in the mode of the ontological gothic seem to imply, or pander to, a sociology of looking detached from causation and consequence. And in this, as guessed—a point we confront once more in the final chapter—they may well solicit acquiescence not only in a culture of fantasy but in a politics of the unreal.

SIX

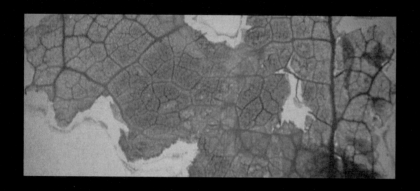

MEDIA ARCHAEOLOGY, HERMENEUTICS, NARRATOGRAPHY

The 2000 appearance of *The Golden Bowl* marks a valedictory summation of the entire filmic substrate in cinema. In the movie's closing film-within-the-film, there is a last-minute backdating of the medium's filmic rudiment to early *actualité* footage. And not all of this footage is simulated or digitally manipulated. History itself shows through in shots of shipboard, harbor, and cityscape. But all of it reads—and I do mean reads—as if narrative were taking its leave, from within the heritage mode, of an entire photomechanical century. An implied genealogy of high-gloss historical melodrama and its period recreations comes through in retrospect as a derivation from the archive itself. In turn, the narrative's thematized ironies of aesthetic mediation come down to a tacit archaeology of imprint and projection. From this perspective, the stratification and sedimentation of representational technologies—across a whole range of nineteenth-century visual practices, from photographic art inventories to museum slide shows to waxwork simulacra—are as much a part of cinema's predigital history as is any directly traced lineage from photograph to projected photofilmic motion. Why shouldn't a single film be able to sense this—sense it productively, showing it as part of the sensuous rhythm of its own execution?

The films whose medial thematics we've been reading since *The Golden Bowl* straddle the same millennial watershed marked by the year 2000. Yet they do so at times as if the fantasies—and anxieties—of their protagonists were those of a postfilmic cinema itself. Among such anxieties, of course, is that of time lost or merely simulated, neither lived nor relived. Converted on the spot to fantasy, we recognize this anxious temporality in the repeated flight, film after film, from something like the biographical (read: historical) present that besieges at another level the character's own medium. It isn't that cinema resists the digital by wanting to turn back media history. Rather, such concepts as real time, random-access memory, selective amnesia, virtuality—all inflecting the drama of consciousness itself for a character-driven cinema—have been brought into new relief against the background of cinema's shifting optical support, whether subtly, bluntly, or even brutally. This relief is the work of a medial hybrid—a no longer exclusively photomechanical cinema—that is inevitably less embedded in material succession than was filmic (or photo-synthetic) cinema, less based in the inelectable before and after of the fixed-image trace. Human presence in and across time, newly dubious in many of these postfilmic narratives, finds its apt and uncertain model, instead, in the metamorphic instability of the digital frame: the process without procession of its mutating field, the displacement of exposure time by compositing.

Reviewing the method that precedes it, the thrust of this closing chapter can be put simply enough. My intent here, first of all, is to doubt out loud that the issues of media archaeology, genealogy, or history so markedly raised by digitization, and so broadly debated lately, stand as little to gain as we are often led to think from textual readings of a given film. And, second, to pose narratography as one mode of such reading. Institutional media study and visuality theory often assume there to be no demonstrable profit in analyzing the actual screen plots that engage ocular mediation either directly, through technofuturist settings, or through their counterparts in the virtualities of psychological or supernatural thrillers. Even when there is no outright polemic against interpretation, silence gives dissent. In much of the new archaeology, instances of the medium in question are not polled about their own operating systems. One is widely if tacitly discouraged from thinking that the narrative work of screen texts might ever break back to their own conditions of visibility, entering thereby into a material history of the visual. Allow me to hope, when even so sketchy a chapter as this is finished, that it will be forgiven for having characterized this widely entrenched (if often latent) premise, up front, in a way that sounds as unlikely as in fact it is.

In closing, then, we return to Deleuze's own endlessly suggestive conclusion to his second cinema book, where he wonders what toll will be taken by the digital on future cinematic figuration. To say that contemporary cinema does not just extend the time-image but stretches it out of shape, distorting it for purposes unknown to modernism, would be one thing. But analysis still wonders about those purposes. How can we know them? What is one's best guess about their cultural work? For all the filmic evidence already behind us, the answer isn't there yet. And may never be. But some tentative conclusions about the sociological valence of such screen narratives is in order. For these are films (as paid entertainments) sent into commercial negotiation with a cultural imaginary that contrasts dramatically with the former modernist regime of temporal figuration. One way to pinpoint this contrast, from within plot, is to take seriously the traces and often erased tracks of optical allusion.

At this point, we might call up Wolfgang Iser's terms from literary anthropology, where the "fictive" stands to the "imaginary" as created object stands to a general cognitive disposition.[1] Fiction, in other words, materializes the imaginary for a given culture. For Deleuze, the time-image of cinematic modernism stands in an analogous relation, as invented figure, to the imaginary base of thought's own virtuality—to that facet of the real which does not require presence for its actualization, including the mind's functions of memory and projection. But that is, or was, modernism's unique way of giving fictive form to a cultural understanding of consciousness. Enter, since then, the postmodern fantastic in its recent narrative turns—and speculative gambits. What if the human imagination, including the tracks and stray traces of its memory effects, were itself fictive, an invention? Or, even if real, nonetheless drastically renarratable at will, just like a fiction—its chronologies suddenly reversible? Or what if, instead of generating a master plot to articulate its desires, human subjectivity were the constructed object of some external or at least impersonal instrumentality (alien electronics, supernatural providence, schizophrenic delusion)? What if Being were the mere interface of the Other? Poststructuralist psychoanalysis thinks so. Postrealist cinema, of the sort we've been canvassing, likes to indulge in the hypothesis as *fantasy*, safe in the "unreality" of its premises.

It does so within circumscribed stories that, again and again, surprise us with a revelation before dismissing it from all urgency within the mechanisms of the unreal: the fact that all is artifice or delusion, posthumous or electronic. That's the extreme Hollywood form of the ontological switch. It has grown so predictable by now that the only real trick is to make you forget you had half guessed it from the start. It's almost a gauntlet thrown down to credulity. You know you should have seen it coming. Partly you did. These plots play games with your own passivity or disavowal, your own acquiescence in the deceptive. They do so in service to the deferred thrill of *dis*illusion. Less dependent on overt reversal, less digital in its procedures and its narrative formulas alike, European cinema of these same years tends to keep us guessing on through the end. Its uneasiness lingers, rather than seeking its terminus ad quem in a dependable shock. For the deferred willingness to be suddenly disabused has a violence all its own: under conditions of suspended disbelief, a kind of padded masochistic charge.

Then, too, another tentative generalization may cover many a base in conclusion as we move now to historicize further a cross-medial genre, the fantastic, both in its recent cinematic revival and in the cultural symptomatology of its variants. As for European psycho thrillers of the erotic or mnemonic uncanny, Being is not in doubt so much as is Self. Alienated self-image, emotional dissociation, a nonidentical relation to one's desires, a nonsimultaneous relation to their objects in a kind of mnemonic mirage of recognitions: these are the mainstay of the European uncanny. Often they involve a kind of rootless, transnational guilt that either fails to speak its name or equivocates its politics, whether postwar or postcolonial, with personal neurosis. Equivocates—or at times allegorizes. That's why we've noted more than once the stress on an ethics of epistemology in this mode of the fantastic, epistemology rather than ontology. Such screen narratives are direct inheritors in this way of the modernist "art film" and its existential problematics, as most recently in the metacinematic videology of Haneke's *Caché*. In Hollywood lately, by contrast, the anomie and anxiety, rather than free floating, tend to be funneled by plot, or complot, into overt scenarios of paranoia or delusion. These are stories in which a nervous distance from anything like self-assurance is likely to mean that the hero is less troubled by life—or by a collective unconscious—than already dead.

Yet any facile dichotomy between epistemology and ontology defines, more precisely, separate points along a spectrum of skepticism. This is a stance of distance itself, of intractable doubt. This is what two of the most important philosophical writers about film have located as central to its medial condition—and its cultural mission, or at least cultural function. Distilling a line of thought more fully worked out in the books of Stanley Cavell (and taken up at the end of this chapter, and then again in the appendix to follow), Deleuze puts it simply enough when he suggests that the purpose of the cinematic virtual is to reconnect us to the real—rather than (one must add), as more often in recent Hollywood films, to reconcile us to a lack of actuality. In his speculations about the fate of the time-image, however, Deleuze didn't foresee, or at least not explicitly, what has become one of Haneke's abiding themes: just how much further from reach, under hypermediation, the real would come to seem. Approached via critique in Haneke, this distancing is addressed by fantasy in Hollywood. Filmic cinema for Deleuze, however transfigured its indexicality, gives us "reasons" to believe in the world again, rather than, as in the latest Hollywood plots, offering an exit route into the entirely unreal, where virtuality is an instrumental tool rather than a heuristic protocol and a mood of mind.

No less with obvious fantasy plots than with other popular genres, it is often difficult to sort out the deeper cultural fantasies to which the spectacle plays, the anxieties elicited by it, the contradictions massaged, the solutions floated. In the elegiac and frequently transnational uncanny of recent European cinema, the violent burden of twentieth-century history finds itself appeased and redistributed, often across national and linguistic borders, in narratives (repeatedly co- or multi-national productions) whose plots indulge in eerie commonalities and odd psychic transports. At times the New Europe seems to be a reconfigured libidinal imaginary, where the anxieties of a biographical past overlap with those of national memory in a complex counter-transference of border-erasing identity formations freed, at least briefly, from the weight of the political—only to refigure it at a different level. Often cut loose from the apparent historical basis of nationhood and its policed limits, uncanny transgression becomes the new mode, if you will, of human interface. I think here, once again, of Žižek's sense of the "interface effect" in Kieślowski (chapter 2), an effect often photographically marked.

In films contextualized within a postphotographic notion of interface, however, even Hollywood thrillers that mask their digitization (or the computer-game prototypes of their plots) seem to draw, nonetheless, on a technologized model of virtuality for their imaginary figurations. This is new, but not entirely. And genre determinations (however far from the canonical literature of fantasy) remain in place to complicate any reception theory of such narratives. So we need to look back at the genre's heyday in order to adjust the cultural anthropology of the fantastic. Historically circumstanced, the efflorescence of literary fantasy in the second half of the nineteenth century drew on a specific cultural context for its internal explanatory poles. Plot solutions residing in the marvelous tapped the residual supernaturalism of Western culture. Explanations tending toward the uncanny anticipated—and Todorov is explicit about this much, at least—the emergence of psychoanalysis and its adjunct in modernist distanciations of the real, where *everything* comes to representation as a species of the uncanny. What this bears witness to, in short, is the rise of the psychoanalytic paradigm as a newly pervasive cultural schema—and this for the travails no longer of the ensouled self but of the embodied subject. A century later, the technofantastic plots of recent Hollywood films, along with their low-tech supernatural variants, indulge in what we might call a postsubjective virtuality. We've seen how. Now why, exactly?

KILLING TIME

Having been videotaped in the presence of death, that ineluctable kernel of the real, the hero of *Caché* flees to the cave of commercial film projection. It is there, implicitly, that what is indexed, even when lethal, remains fictive, and where time is recursive and unreal. In Almodóvar's *Bad Education*, the irony has been more directly voiced. The impulse to "kill time" at the movies is a sick pun in the murder plot of a retrieved and impersonated past. In contrast with either of these extreme cases, Hollywood's *chronocides*, whether dystopian or supernatural, are more explicitly fantastic. So, again, why? In these murky areas of cultural mythology, not being able to say for sure is the best reason for trying. One might suspect, at least with supernatural thrillers like *The Sixth Sense* or *The Others*, that these plots were playing to the religious Right, recovering from genre history a form in which supernatural explanations might still feel viable. Even limbo suggests divine intervention—and even when the light at the other end of the tunnel, the divine refulgence, as in *Brainstorm* (Douglas Trumbull, 1983), *Jacob's Ladder*, or (below) *The Jacket*, may be entirely subjective: a final migrainelike breakthrough to unconsciousness. More than this, even sheer digitization of the human world by alien overlords or robots (from *Dark City* to *The Matrix*)—since it is just a fantasy, and a human one at that—tends to confirm, by contrast, a comfort zone of humanity outside the machine. Such plots could return us, that is, if an audience were so inclined, to a sense of the species as created rather than made, not just emplaced but actually redeemed within a differently apprehended totality.

One could, therefore, see the recent return of fantasy as a throwback to an age of eroding supernatural belief. One might thus identify a genre stationed to shore up again the intuition of a spatiotemporal *beyond* that tastes of the spiritual as well as the cosmic. One might. Even the scarified punks picketing the EYEtech Corporation in *The Final Cut* could in this sense seem to have been commandeered by an evangelical cause, more right than left, in their fervid objection not to cloning directly, but to its psychodigital equivalent. Yet the associational crossfire of pop-cultural issues is likely to confound any such clear-cut determination, there or elsewhere: to hold sociological checkpoints in suspension instead, tolerant of contradiction because availed by it in the spread of its audience appeal. One might, in short, think of fantasy as quintessentially right wing of late. But I think not. That wouldn't be too easy but too hard.

And these films are ultimately relaxing. They induce, that is, the slackening of all vigilance for the fun of impertinent surprise. Impertinent in every sense. It is neither the ontological nor the theological dimension of these stories that bears the weight of their affect, then. It is the fantastic itself that is their appeal. A genre that was once culturally heuristic—as a residuum of belief and a test of emergent models from alternative explanatory systems—has become further commodified in its own right. It is superannuated by sociohistorical reality and retained vestigially as an entertainment trope alone. Coming with a twist rather than the momentum of a quest, or even a speculative inquest, the disclosed afterlife narrated by any one film seems no more apocalyptic than the forcibly reconstituted past visualized by another. Each testifies to the fragile, unactualized status of virtuality (in a histrionic rather than philosophical sense). This is the virtuality of a narrated present whose past is always relative and revisable. It isn't what the world turns out to be that counts; it is the fact that it isn't what you thought it was. The aesthetics of suspended disbelief, the earlier dispensation of genre response, has become a suspended belief in the world itself. This, then, is the fantasy played to by these films: not one of supernatural or psychosomatic empowerment after all, but of passive surprise and acquiescence.

Sure, it is comforting enough that outside the theater we're a good deal more certain of the world than we are within. Art and life are more than ever separate spheres—or so audiences must enjoy thinking. Media on the one hand, reality on the other. Sure. And if there should ever be anxieties about the mediated nature of one's interface with the real world, the virtuality of one's place in it, including anxieties about one's relation to commercial media themselves in the manipulation—and surveillance—of desire, these films show such an extreme and hypertrophic form of this vexing ontological nervousness that the outside world would have to seem real again by contrast. The ontological gothic reconciles us to the world not by catharsis but by default. It is thus an anodyne genre—and not only in final affect but ultimately in its narrative form. Films whose teleological narratives concern life beyond the grave, purgatories renegotiated, backlit heavens entered at last, especially when the onset of otherworldliness upends the spatiotemporal universe we thought the characters were immersed in: such films do not vouchsafe a revelation so much as undermine a confidence—and confirm a skeptical distance. Just for fun.

As a result, these films don't play to the religious fringe so much as further yet from center: to a deeper political disengagement, as we've seen, from reality itself. This must happen of course, if happen it does, against the frequently heroic drift of their avowed plots. The hero insisting on his human rather than cyborg status, even if he's wrong; or on his organic rather than digital status; or demanding to reclaim his own memories, even if they are artificial; or hoping to take charge of his past by either minimizing or deleting it; or refusing to acknowledge his death until he has cleared away the burdens of unfinished business: these might be seen as fantasies of *re*enfranchisement. In a world of global political abstraction, one at least yearns for some negotiation with existence where a difference might be made, even if only a difference within. Such would be one way to take the *affect* of these films: as wish-fulfillment fantasies of recovered control over one's time in the world.

Maybe, but again one doubts it. One doubts, in other words, that such is the deepest appeal of the ontological (or mnemonic) distortions in these screen stories—stories of what we might ultimately want to call the narratological gothic. Defying the given, the heroes of these films are narrative actants transformed into fantasy agents whose true cultural discourse in VR is, as we've seen, repeatedly staved off by plot or *mise-en-scène*. This intertext in electronic virtuality comes to light only by a narratographic attention—and here I mean on an audience's part, not just a critic's—to the image manipulations that transfigure the character's bodily inhabitation of space. These are tamperings by which a hero, in the end, often feels not empowered so much as willfully inactivated. In response to the system, whether biographical or technological, often in response to lived time itself, human character opts out. Heroes may access improbable powers to do so, but the imaging of such untoward force is only called up to open an escape hatch. For the audience, too. In a world where the very concept of the worldwide grows increasingly mediated and impalpable, a spatial Webwork without Euclidean coordinates, the fantasy of temportation—that most extreme among the recent plot options—seems just that: a fantasy. And whole films with it: not parables of enablement but escapist games for looking (along with their heroes) the other way—whether back or forward. It is in this light that one is tempted to see such films as vehicles for little more—yet nothing less—than killing time: abandoning the imperatives of lived duration in a new theater of the unreal.

BODY TIME AND TEXTUAL LIBIDO

Johnny Mnemonic is the first in a decade of heroes who, if they seem to incarnate a synecdoche for the "spiritual automaton" of cinema,[2] or in other words its thinking machine, do so at a real but elusive risk to their body's lived relation to time. Strictly mechanical automata are easier to contemplate. Psychomechanics and automatism: the composite artifice of filmic cinema is neatly localized by either the externally willed moving body and its mnemonic transports or the robot, from Cesare in *Caligari* to David of *A.I.* In all such cases, as we've seen, the cognitive authenticity of the virtual has been usurped or manufactured.

   "Spirit borrows from matter the perceptions on which it feeds and restores them to matter in the form of movement which it has stamped with its own freedom," writes Henri Bergson in the closing sentence of *Matter and Memory*.[3] Out of context, that rhapsody on motor impulse might seem like a high-end motto for the films of temportation we have been considering. Not so, however, for Bergson's advance proviso, two sentences back, is that all such freedom to stamp the world with the body's own motion depends on passing through the "meshes of necessity" (249). In recent trick plots, these meshes have been strained wide, becoming fantastic loopholes instead, whether by digital download or schizophrenic detour. Films like *The*

6.1

Thirteenth Floor or *The Matrix* on the one hand, like *Donnie Darko* or *The Butterfly Effect* on the other, offer narratives that, rather than project memory into matter, reduce the virtual life of memory to sheer planar space. This is a psychic space that can then be inverted or removed at will, dematerialized, violently remade. In such narratives, the body is no longer a Bergsonian "center of indetermination" (that concept so crucially borrowed by Deleuze), awaiting from without the very change that also reconstitutes it within. Rather, the body is a vector of overdetermination, willfully motoring its way to a time no longer its own.

A decisive sensorimotor gimmick of *Minority Report* (Steven Spielberg, 2002)—a central *trucage* in the futurist ambience of its design—takes its suggestive place in this syndrome. Detached from their own vestigial bodies almost as completely as is the free-floating cranium in *City of Lost Children*, the brains of three female psychics foresee the violent futures which the technologically supplemented brawn of the detective-hero is meant to intercept. To this end, "precognized" fragments of such premonition can be literally *manipulated* into a forensic montage on a huge transparent screen by the detective's manual dexterity. Spielberg's "precrime" hero often selects his evidence from double-tracked parafilmic strips—in the opening segment even clipping out, as it were, a punning shot of the very scissors that the new digital editor doesn't need but that the would-be murderer will (fig. 6.1). To foil this eventuality, a coherent narrative of the future must unfold before the detective by ad-hoc montage. This would offer a *lectosign* (a readable image) of cause retroactively fitted together from flash-forward effects. Requiring no prosthetic controls with which to click and drag the images, the hero's skill is brandished as a kind of punning upgrade of the Palm Pilot. Here, by hand as it were, is the postcinematic ability to materialize and reshuffle time rather than to memorize it. This is all part of a "forking-path" narrative that is also a "garden-path" variety as well, with its false lead and trick ending (the villain pulling the trigger in premonitory close-up not on the hero, as we've been proleptically led to think, but on himself). In Spielberg's film, time has been somatically mediated as never before. It is accessed not only through the medium of the precogs' enciphering brains but through the answering medium of the detective's deft touch. According to this one plot turn, telepathy has become as much haptic as optic.

And in a cinema of tactile as well as ocular response, this question of a haptic optic is inevitably archaeological in its horizons. Hence Mark Hansen's debate with Lev Manovich. Hansen, who thinks Deleuze minimizes the place of the body in visual *response*, stressing instead its on-screen role in the affection-*image*, also finds Manovich drawn too exclusively to the role of the haptic or manual at the *production* end of visual artifacts. As Hansen sees it, Manovich thus underestimates the place of tactile dexterity in the *performance*—and hence the generation and reception, rather than just the initial construction—of protocinematic motion effects like flip books and lantern slides.[4] The point of departure for each recent commentator is, nonetheless, similar. Both Hansen and Manovich are quick to stress the fact that film is an affair of indexical capture. By contrast, the digital image can be generated from within, implying no real-world referent and entailing no necessary succession from one arbitrarily framed field of pixilation to the next, a next that (by infinitesimal shifts) replaces it rather than displacing it wholesale. In just this respect, the missing third term at the historical switch point between coherent frame and constructivist pixel, comparably missing in each theorist, is the filmic photogram. Overlooked or undervalued, nonetheless the subliminal flicker of the celluloid frame cell has consequences for reception (even when not for direct perception) that go entirely unexamined in each account—if according to opposite prejudgments: Manovich (as noted earlier) because the photogram, like the dot matrix of a photograph, can't be seen; Hansen because only experimental and interactive video can lead us, he thinks, to a similar materialist level beneath routinized perception. It is in his attempt at a broader archaeological base, then, that Hansen faults Manovich for not stressing a precedent context for the haptic in cinema: namely, the manual operation of image systems, and especially the bodily engagement with them, that defines a common thread of somatic mediation from, say, panoramas to stereoscopes. These image systems incur points of definitive bodily interface that emerge long before, and come into determinant contrast with, the advent of cinema's passive body as stressed by Manovich: a body immobilized before the screen spectacle. According to Hansen, and despite his argument with Deleuze on film viewing itself, it is only after the long cinematic hiatus or digression that interface technology can fully return the body, rather than just the retina, to the focal point of vision.

Yet it is in full view of such an archaeology that a complementary narratography of film, well before digital or virtual installations, calls out the moments when the "passive" receipt of the celluloid is bothered or confounded. These are points where (in Lyotard's terms for "acinema") the representational order gives out—and must be actively reabsorbed, via material or medial conversion, into the narrative order. Or where (in Benjamin's related terms for photography) the order of the "optical unconscious" returns from suppression to be felt along the bodily pulse of *filmic* reception.[5] For Hansen, instead, it is only the contemporary avant-garde of installation art that installs the spectator before the true material conditions of her optic experience. Only by confronting in this way "the synthetic interval comprising duration" does one gain a newly affective "perception of the flux itself, time-consciousness."[6] But cinema as film, cinema as a writing with movement rather than a simulation of motion, has been there first. Hansen doesn't need recourse to Douglas Gordon's slowing down of *Psycho* to two frames per second in order to discomfit one level of suspense with another. He has on the commercial screen, to begin with, the foregrounded frameline of Hitchcock's own film in the scissored jump cuts, say, of the famous shower stabbing: the interstices of time laid open to jagged (if still invisible) view in the moment of fatal penetration and arrest. What Lyotard would call the representational "arrhythmies"—with all their shock to the viewer's system, all their somatic disorientation—are certainly recuperated yet again at the narrative level, in Hitchcock and elsewhere, indeed almost everywhere. But that isn't to deny their ocular force. Nor does it deny the place of the spectatorial body in the affect of response.

Between apparatus and apparition, the body is a medial term in its own right, registering on its pulse all generated shocks to the celluloid system. Responding to the filmic as well as to the fictive is, in other words, a way of proving your existence apart from the identifications induced by screen narrative. A phenomenology of viewing comes at this issue in complementary terms. In the cinematic bypass of epistemology suggested by Cavell, we have use for *a* world rather than *the* world, a *World Viewed* (according to his title) and renarrated for us. This is for reasons very close to the heuristics—and the ethics—of virtuality in Deleuze, where "belief" must always be freely willed, never just automatic.[7] Here is where it must be obvious—or, if not, where it should be made apparent—that a response to narrative textuality, whether verbal or pictorial, has no inherent tendency to forego the somatic impact of the medium, or even to occlude an archaeological long view of its palpable force in reception. Entirely apart from the world represented, then, and the special effects that may thereby be enlisted, the *affect* of fiction becomes a testing ground of the real.

But "belief in the world," in the time of the world, is precisely what the heroes of recent Hollywood films, when uncanny disorientations turn supernatural, must be desperate enough to see past or sometimes (as in *The Matrix* or *Vanilla Sky*) brave enough to recommit to. In these and other films of temportation, time is itself submitted to a kind of malleable holistic reframing. As such, lived time is made available to something not unlike the metaphysical equivalent of computer memory, with its exemplary logic (and textual metaphorics) of "cutting" and "pasting." In the heavily narrated and often overtly tricked plots of temportation that have concerned us in these last two chapters, the human agent is regularly converted to a scopophiliac of his own reread past. Exaggerating Deleuze's sense of the modernist actor turned "spectator" of his own duration in a kind of postmodernist travesty, the hero empties out both relevant temporalities, active and passive: that of endurance and that of contemplation. Rather than submitting to some exfoliation of the "*materialized infrastructure of the enlarged now itself*" (the privileged site of digital art according to Hansen's italics, 260), heroism is redefined in fantastic terms. It emerges as the effort to cancel out an elapsed past in vaulting across an alternate present to a remade future. Here, once again, is where a kind of medial allegory inevitably sets in, inflected by digital allusions that may or may not find technological manifestation within the plot. Narratives of this kind run counter to the temporal model provided at base by the flux of the photogram. They break stride with all norms of temporal consciousness as differentiated and subdivided in passing. In their magical and paradigital temportations, the material "now," further materialized as space per se, is foreclosed, minimized, exited, logged off.

The postfilmic condition thus seems to have shaped these postrealist plots from the inside out. In such films the body's once normative place as passive and immobilized spectator of a screened world (Manovich) has been shifted, but not to the register of a more aesthetically probing response, as in the haptic vision of digital art (Hansen). Scarcely given new access to the intervals of change in process, instead the body of temportation has merely been converted to a mobile rather than an immobile spectator—an active seer of the subject's own for(e)gone world, including its own "screen memories." We are by now well familiar with the mixed blessing of this dispensation: the ability to dispense with the constraints of sequenced duration. Such protagonists are allowed to cut and dissolve at will through a paradigital editing process that requires no return to the indexical record of actually lived time. In the hallucinatory ontological montage that results, with the body serving to anchor nothing in its haptic transfigurations, time is itself reformatted at will.

As we've repeatedly noticed, one analytic consequence of all this is hard to avoid. At the level where narrative and representation are processed together as twin facets of textuality, any narratographic response is inevitably a kind of microhermeneutics. Why—and why not? I mean: how so, and according to what objections should this be ruled out? Here a return to Lyotard has much to recommend it—Lyotard, whose dovetailing of the figure/discourse dichotomy is an insight inseparable from his sense of mediated storytelling as a self-regulatory system that routes representation through narration. Merely to identify this as one form of "libidinal economy" is to summon back the body's place in the performance of text, and thus to answer complaints against the supposed immateriality of this textualist paradigm, so alien to media and visuality studies of late. Such media archaeology or visual theory often seems inclined to keep a clarifying distance from textual hermeneutics in order to keep pure the object of study as vision or mediation per se (or information processing) rather than any given instance thereof. As suggested at the start, this is one of the axioms, spelled out or otherwise, that a book like this is bound to be least convinced by.

Concerning narrative cinema, a more productive assumption is instead that something resembling hermeneutics—here in the tight-grained form of narratography—is doubly valuable as an approach to just the questions raised by media study. First, and by no means insignificantly, narratography helps situate for film investigation the cinematic experience (whether purely filmic or entirely digitized or everywhere in between) as a cultural object, with its own stance to the ambient media it recruits or alludes to, incorporates or suppresses. Second, and at least as important, only an intensive reaction to the visual object as text is capable of experiencing the continuous tension—continuous and *defining*—between an imaging medium and its phenomenological *disappearance* into a viewed scene. When that vanishing is resisted in certain aggressive returns of frameline against screen frame, or more recently of the pixilated against the pictorial, it is only a textual response that can process these ruptures as refigurations within narrative discourse.

Again, it is Lyotard who clarifies this operation, and with memorable concision, in his brief comments on the somatic affect of violence (violence within and to representation) in the paired death scenes of the film *Joe*. Certainly, any discernible "arrhythmies" of the track in these sequences, breaking with normative movement, are to be seen as inferring the most obvious of time-images. They constitute in effect the return of a repressed motion underlying pictured movement: the photogrammatic series beneath the photogenic screen frame. They are in this sense, from within an antithetical Bergsonian resistance to the movement-image per se, the irruption of the time-image degree zero. Had Deleuze been interested at all in the on-screen materialization of the strip (apart from his remarkable pages on Dziga Vertov), he might have seen the time-image as produced in just such an exemplary fashion by the gaps between, in Lyotard's terms, the representational and the narrational orders. These are gaps often registered by what we have further isolated as the celluloid order. Isolated, first—and then surveilled, in all its covert operations, as it colludes with the digital under the aegis of time itself.

If he had done so, Deleuze would have placed his elaborate semiotics of *opsigns, sonsigns, lectosigns*, and the rest in a more direct contact, where they belong, with the larger implications of the textual turn. From such an understanding, cinema emerges—emerges split second by split second from the photograms it elides—as nothing less than a "writing" with and in movement/s (Lyotard again): perceptual shifts that are inscriptive before phenomenal, differential before directional, serial before vectoral. Hence the usefulness of renewing Lyotard's simple-enough distinction between representation and narration. Here is yet another version of story versus discourse, *fabula* versus *syuzhet*—but not quite, for his distinction is one without a stable difference, one in which the line between picture and its description is everywhere permeable. This should allow us to say, with equal simplicity, that textuality manifests itself as that very condition of discourse (visual or verbal) operating across the ordinary correlation, and occasional frayage, between representation and narration.

Or say that textuality is what narratography discovers in narrative. The more hermeneutic in certain cases, the more medial; the more caught up in the texture of image, the more archaeological in inference. So let's move ahead one last time, in both terminological and historical review, from the frameline violence of *Joe*—sampling as it does a sixties cycle of death-drive plots and those freeze-frame finishes first discussed by Cavell—to the evidence of the last three chapters.[8] Including such disparate temportation plots of a revis(it)able life as *Donnie Darko* through *The Sixth Sense* to *The Final Cut*, this is an ongoing cycle whose narrative formats ultimately depend on, repress, or mostly ignore the model of cybernetic memory—and which thus stand in a variable relation to the technoculture of their production. In any event, to see them as part of a Hollywood "cycle" of ontological ambiguity is not to use the term in the more specialized sense employed by Rick Altman, since they are identified with no particular studio or star (even though they sometimes do invoke, in their repeated casting of a Robin Williams or a Keanu Reeves, a certain set of characterological expectations across plots).[9] As successive popular instances of a continuously evolving prototype bearing no particular industry signature or corporate logo, however, these narratives of temportation—crossing as they do between more established genres like sci-fi and supernatural thriller—are best seen, I suppose, as an intertextual cluster.

In any case, hitting on the right term, whether of genus or species, isn't as important as the categorical thinking to which such vocabulary may further lead, as Altman's own widening beyond genre to genealogy at one point makes clear. Typological study has a way of penetrating to the level of inscription even when not pursued by the attentions of a given genre theorist. The contemporary trend we've been isolating shows one way in which this can happen—and with what results. However much the genre contract is differently at work in the revitalized fantastic of both European and Hollywood production, mediation looms large, and temporal mediation in particular—from the uncanny photograph to the wholesale digital surround. As a new screen cliché and cultural symptom, for instance, the hero's narrated potential for a return into an elapsed past (including an entire life as lived and died), a return for the purposes either of clarification or of rectification, invites a shift in narratological parameters, to be sure. But at the level of narratography, the recurrent move to a particular kind of remediation in the closing moments of these films, with the shifts in actual graphic texture this precipitates, comes markedly to the fore. To appreciate such effects in action, we need to consider the fuller metahistorical context that such medial deviance brings suggestively in tow.

MEDIATION BROUGHT HOME

As Altman reminds us, following Tom Gunning on this point, the film medium as we know it didn't see itself coming. Long before its narrative options were consolidated and its commercial distribution maximized, moving images were conceived by Lumière to be an extension not just, in technological terms, of instantaneous photography but, in sociological terms, of the family snapshot. As both recorder and projector in the same alternating mechanism, the motion-picture apparatus was in effect designed first of all for home movies, for personal record rather than fictive narrative. Its invention didn't foresee what Altman calls the "remote consumption" (186) of film as mass cultural object. Genealogically conceived, one could say that the evolution of cinema as institutional medium is a displacement from the home movie to the movie house. This is a consumerist turn whose lost immediacy had to be compensated for, in part, by interpretive communities reconstellated around the targeted appeal of separate genres. A full century down the road, we thus arrive, by medial reversal, at a climactic throwback to home movie footage—or its upgrade in video or even digital implant—in such films as *Johnny Mnemonic, The Butterfly Effect*, or even *The Sixth Sense*, with its death-moment reprise of the wedding video. And on the European end of the spectrum, there is the traumatic estrangement of "home video" itself for Haneke's *Caché*, where, to keep the domestic irony in play, the protagonist's isolated widowed mother speaks of her "remote-control friends" on TV.

What does the larger sense of media archaeology tell us that the films themselves cannot quite say? In the Hollywood case, with human ontology more feverishly at stake, the closural reversion to home movies taps into the conceptual prehistory of filmic device as an intimate prosthesis of domestic emplacement, a face-to-face index of origin. It harks back in this manner to a prenarrative imaging in the evolution of media devices. Such moments thus become the kinetic equivalents, in historical as well as structural terms, of the inset and reframed photograph or the freeze-frame. Like a still shot breaking with the flux of motion, so, too, a private document of motion capture embedded in a fictional film is not only an interruption but a recapitulation of the medium's own turn-of-the-century advent. So closely tied is all this to the intimacy of family record and its organic and mortal chronologies that one may sense Lumière's intentions preceding by exactly eleven decades the home-implant funerary nightmare plotted out in the satellite-released *The Final Cut* (Altman's "remote consumption" at an extreme distance). At which point, yet again, one cannot help but recall the strangely prescient overtones of such coinages—for the early cinematic apparatus—as Biograph and Vitagraph, to say nothing of the Zoetrope before them in the archaeological past.

When Altman speaks of "genre in the age of remote consumption," he is, in effect, highlighting the way in which film history can be charted as a quest for lost presence, first of the profilmic body of a loved one, then of the image's tight-knit communal (before its commercial, mass-marketed) reception. "Condemned by its mechanical nature to substitute spatial and temporal absence for an originary presence, cinema began with the implicit charge to restore presence, at first through familiar images of everyday subject matter, later through the familiarizing devices of narrative organization and genre formats" (186). What we began by examining in the metafilmic coda of *The Golden Bowl* is the condensed rehearsal of this transition from private to public record. For what begins as a kind of home-movie travel footage on shipboard, in the Ververs' steaming toward America, becomes a full-scale commercial exposé in the ensuing newsreel document, including its reframed headline photographs of a thereby twice-removed presence. Though appearing decades after the narrative stabilization and commercial distribution that led, for instance, to latter-day heritage film itself (and the latest Henry James subcycle in particular), this one vintage plot ends, as we know, by reverting to the medium's own prenarrative roots in a cinema of *réalité*.

In the films that have preoccupied us since the first chapter, it is exactly this metafilmic dimension that has been visited upon the heroes rather than merely the spectator. In the process, so we are now seeing, their closural reversions serve to factor in a larger spectatorial history. For they often appear cast back even beyond primitive documentary footage, back to the domestic expectations for the apparatus in its developmental stages. To understand the rehabilitation of "presence" (or call it a phenomenological conviction in the image) as being one mission of genre's familiarizing work in the decades to follow: this is to find genre subsumed to exhibition as a facet in itself of both the praxis and the history of mediality. In a given film, however, a genealogy of the medium, or a more broadly-cast archaeology of its visual prehistory, can operate with full intertextual charge only through specific permutations of the image plane. These locate, therefore, the real methodological breakpoint at the medial interface between genre-based analysis and (post)filmic scrutiny. Such complementary operations can, of course, be rephrased in a dichotomy long familiar to these chapters. For where narratology is disposed to sort out the plot complications channeled by genre and its baroque recent variants, it is nar-ratography that registers in these same films the graphic fields of their remediation. These are the planar fields by which an archaeology of the moving image itself gets cycled through the visual configurations of its storytelling function. POV shots in the Hollywood time tunnel are as crucial a determinant in this respect as is, for instance, the return of the repressed in the hypermediated fixed-frame "setups" (both senses) of *Caché*.

Presence remains a notable crux, but only as transferred from medial to dramatic crisis. Rerouted from its intended extension of the private domestic archive, cinema as film has thereby come full circle. A marketable generic expectation is met now by film plots of robbed or squandered or virtually recouped "presence" rather than its passive acceptance. For narrative agents themselves, this is an ontological presence erased, remade, or revised by fantastic transits often digitally implemented—or merely troped as digital: in other words, by various forms of temportation from time now to time when. And this would include not only the returns to home movies or their video equivalents from *Johnny Mnemonic* through *The Sixth Sense* and *Vanilla Sky* to *The Butterfly Effect*, but a further throwback yet. As registered in its emergence from the bared celluloid order of a last uncanny lap dissolve in *One Hour Photo*, this is the closural reversion to just before Lumière's moving-image machine altogether: the private circulation of separate family snapshots, whose indexicality is simulated in this case for a reparative fantasy.

The latest Hollywood instance happens to embed the family photograph in reprinted book form, but with no less intimate and restorative an intended charge. And this most recent example of a defanged and sentimentalized ontological gothic also involves the clearest possible case of a digital intertext in the upgrading of photography as a spatiotemporal communica-tions medium. Photographic memorial, digital *trucage*, time travel, and the ontological trick ending all converge once again in a dreary cross between *You've Got Mail* (Nora Ephron, 1998) and the Korean ghost film *Il Mare* (Lee Hyun-Seung, 2000), on which the new film is more closely based. The parallel-universe plot of *The Lake House* (Alejandro Agresti, 2006) is divided between the architect-hero in 2004, the doctor-heroine in 2006: namely, Keanu Reeves and Sandra Bullock, reunited here across time (from 1994's *Speed* [Jon de Bont] and their subsequent "separation" when Reeves turned down the role opposite her in the 1997 sequel) not by either magic or chemistry but by self-conscious Hollywood casting.

Initiated by the heroine's "Dear New Tenant" letter (addressed from her eventual—rather than past—occupancy of the lake-reflected and hence optically doubled glass house, and request-ing simply the forwarding of her mail), the characters continue inexplicably to communicate with, and thus "fantasize" about, each other by an instantaneous exchange of notes in the mailbox they share two years apart. The admitted "lack of privacy" of the house has to do, it would seem, with its transparency across time even more than space. Almost by formula, the temporal touchstone of photography seems mandated. In the climactic phase of the film, as a technological emblem of healing time travel, an image of his earlier past becomes the first of two gifts the hero receives from his future lover. Appearing in a posthumously published book on his father's architectural career, this is a solacing photograph of the harsh genius with his arm around the hero's boyhood self in front of the same lake house.

By such paranormal special delivery, the hero's greater gift yet from the future, beyond the image of a past presence, is the quasi-telepathic heads-up about his own pending death. For just in time the heroine realizes—along with the audience—that their contact has in fact been so supernatural, all along, as to be actually achieved from beyond the grave. Narrative scrambles to explain—and hastily unscramble—this twist as an implausible time loop. For it is up to the doctor now to perform a paramedical intervention. Unless her infatuated correspondent heeds her final message from the future and stays put, he will become the anonymous man who *had been killed* in a car accident back in 2004. On her initial assignment out of med school, he is the first patient who died, as it were, in her arms. The whole plot can now be seen as her wish-fulfillment fantasy, come impossibly true, of having saved him after all, bringing him back into her ongoing world, making the future itself possible for him. It is as if *The Sixth Sense* were unfolded as plot from the wife's point of view rather than the slain hero's—and with a happy ending at that.

Even technique in *The Lake House* is recognized in retrospect as having been doubly tricked. Not just figurative for the premising temporal gap between their lives but literal for disembodiment, the hero's "ghostly" image has several times materialized and faded away from the heroine's presence: a metafilmic clue about not just the evanescence of their verbal intercourse but also the phantasmal nature of his whole equivocal existence. At one point, as if in some conceptual video installation, the two long-distance lovers are spliced onto adjacent park benches, with passers-by laterally dissolving (into others like them) across the marked gap of years. What all such lyrical and seamless digital manipulation suggests, finally—after the plot takes its ultimate narratological (and ontological) turn—is that the evocation of overlapping time frames has been given a further narratographic spin. As part of a reflexive history of optical superimposition as *trucage*, we are returned to the original magic of cinematic spectrality in the materialized undead: he lost to her by fatality, her future existence merely virtual to him in the first place.

But beyond this, the contemporary technocultural matrix. Though with no bodily temportation, once again the "wonder" of audiovisual technology is quietly transposed into the marvelous time frames of plot. In yet another deflected digital intertext in postfilmic cinema, that is, we note that communication has been achieved between the decoupled pair, fantastically enough, through the real-time immediacy one ordinarily associates with electronic relay—so that increasingly, as plot conventions settle in, the red flag on the preternatural mailbox flips up and down with no letters visible, all text elided instead into voice-over dialogue. Though the Asian ghost genre from which the source text, *Il Mare*, derives tends to embed the archaic within the modern as a kind of cultural ghosting, the Hollywood version of such fantasy is more likely to magicalize technological innovation itself in disguise. Where, for instance, wireless short-wave radio cloaks and refigures instantaneous digital communication in *Frequency*, we come to recognize that the transcendental sentimentalities of *The Lake House* have also been, under cover of a haunted epistolarity, as much an allegory as a mere fantasy of intersecting time frames. The supernaturalism is in this sense normalized. To some extent we identify, all too readily, with the uncanniness of this spooky communication.

Contemporary microchip culture actually knows how it would, how it does, feel, since wireless communication has achieved in the last decade a new level of domesticated tele-pho-nic magic. Indeed, cell phones make their now routine appearance in the plot—twice when the heroine's old boyfriend tries to "reconnect" with her (the first of these calls when she is half hoping that it might be her unseen lover from the recent technological past instead), and once when the hero gets a call from the hospital about his father's death. All of this descends as subtext, however, from the failed early appearance of such technological mediation after the fatal car crash, when the novice doctor dials for an ambulance and then drops the unavailing cell phone alongside the dying body. From here out, in their reciprocal reach into each other's time frames, we are to believe that these supernaturally attuned lovers have reverted to an ancient and founding means of communication technology—the earliest private "medium" in the hand-scribed letter. Yet given the way their rapid voice-over exchanges come to condense and refigure this scriptive pretext of the plot, this film too, like *Frequency*, ends up evoking nothing so much as spectral instant messaging with the dead—or call it the always annihilated presence of electronic "contact."

Even the hypermediated pyrotechnical futurism of Spielberg's *Minority Report* takes a turn into the medial reflex of nostalgic imaging rooted, like *The Lake House*, in a posthumous photograph; indeed twice over in adjacent sequences. We have seen the plotted maestro of orchestrated image fragments trying to convert the projected flash-forwards of a violent future effect to the coherent *lectosigns* of present causation. Compared with other Hollywood films of existential self-exile and psychic temportation, one might well think that Spielberg's *Minority Report* stops short of this ultimate perversion of the Deleuzian virtual, whereby your past can be

6.2

6.3

territorialized and repotentiated at will. Just short: by inverting the logic to a future that can be nipped in the bud—and precisely by the spatialization of time as shuffled prefigurative frames. From the marvels of electronic prophecy, however, we soon traverse, by rehearsing, a precedent media history, archaeologically reconceived. This takes us, in the very next scene, from the transitional uncanny of photographic record to, yet again, its upgrade by the protocinematic (as well as, in this case, postfilmic) nostalgia of home "movies." After envisioning the future from an intermittent barrage of moving-image traces, the hero returns home to a rightward pan across his dead son's more neatly arranged and conventionally elegiac photos: the same camera movement, though reversed, that revealed the scientist's dead son as robot look-alike (and model) in Spielberg's earlier *A.I.* (fig. 6.2; compare fig. 3.17). This memorial drift of the camera is now followed in *Minority Report* by a holographic digital video of the mourned child, dialed up by his seated father (fig. 6.3) from one among several electronic monitors on his desk. The boy's image is running in place as if it has emerged into electronic apparition from a two-dimensional black hole on the wall-sized home screen: a kind of reverse indexical shadow from which his present virtual incarnation—occupying the same "diegetic" space—seems spectrally peeled away. Whatever the exact Deleuzian valences of premonition might be within the logic of the time-image, certainly few films could make it clearer—clearer than in this segue from photographic to holographic memorial afer the earlier scene of optical prefiguration—that time past is just as virtual as time future. And thus no less real.

And *Minority Report* adds another turn to our consideration as well. We have speculated already on the relation of the DVD supplement of "deleted scenes" or "alternate endings" to the parallel-universe mentality of the new cryptodigital subgenre of film fantasy. We have sensed that these new plots of lives otherwise lived or elided, or of deaths undied, are in a

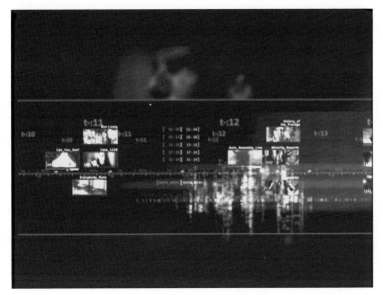

6.4

sense thematizing, by fantastic extrapolation, their own networks of commercial distribution and private exhibition. But the process works more obviously in reverse: when the DVD apparatus imitates by reprise the visual layout or logic of the film itself, especially in the ingenuity of its search engineering. In *The Final Cut*, as mentioned, the remote-control viewer goes to a scene (presumably the second time around) by way of a push-button "rememory." With a yet more kinetic visual design, in *Minority Report* the viewer picks out the chapter span (1–4, say) from a lateral cascade of images: an unmanageable shunting of frames, stabilized only when the numbered spread is selected (fig. 6.4). Otherwise, these sample images from the individual episodes float and skim elusively across the screen field of the user more like the precogs' forevisions than like our own favorites from among remembered segments of the narrative.

This effect is not just graphically thematic but almost psychosomatically so. For with its remote interface—and the agile interactivity it at first seems to require—this menu effect puts the viewer in the hero's heady place as narrative operator, finessing out of discrete and fleeting images an actionable (in our case recognizable) sequence. As viewer identification is extended in this way from narrative into "paratext," the latter is converted again to metatext. And in the process, the remote body of cognition is somatically enlisted as if to make, or remake, a pertinent narrative succession. (Indeed, as we will see below, such paratextual interaction is part of what Thomas Elsaesser would include in an expanded new definition of diegesis, or narrative space, itself.) With *Minority Report*, to search and recover the elapsed in this way has a media-historical charge—as well as an intended thematic kick. For these are newly facilitated haptic functions that were once quite unavailable to, because already preformatted for, the passive spectator—the spectator first of photography, then of film. In Manovich's terms, one supposes, as in Laura Mulvey's related sense of "delaying cinema" by electronic intervention, the DVD operating system returns us to the manual prehistory of motion-picture technology, with its automatically sampled and extracted time.

Short of this, which is to say even in the film's theatrical release, *Minority Report* circulates its own digital marvels back through plot, as we've seen, against the yardstick of photography's static imaging. It is in just this way that, when viewed against the backdrop of other recent Hollywood films in the fantasy or sci-fi mode, *Minority Report* appears quintessentially hinged around a bidirectional logic of the time-image. This is an image that waxes, once again, not just nostalgic but metahistorical. The digital film editing of a death coming is, in fact, a kind of trick beginning—in the sense that the hero himself will later be part of the incriminating disembodied future in which he traffics. Plot moves next through the memorial photographic record—and on to the holographic projection—of a death long past rather than still impending, the son's rather than the father's. What all this concatenated mourning work in *Minority Report* helps bring into focus, for one thing, is a heightened self-consciousness, here and elsewhere, about the photomechanics of cinema from the brink of its supersession by the digital.

And, for another thing, it instances a widespread ontological loosening in the relation of lived existence to both human futures and human pasts. More obviously yet in other recent fantasy scenarios, cinema history, even while overstepping the limits of film history, has moved in its narrative ingenuities from the presencing work of genre at large (in Rick Altman's terms) to a particular (if flexible) paradigm of *non*presence. In the Hollywood narratives we've been contemplating, moments of climactic alleviation or revelation often round back not just to the plot's trick beginning, with its buried narrative matrix, but to ocular rudiments lodged more deeply yet in the archaeological sedimentation of the medium itself, its history and prehistory. Narratologically understood, characters are returned by *syuzhet* or discourse to the alternate replay of their own *fabula*—replay or, in the more extreme gothics of nonpresence, *summary* cancellation, in one sense or other of that last adjective. Narratographically seen, however, recent plots of fantasy or technofuturism turn on themselves as visual systems. They repeatedly come round (as in the photographs or hologram of *Minority Report*) or ultimately wind down (as in the closing videos in *Johnny Mnemonic* and *The Sixth Sense*) to a kind of explicit medial nostalgia that is elsewhere shrouded in either supernatural or high-tech disguise.

A medial nostalgia—and sometimes a specific genealogical irony. This is where the new genre of the ontological gothic performs its by now familiar reflex action in the remediated sneak ending (or, alternately, in the move from forensic visual telepathy to the domestic use of both photography and holography at the start of *Minority Report*). As if recovering the original market forecast for the apparatus of moving pictures—as a tool of private record and family memorial—human desire reverts to a cinematic imaginary purged of spectacle. In this respect, the ultimate home movie may not be the edited rememory screening of *The Final Cut*, or even the son's possible complicity in the domestic surveillance of the father in *Caché*. It might come from *The Sixth Sense*, instead—and not from the wedding tape in the main plot, either. For there is, as well, the hidden surveillance video that is posthumously exposed by the displaced agency of the daughter at her own funeral. Here we have what amounts to the trick ending of a subplot, designed to reveal the mother's gradual poisoning of her own daughter, week after week, in her domestic sickroom. Narrative is thus yet again bent round toward its medium's own outdistanced commercial intention, where diegesis would have once opened out only as far as the viewer's own domestic habitus.

Visible in such narrative turns are ironies coded almost inevitably as a matter of visual grain and scale, of optic materiality per se. From amid all the flourished electronic virtuality of such ontological and existential thrillers alike, their postrealist plots, at least in their more sentimental variety, often reach back in fantasy for some originary filmic node of the *heimlich* itself. Thus are they found graphically as well as biographically *homing in* on the intimacies of indexical rather than virtual interface. In the broadest terms, one might want to say that at such a crux of remediation, cinephilia has blurred into epistemophilia in the very course of their twin effacement by the postrealist narrative twist. Despite the politics of the unreal that permeates this genre of *non*presence, in the last minutes of many a plot something more authentically cinematic (or at least analogic, maybe video-cam footage) is held out—and clutched at—even if immediately negated, as for example by ultrasound self-infanticide (within a home movie) in *The Butterfly Effect*. To put it again in terms shared silently by Cavell and Deleuze, in these films a restored indexical nexus is often clung to by human agents as proof of having once been in, and even believed in, the world. A crisis of "presence" is thereby manifested not just *in* their plots, for characters in action, but by the films themselves as image systems—in their fullest cultural associations.

More than the rudimentary trick effects such plot turns may involve, the logic itself may be tricky. It shouldn't be too hard to sum up for the reader, though, since every viewer knows the feeling. In this book, I have tried standing far enough back from these Hollywood thrillers in the subgenre of contingent temporality or downright "nonpresence" to find them aligned with counterparts in the contemporary European cinema of mnemonic hallucination and psychic transference, as well as with a related visual "excess" in Asian films of the historical fantastic. From that distance, we note another perspective opening up within the overarching genre frame. In the media reversions of such Hollywood plotting—and the retrogradients of visuality that signal this regression—narratography finds a further textu(r)al irony. Such films tuck their fantasies back into the folds of mediation per se. The archaeological reflex is by now familiar. From amid the technological bravura of pervasive electronic effects can emerge, and against all odds, the recovered innocence of industrial cinema itself. For these plots often turn on the fantasy of their narrative medium's own fantasy alternative: its lost origin, the surrendered intimacy of its function as a prosthesis of family memory.

But even here there is room for bad faith. The futuristic hologram of the hero's dead son in *Minority Report*, though a narrated "marvel," is also rather a mess. We can barely recognize the boy's image amid the photon streaks that materialize it. That's the very ingenuity of its *trucage*: that it must mark itself off, precisely as digital illusion, from the entirely unmarked digital illusions that bombard us in the rest of the film's futurist *mise-en-scène*. Sci-fi is full of such moments, always has been. These are moments where an audiovisual prognostication can only be shown as vanguard optical technology by being called out as "fake," by breaking with the neutral wonders (the pervasive *trucage*) of the cinematic surround.[10] The plots we've been considering offer a special twist on this general tendency. High-tech cinema may often wax nostalgic for home movies via the upgrades of such a domestic appliance. But execution itself, certainly in Spielberg's film, works otherwise. As the posited future of precinematic home imaging, holograms make the commercial turn to mass-market narrative look great by comparison. For all its high-tech flamboyance, the case is typical enough as a flash point of response. Stationed always at this intersection of plot and picturing, narratography is there to gauge, and only by closely engaging, any such graphic rift induced by visual discourse in its channeling of medial image as narrative text.

Textuality: an abstract noun that need seem no more immaterial than many others, including an abstract noun like *mediality*. Instead of being put to continued use as an interdisciplinary lens for the recognition of systemic interconnections, however, the textual turn in critical theory has fallen out of urgency into disuse. It tends now to be ignored or denigrated—both by genealogists of cinema and by archaeologists of visual mediation—as a flight from the somatic apparatus of reception, historically embedded, into the structuralist bias of an immutable *langue*. One way to resist this resistance, this dismissal, is to rethink filmic textuality (within a longer view of cinema) from the photogrammatic ground up, where recent differences between the binary and the serial "undertext" of motion picturing, between the digital and the plastic substrate, might take narrative shape. And one mode of this rethinking is, in fact, interpretation—by any other name.

Unlike narratology in its approach to a plotting loosened from the details of representation, there are reading procedures in which mediation would instead show forth as a direct means. Intermedial in ultimate scope rather than more broadly transmedial, narratography offers a monitoring in progress of the body's response, whether to lexical or indexical traces—a response local, oscillatory, and constitutive. This is at least as clear for filmic cinema as for alphabetic prose. With style understood as the mode of adequation between plot and story (our adjustment of Bordwell's tripartite division), the celluloid order can image (and refigure) the rifts, recursions, and overlays of time within the gap between representational and narrational orders. Even Deleuze's writing about cinema seems to invoke this in the very gesture of revoking its relevance. In his favored examples of superimposed "sheets" and irrational "cuts," layers and breaks alike, Deleuze may thus be seen (recalling Riffaterre) to have founded his models of temporarily, almost despite himself, on the suppressed matrix of the frameline per se—and its interstitial generation of movement. The frameline: where out of flux and its perceptual frictions is constructed the image of a motion other than its own, a movement on-screen rather than of the strip. But the medial duration that founds such effects must, whether or not it is converted to a figure of time, undergo a decisive rethinking under the reign of binarity rather than seriality.

This is why the narrative analysis of screen plots engaged with the digital, either by inference or by explicit futurist setting, cannot be deemed tangential to the history, the genealogy, the transnational cultural geography, you name it—let alone the medial archaeology—of the screen medium thus textualized by narrative. One reads the image, narratively construed, even if one doesn't invite, much less strive for, the freestanding nominalized result of *a* reading. Advocated here is an openness to hermeneutics that does not, in short, seek interpretations in the reified and pluralizable form. It is instead a mode of graphic apperception that is ongoing, provisional, always speculative. In it, the famous hermeneutic spiral is not found working round, as usual, to ever more discriminating and exhaustive accounts of the same object. Instead, the spiraling of response rounds back continuously from object to its conditioning ingredients, from plot to medium, in a mutual adjustment that finds no complacent plateau for the work of signification.

In regard to all this, at the draft stage of this book an unintended and rather downbeat serendipity in its table of contents occurred to me only after the fact: namely, that in closing with Spielberg's interface fantasy of haptic mediation as a last named example (before other and more recent releases came to my attention), I was actually calling out a potential subtitle for this last chapter. In point of fact, any case for reading as a mode of, rather than an alternative to, contextualizing—for reading, in sum, as our most direct response to the narrative signals of mediality—is indeed a case likely to be filed these days as a minority report. This does not mean that narrative interpretation has no visible practitioners. Far from it. By not being stressed as *textual* analysis in the same way it once was, though, such analysis actually elicits fewer visual bearings—fewer explicit images—than it might. Yet hermeneutics has its prominent adherents still. Though without sharing Edward Branigan's interest in a cognitive parsing of the actual visual frame, Thomas Elsaesser and Warren Buckland have followed his lead in finding analysis and interpretation merging into each other, if not always joining forces, on a fairly continuous spectrum of response. Such a spectrum runs from objective description through aesthetic or social evaluation—with analysis attached more directly to the internal working of a narrative object, interpretation to its larger contextual ramifications.[11] Disciplinary rifts are discernible even here, however. For when Elsaesser moves in a more recent position paper (discussed below) to a revisionary account of media archaeology in terms of "diegetic space," he leaves behind any concern for the narrative structure of a given diegesis. Such divisions of labor between visual history and hermeneutics are all but universal—yet anything but inevitable.

In their earlier commentary, as Elsaesser and Buckland assess the changing academic climate of film discussion, they note how interpretive work is represented most often in contemporary film studies by a "hermeneutics of suspicion," where the film is analyzed for "symptoms" of its implied ideological burden, its "structuring absence or blind spot" (231). Here we have been keeping in mind that this blind spot may be medium deep, and hence archaeologically blinkering, as with the elided play of commercial electronics in fantastic narratives of transtemporal interiority and mortal virtuality. It is for this reason that I began by asking, why, in Hollywood cinema, are there so many plots of *non*electronic virtuality in the age of electrographic imaging? With such a question still in mind, it will soon be time, on the occasion of a recent screen example, *The Jacket*, to isolate for closing attention the exacerbated virtuality of death: that asymptote and parody of the Deleuzian time-image (as well as of Bergson's *dureé* before him). It should be no surprise at this late point to add that, in doing so, one puts this matter—this issue of materiality—in its fullest archaeological context only by putting it to the hermeneutic (which is to say here a narratographic) test.

WORLD ENOUGH AND TIME: DIEGESIS, INDEX, TEXT

Recent essays by two leading film theorists, Elsaesser again and Paul Willemen, may be said to approach the issue of temporality and screen fantasy, in their material constitution, from opposite orientations of the projected image. Stressing visual narrative in reception on the one side, the process of optical generation on the other, each approach factors in digital mediation as the new conceptual challenge. Together, they widen the very definitions of diegesis (telling) and indexicality (record) in complementary ways, each with an eye to an archaeology of spectacle. Building on Noel Burch, Elsaesser finds a revised concept of narrative space to be necessary for the practice of "cinema history as media archaeology."[12] For him, diegesis becomes, in this expansive sense, the space of narration (the public or private event of the screening) as well as the space in narrative (the screening of event). In ways closely related, therefore, to Altman's emphasis on the original closed circuit of domestic imaging foreseen for the new medium, the parameters of diegesis—and ultimately of genre—thus reach outward into the very field (and stance) of reception.[13] In this sense, as Elsaesser wants to insist, diegetization is more important than digitization in assessing cinematic transformations over the last quarter century. Only diegesis, that is, can encompass the work of narrative action and narrative interaction, a distinction we have seen Elsaesser make more sharply in an earlier essay (noted in chapter 3) as the difference between "narrative" and "navigation." His purpose now is to transcend this duality and hence to escape the latest (and lazy) truism that high-tech cinema has returned from the epoch of "narrative integration" to the inaugural mode of sheer "attraction."

His new thinking seems to have begun roughly where this book does as well: "Why should fantasy have become the preferred mode since the 1980s?" (90). Though not exactly rhetorical, the question is neither answered nor often returned to, however, in his extensive essay.[14] Since a resolution of the issue wouldn't rest for him with the marketing of the digital "attraction," it must again have to do with the field of interaction (in the everyday as well as the technical sense). Our own version of the question has concerned, of course, what Elsaesser mentions merely in passing (and cryptically at that) as the differing "grammatologies of time" (111) now at play on the narrative screen. Such, one assumes, are the self-constitutive differentials of time-consciousness implicated, in turn, by the shift from filmic to digital imaging, where the mechanically advanced photogram finds its alternative in the flux and backwash of the numerical algorithm. Across the uneven developments of this transition, as explored in these chapters, optical allusion has kept the act of viewing alert not just to the "event character" of cinema (Elsaesser's broader diegetic zone) but to character and event as pictured—even if ontologically undermined—by a given mediation and a given plot.

Willemen's response to current screen practices is, in the long run, closer to our concerns than is Elsaesser's. Not only does Willemen approach more nearly to an account of the contemporary fantastic, but his method of "allegorical" hermeneutics (though scarcely ruled out in Elsaesser's previous thinking) is more closely married in his latest work to certain specific valences of media archaeology. Further, where Elsaesser minimizes narrativity in delimiting diegetic space, Willemen sees certain technical devices as "inflating" the narrational apparatus, extending its discursive reach. In one of his essays, an exemplary answer in its own right to Elsaesser's call for "cinema history as media archaeology," he sketches the operation of the zoom from its early uses in 1920s cinema.[15] The zoom is a lens effect. Its camerawork is not a traverse of inhabited space but a strictly ocular adjustment within the ratio of diegesis to display. The zoom *is* a visual event, as much as a closing-in upon one. Establishing this device as "an emphatic, gestural marker of narrational performance" (12), Willemen is thus able to track its lineage not just to novelistic focalizations of various sorts but all the way back to the audience address of oral storytelling and its phatic gestures.[16]

What, then, about the so-called zoom mechanism of still photography as an optical allusion within cinematography: a selective close-up played off against the possibilities of film's motorized closing-in? Even the most domestic example of the predigital zoom, as a feature of 35 mm automatic cameras, can be installed as an emblem for film's time-loop plotting in the spate of such turn-of-the-millennium narratives, especially at their least digital. It can operate in this way almost like an in-joke on the more high-tech manifestations of similar narrative material elsewhere. As with near or distant futures coming to the rescue of a past death in *The Lake House* and *Frequency*, but prolonged in this case by a plot-length equivocation in the classic genre mold of the fantastic, the disoriented hero of *Happy Accidents* (Brad Anderson, 2000) uses the title phrase, in the singular, to describe how he came upon the contemporary New York heroine's date-inscribed photo, crumbling in a junk shop, four centuries into the future—and came back smitten to seek her out. But only a conventional "zoom" button will come to the rescue as explanation.

Until then, we waver between crediting his narrative of supernatural time travel or accepting the diagnosis offered by the heroine's female psychiatrist: that his delusions are caused by "temporal lobe epilepsy," a condition suffered (we're told) by Lewis Carroll, van Gogh, and others, which can manifest itself, just as the hero says time travel does, in spasms of disjunctive temporal consciousness—as already coded intermittently by the film in fast-forward, reverse-action, and freeze-frame distortions of the track. Who's to say? Between the psychological and the supernatural explanation we hover until the climactic moment, where we find that the hero has indeed arrived from the future for the clairvoyant rescue of his beloved from a fatal taxi accident. This revelation is itself conveniently intercut with our discovery that the psychiatrist has been planting false explanations derived from the uncanny effects of brain disorder and is herself a time traveler hoping to sustain a cover-up. For the heroine, the confirmation of her lover's unbelievable narrative comes when she finds a photo of herself from the present, taken by a girlfriend, that was intended to be a shot of the couple together at the beach, but that has accidentally excised him (the man who has only invaded that present place at her side from his own future anyway) with an on-camera slip, recalled now in flashback, of the "stupid zoom"—and closed in on the heroine's face instead.

Proof positive, this image is shown to her in a print inscribed on the back by the photographer as a "happy accident"—the seed phrase of the hero's own until-now fantastic explanation of the found photo. Instead of the viewer witnessing by special effect some graphic chute or cosmic zoom of what he calls his "backtravel," the "long-distance" refocusing of desire is referred away to everyday optical science as its technological intertext, where the fetish power of photography becomes a more general touchstone for that instrumentalized recovery of the past that can take material (if not, as a rule, bodily) form. Like the protagonist of *One Hour Photo*, the hero's fantasy here too—magically activated in this case by the time-loop plot—is to put himself *in the picture* of his own desire. But the norm remains. Photography, in short, *is* motionless "backtravel": ocular temportation without the preternatural element of spatial transit. And film, too, begins in such a fixed-frame index and its manipulation. So that this film, in a kind of archaeological self-consciousness about cinematic temporality, chooses to open with a double liminal moment: a first peopled shot of narratively extraneous but thematically cued reverse-action footage of a Manhattan street, as preceded, even before the credits, and with even less immediate pertinence, by a cinematic image of the beached log caught just before the lovers (will later) have entered the frame to get their picture taken leaning against it—an image still in search of any and all plot; in this case, a pregnant photographic moment lying in wait for narrative's motivating cinematic duration. By contrast, other recent zoom effects, when digitized, mark cinema's more aggressive distance from any thematization of its former filmic base.

Beyond the narratively self-referencing cinematic zooms stressed by Willeman, another of his essays takes up a more recent device of digitally enhanced narration that similarly oversteps, like the zoom, the bounds of normal diegesis. From his Marxist orientation, he is interested in the way the stylistic excess of certain computerized screen effects—like the 360-degree rotoscope of Asian action cinema and its Hollywood derivatives—serves to "allegorize" the unchecked resources of the new venture capitalism, so that we marvel not at the craft but at the investment.[17] Its deployment for high-concept plots in the mode of the fantastic thus displaces the institutional fantasy of global market appeal into the local dazzle of the "attraction," where technique seems to emerge from the various "social imaginaries" (19) of "corporate dreaming" (18) as they are linked inexorably to industrial capitalism.[18] In the process, Willemen's own method returns him (and he is richly explicit about this) to a founding move of structuralist semiotics, where the substance of the form makes for a new content of the "inflated" expression. Here is where Willemen's signal contribution takes a different turn from Elsaesser's emphasis on diegesis to concentrate instead on indexicality. Where Elsaesser wants to break past what he sees as the conceptual deadlock between digital spectacle and cinematic narration, Willemen seeks to resist the comparably cemented antinomy between index and icon.[19] It is the latter dichotomy that would, in his view, too quickly segregate photomechanical record from digital fabrication, or in other words—for Willemen himself specifies the matter in pointedly "temporal" terms[20]—too rigidly distinguish filmed time (recorded index) from strictly screen time (projected icon).

Hence the crux of fantasy's digital manifestations. In them, the "real" is more easily dismissed than otherwise—without that "resistance" one finds "still operating in chemical indexicality" (19). In the deviant "temporality" of such new imaging, the drift to the "virtual" allows the desiring machine of venture capitalism to speak more clearly from its own unconscious will to astonish. For Willemen, this is anything but a vitiation of the index. The digital spectacle is not just an icon of the unreal, the unfilmed, but an index of the targeted ingenuity—as well as the broader technological imaginary—that produces it. Its lavish and transparent expense is cashed out as fascination, its expended labor converted to surplus value and reinvested less within a diegetic continuum of response (which would be Elsaesser's claim) than as symptomizing display. Under Willemen's allegorical hermeneutic, again, the form is its own content. Another sense of the digital index in such *trucage* should be clear from the preceding chapters. Often enough, what is traced by narratives of digitime is not only the money that went into them as screen spectacles, with their risk-hedged budgets, but the more widespread commercialization of electronic fantasy in other mass-market and more interactive venues.

Either way, we are—need it be said?—on familiar enough methodological ground. In response to traditional cinema, narratography attends to the diegesis in view of the cinematographic strip whose in-camera original was once filmically present to the performance of the narrative, in whatever fragmentary fashion the story got constructed. Similarly with postfilmic cinema, narratography attends to the diegesis, fantastic or otherwise, in view of the indexed traces of its digital mediation per se (themselves nonindexical). It is there, in such attention, that the move from frame time to framed time, along with the "grammatology" (Elsaesser's brief hint) of each, can be registered. World, narrative time, and their processing by the spectator: that's the "interactive" zone of attendance (by any other name) involved in anchoring the narrative and realizing its inference. Story space, screen imaging, and their discerned ocular ratios: that's the projected scene as medial function. The meld is inextricable: diegesis, index, text.

Whether we conclude that what is indexed by the digital icon in recent cinema is venture capitalism in a post-studio system or a broader erosion of embodied presence under a regime of the image, narratography has only one way to decide. It must specify the often imperceptible slippage between world and its manifestation over time—let us say, between "real" movement and optical (com)motion. Willemen's two essays, brought together around a single high-profile instance, should help clarify. For one could well attempt an allegorical archaeology of the new *digital* rather than cinematographic zoom, the work of the coded grid rather than the lens. In fact we have. This is the very effect that has characterized so many of the temporal plummets in the films we've been canvassing—from *The Thirteenth Floor* to *The Butterfly Effect*—as present-tense space contracts to the portal and chute of willed displacement along the ocular time tunnel of desire. What, after all, have we been working to graph in response to these moments if not the temporal content of such spatial form: the cybernetic or schizophrenic morphing of time itself, its biographical transmogrification, often the black magic of a self-annihilating escapism? Unlike the rotoscope process, for instance—in its normal use for stylistic inflection alone—such equally bravura devices actually swell the diegetic space (or spacetime) in the process of vacuuming out its human presence. In their warping of temporality, these effects of digital rack focus (so to say) or swish tracking (rather than swish panning) include, of course, the so-called cosmic zoom itself. In a tacit conflation of historical categories, they thus offer the reintegration of display, turning it psychosomatic within plot, subjectivized. Film history's reigning dichotomy between visual attraction and plot action is thus renegotiated from within the latter. At moments of plot crisis like these, the mind's eye has become, within the diegesis, its own spectacular screen, at once for memories, their evasion, and their transformation.

A comparison with a more traditional zoom effect, even in phantom form, is instructive. The last camera position of *Brokeback Mountain* (Ang Lee, 2005) begins with the titular shot of a dated postcard that recalls the Panavision expanse of the film's liminal shot—and the scene, somewhere far in the distance, of its eventual homoerotic Eden. That mountain space is by the end only a memory: only a time-image in the field of the virtual, as Deleuze would have it. But in the mourning of the surviving lover, and through a POV shot while his bereft stare lingers on the thumbtacked card he has just gently put "straight," there is the momentary hint that this lost scene might be undergoing reactivation as a phantasmal locus of habitation. Into its iconic terrain, that is, the camera abruptly seems to zoom—as if its lens action were traversing a real space. For just a split second, the viewer might assume the kind of optical plummet familiar from recent time-travel plots—or a strictly graphic metaphor for the same: a figurative dive into a suddenly immersive past, entirely illusory and subjective, momentarily appeasing. Not so, since the motion is not in the camera work at all but in the diegesis. Movement comes only from the final closing, no less, of the closet door—on whose inside the postcard is pinned—a door that rushes toward us for a brief moment before being hinged away forever. Beyond it, echoing the frame of the vanished postcard, is no represented and idealized place now but only a stark window on the present flatland, its irregular caulked seal repeating the scalloped bottom edge of the hokey postcard photo: the world as dimly viewed from a stationary mobile home in a life going nowhere.

So let us turn to one more (and quite minimalist) case of digital zoom as example—and this within a committed effort by the director to avoid or at least revamp this computer-graphic cliché. The effect is held this time mostly to the digitized iridescence of a character's "visionary" eyes in close-up rather than released to a full digital careening into a past space. This happens in the most recent of the time-loop plots from postmillennial Hollywood, a narrative that looks back across more than a decade's worth of derivative experiment to the death-moment fantasy of *Jacob's Ladder*, where an irradiant lyricism rescues plot from a gruesome nightmare. This new film is *The Jacket* (2005), directed as his first mainstream feature by experimental filmmaker John Maybury. As we realize only after the diegesis has run its course, the nonphotographic final credit sequence—celluloid without chemistry—casts us back in contrast, like a bookend, to the easily forgotten digital graphics of the opening. Indeed, it is not far-fetched to see the hero as a kind of scapegoat for the tension, or at least difference, between film's plastic substrate—real and material, including the indexical precursors of photochemical imprint—and the new digital phantasmagoria.

Shot dead in the early days of the first Gulf War in a melange of video jump cuts and color bleeds ("I was twenty-seven years old the first time I died," we hear in voice-over), and then slain again stateside with another bullet by a psychopath after unaccountably surviving the battle wound, the hero lives on (out of his body, we later realize) to make something out of a life already lost. Time travel becomes a metaphor for an appeasing deathbed projection. From five minutes in, the entire narrative is a plotting of the sheerly virtual. This process allows the hero to discover, far into his own unlived and ultimately cancelled future, a photograph of the same young girl, in the keep of her grown alcoholic self, to whom he had given his dog tags years earlier, just before his second gunshot death—evidence which he notices alongside the photo album. The hero's ability to join the grown girl in this revisable future, and to move back with her into her own past to correct its errors, is facilitated—and, as we later recognize, *figured*—by his being strapped into the titular straitjacket. It is under this constraint, yet with a drug-induced release into fantasy, that he is entombed in a wall of coffins under surveillance by a sadistic psychiatrist who is trying to cure him of his delusions of rebirth.

Only at the end does plot revert to the trick beginning, letting us discern finally (at least this is one probable reading of the film's complicated but muted switch finish) that the real *fabula* behind the film's achronic *syuzhet* is that the protagonist (we find through flashback confirmation) actually had been not just fatally but *finally* wounded the first time around. Having been shot point blank by an Arab boy he had tried to help, he subsequently fantasized his survival of not one but two deaths so that he could imagine some purpose for his existence. By generic mandate, of course, given the romance subplot, this search for purpose requires some selfless but autobiographically redemptive gesture toward a girlfriend or lover. Libido is silenced by the larger arc of an Eros shadowed by Thanatos. Unlike the hero of *The Butterfly Effect*, however, Jack in *The Jacket* doesn't have to arrange his own death in order to accomplish this. Like Donnie Darko, he has merely to accept its ex post facto revelation. As we in the audience come to do, he has to recognize that the straitening constraint of his nightly crypt, also his only source of liberation into the power of time travel, is exactly the morgue it looks like, each coffin awaiting its corpse—and what dreams of resurrection may come. The whole plot has thus been laid out, suspended with him, in an interment from which only fantasy can extricate the mind's eye. And if, by compounded association with the titular apparatus and the hero's first name, all this seems to be a thinly veiled supernatural variant of a sci-fi hero's jacking in to an electronic virtuality, it is only by digital intertext—rather than by plot turn—that this association is conjured.

Unique to this otherwise formulaic false-survivor plot is what follows after the de rigueur special effects of his time travel. But even these effects are designed to resist the predictable equation between electronic ingenuity and existential transport. In the director's allergy to the typical "tunnel shot" of a film like *The Thirteenth Floor* (he and his technicians are explicit about this in an interview on the DVD supplement), everything is played out instead across the surface of the hero's iris. The requisite boring through time is thus rendered ocular (visionary) rather than strictly marvelous (or technological). As the hero tries to break free from his nocturnal coffin, the camera enters the veined avenues of his vision per se. This time, in echo of the blood that gushed from his eye sockets on the battlefield, it is a red veil that fills his field of vision: more as a color filter for his retrospect than as the draining of life itself. What ensues is a free fall into hallucinatory subjectivity rather than into a fully spatialized past. At one point a close-up of Jack's own eye discovers the reflection of himself superimposed over his fantasized future lover (fig. 6.5), two time frames thus held in laminated suspension. But this kind of sensuous lyricism is continually inter- and under-cut with Gulf War carnage in the same startled frame of the hero's literalized "iris" shot, sometimes explicit battle scenes, sometimes merely a graphic association with the film's true liminal shot. For well in advance of his first death, *The Jacket* has opened with infrared digital telescopy through the crosshairs of a bomber's targeting system. And it is this

6.5

same gun sight that appears faintly super-imposed over numerous iris close-ups in the subsequent fantasy transitions.[21]

The pyrotechnics of those opening digital sequences of *The Jacket*, both air-borne and on the ground, await their final overthrow, however, by the indexicality of the evoked photo cell on the flickering strip—and even its nineteenth-century precursor, the isolated nonmechanical "photogram." Having exploited the digital to figure the hero's posthumous desire, Maybury thus returns, in marking Jack's death, to cinema's own origins, not in spectacle but in visibility per se. After all the coruscating effects rippling in maelstrom across the optical lozenge of the dead or dying mind's eye—those localized visionary equivalents of the temporal zoom, by which the subject's will to live is catapulted by turns back into his own impossibly outlived past and into his own unachievable future—after all this, some ocular relief is needed. Throwing over entirely its digital bravura, *The Jacket*'s final credits do nothing but project the very things of the world that death leaves behind. Unlike a host of sci-fi time travel-ers since Johnny Mnemonic, though much in the vein of fantastic temportation, Jack hasn't, we know, been plugging in to an electronically aided fantasy. The computerized virtuality of digital editing has instead emblemized, rather than actually staged, his mind's own electrochemistry. Far from unprecedented, the optical tropes, even though more sophisticated, can't help but remind us of *Donnie Darko* or *The Butterfly Effect*. As if finally to disinfect its time-images from all taint of VR, however, even by cultural association, Maybury's film drives in closure all the way back to the prenarrative logic of optical impress itself. Apart from signification altogether, and even before photomechanical transcription, his touchstone is the precinematic sense of an image traced by luminescence itself in a fixed shadow. André Bazin's sense of photography as the "mummy complex" is trumped here by a world itself fossilized by light.

What results is the director's explicit optical allusion, in turn, to a kind of materialist film practice that breaks from all vestiges of a realist aesthetic into the real itself: pasting bits and pieces of the world, the world before image, onto a transparent strip and giving it up to the projector's light. This is to say that the material object, unlike the character on photomechanical film, becomes its own disembodied shadow, takes place in its own surviving trace, participates, if you will, in its own immortalization. After the hero's belatedly accepted death is symbolized by a blinding light flooding from behind him—available to him only in the symbolic rearview

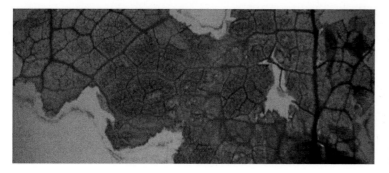

6.6

mirror of his getaway car—the equally backlit but no longer representational end titles begin flashing by. In their flickering scroll of separate handmade frames, they offer the ultimate serialized homage to real rather than virtual presence. Associations rapidly coalesce around these remains of the material day. For what has been designed for this exit footage is a rolling epiphany of the organic and the found: the real. The artisanal work involved is closer to optical entomology than to cinematography, and is indeed called an "insect leader" by the designer in the DVD interview. In it, filigreed botanical and inorganic shapes are flattened directly onto celluloid: a running index of the world materialized by illumination alone (fig. 6.6).

In an archaeological sense, again, what is effected by this is a return toward the original "photogram" of light-writing—before the chemical print, and well before the redubbed photogram cells of mechanical cinema. This final leader would seem to build on that nineteenth-century process where, for instance (and with a nonetymological pun fully invited), a *leaf leaves* its tracery, immediate, contiguous, on a treated surface. And here this effect is inscribed at just the moment when narratology's work, like that of plot, has come to an end with death's vanishing point. Yet Maybury's closing device offers a quite explicit homage, not to prephotographic tracing in general, but to Stan Brakhage's *Mothlight* (1963) as a benchmark text of optical experimentation in the American avant-garde, its title evoking the fluttery on-screen shadows cast by any diaphanous tracery, moth wing or otherwise. Such an optical allusion to Brakhage's renowned spool of nonphotographic projection also stands in for a broader film-historical motive—nostalgic, ironic, or both—on Maybury's part. Long before the postfilmic advent of digitization, and in an antithetical sense, Brakhage's "movie" was cinema without film. In closing off this one fantasy plot of strictly virtual existence and its digital evocations, the final prephotographic strip of *The Jacket* looks even farther back, however, than to the American avant-garde—or even to nineteenth-century "writing in light." Inviting a recognition of the postnarrative suspension of all mediation, these final credits thus disclose, alongside the tacit motif of elegy, their deepest archaeological gambit. From the cusp of the postfilmic, their visual kinesis calls up the late-coming place of cinematography itself in a longer history of indexical tracing and its projected imaging that runs back through magic lanterns and shadow puppets to the metaphysical paradigm of Plato's cave.

Following Paul Willemen, we can see that what is indexed by any "cinema" is not only the world projected but the work of the world's framing, whether digital, filmic, or in this case an unmediated bricolage of the world's debris. The diaphanous final leader of *The Jacket* animates found objects and fragments into a choreography of graphic shapes. Narratology, armed by genre templates, is well enough poised to unravel the time-loop trick of the preceding plot: a diegesis of the death moment "inflated" twice over by the digitized metaphorics of wish fulfillment. More immediately, narratography is caught up short by the flight from photography altogether. In this story of a life twice died, the hero is mourned on exit by the traces of what he has left behind—rather than by any further simulation of the fantasies shored against that doom. It is a gesture of elegy without melancholia. In this final succumbing of the movement-image to sheer moving things, we find, then, the quietest of time-images. With index triumphing over icon, backlighting over photon emission, matter over electronic mimesis, we come upon a vestigial space (no longer story space) that offers world enough but no lived time left, just frame after flickering frame of its aftermath. Maybury's narrative—by reverting from digital mirage through cinematographic suffusion (in the final seraphic flood of light) to prephotographic impress—has performed, in short, its own hermeneutic archaeology of a life died, and died again, along the enchainment of discrete image cells.

In stark contrast with this closing leader in *The Jacket*, we may recall the hyperbolic "flicker effect" that choked off Gaspar Noé's *Irreversible* three years earlier. Indulged there, at the overall plot level, is a play with temporality that is closer, as is *Memento* two years before it, to the strictly structural anomalies of such European experiments as *Run Lola Run* or *He Loves Me, He Loves Me Not*. For all its marginal fantasy elements concerned with premonition and second sight, Noé's film bears little resemblance to the Hollywood thrillers of temportation in the mode of the ontological gothic. Its time-images operate predominantly at the level of *syuzhet*, not *fabula*. It is a discourse about time, while merely a story about death and aborted futurity. As implied in *Memento* as well, Noé's whole point, captured by title, is that there is a kind of violence that cannot be fled from or reversed. Escape is just as impossible when a plot like that of *Irreversible* backs away from such violence (giving us the savage murder, then the preceding lethal violence for which it is a revenge, then the erotic happiness of the couple that preceded it) as when it is found sneaking up on it (Nolan's alternate trick) in inelectable recovery of the already elapsed memory traces. Noé's film ends with the tagline "Time Destroys Everything" (*Le Temps Detruit Tout*) after we have seen the "good times" we know always and already to be doomed. In the penultimate (hence earlier) moment of the film's reverse chronology, we watch the heroine, who has just discovered herself to be pregnant, hugging her belly below a poster of the Star Child from Kubrick's *2001: A Space Odyssey*, with its own tagline: "The Ultimate Trip." Destiny's astral fetus is the obvious node of allusion, but it offers at best a partial lead. The real intertext is a cosmic view of temporality in which history's unfolding moments are already written in advance in some monolithic master script.

Kubrick's evolutionary time travel, in its ultimacy, becomes figured by Noé's film as the vicious circle of the inevitable. Rotating on its axis for the next several minutes, the camera first frames the heroine upside down in a park, where she is (or may be) reading the same book she had been boasting of earlier (diegetically later), with its thesis about the future already written but legible only in prophetic dreams. (In her case, these dreams occur through violent flash inserts of the subway underpass where she is [to be] raped and slain.) Lifting above the children playing near her in the sweep of a revolving lawn sprinkler, the withdrawing overhead shot in the last scene is rotating clockwise along with the spray of water in this vertiginously distanced Edenic scene. Its gradual removal from us, in our optical liftoff, hovers there as a pure time-image of inevitability. The substance of the expression, as Willemen might want it, emerges as the symbolic motion of clock time itself. Eventually, the still spinning camera tilts skyward to halt upon a blanched daytime sky, fading next into a blur of darkness punctuated by galactic forms almost indiscernible in their Kubrickian abstraction—especially because they are interrupted, for the last endless, piercing minute or so, by blinding strobe lights so stark as to induce mild ocular trauma. Out of the blue, or speckled cosmic black, suddenly this movie, stripping bare its own pretext in happier days, has become again just what it was to begin with in its violent manhunt and murder, if in a different key: a rush of images literally intolerable to watch.

In interviews, the director has said that he meant this pounding registration of pure blinding duration to suggest that time is itself frightening. The effect is certainly effacing, blanking out any clarity in the diegetic image. Even with its digital intensification of flicker's rudimentary optic shock, however, this stroboscopic effect serves indirectly to recall (in yet another of those uneven screen developments we have been noting) the primal time frames of the photomechanical strip—or their experimental equivalent in Brakhage's projected intermittent materialities. These flash frames—losing all ocular discreteness, indeed their entire content, in a merely punishing throb of duration—are called upon to figure the preinscribed fate of an impending death. Time frames again become framed time, isolated, denatured, reified, and traumatic. Hence the contrast between such digitally applied or amplified scopic violence and the primitive flicker effect: a hyperbolic prevention of the seen versus its facilitating intermittence.

By this route we get a glimpse into what Bernard Stiegler, drawing from Heidegger's *Being and Time*, sees as the technics of duration itself, leading as it does to film.[22] To capture this insight in the most compressed fashion, one would want to retain, as his translators do, his typographic deconstructions of two key Heideggerian terms. This is a vocabulary concerned with exactly that oscillation of retention and protention that delineates time itself as the differential now. For Stiegler, the conceptual terms are in place well before he moves on, in his (untranslated) third volume, *Le Temps du Cinema*, to the sweeping modern paradigm of temporality afforded by the discrete succession of the filmic chain. This is to say that *Irreversible*, in the mechanical brutality of its plotting, as well as *The Jacket*, in its lyricism, invites the narratography of a post-narrative foreclosure. One recoils back in time from a telos of violence, one projects forward from slaughter into a fantasy of reparation, and both end in an oblique photogrammatic parable. Screened there is a metaphysical axiom that other films, approached narratographically, have also imaged for us. This is a pervasive sense that Dasein, being in time, is manifested nowhere more cogently (*pace* Bergson) than in the filmic support of cinematic projection. For this is where, in Stiegler's etymology and orthography both, the temporal "pro(gram)" is always one that "pre-cedes Death" (216), giving in to it before it arrives, materializing it on the cusp of a previous disappearance. So with the photo(gram) and its elided lethal fixity. Here once more, in screen projection, is time embalmed, change mummified, death captured in execution at twenty-four guillotine strokes per second.

With a film like *Irreversible*, narratology can accommodate the closing irony well enough, even if flinching from it emotionally, for here is a film narrative whose every hope, just as whose violent end, is already and all behind it. Narratography takes the actual pulse of the closing visual melodrama: both in the rapid diurnal camera movement, wheeling the image round like a sped-up sundial, and in the subsequent lacerating gaps of image, their usurpation by pure glare. Whether in the hammering stabs of *Irreversible*'s strobe finish, like the onset of a migraine, or the gentle flutter of the world's residue on celluloid at the end of *The Jacket*, optical allusion mediates both diegesis and index on the way to textual recognition. The stark aggravation of the strobe flashes, the splutter of the prefilmic frame: each, in passing altogether beyond narrative, invites not just an archaeology of intermittence but a grammatography, whether digitized or sprocketed, of time in process.

TIME OUT OF MIND

By such and other means, if these latest releases are anything to go by, we might—though there is no predicting for sure—continue to see screen narratives in which, in order to be figured, time must be either equivocated or eccentrically traversed, as in so many recent European films, or travestied in its normal duration, as by so much Hollywood sci-fi and fantasy. In rendering such material, the filmic frameline (to whatever extent it becomes digitally "compromised") will no doubt continue to make its medial inroads into theme. This is because, at material as well as phenomenological levels, cinema's ability to image the pure virtuality of time and memory must ultimately rest with its capacity for visualizing the way things are that are no more—or aren't quite yet. We remember Arthur Symonds's sense of the choreographic flicker of dance on the Victorian stage. So too with the transit from the filmic to the cinematic frame, the image lasting only long enough "to have been . . ."—to have "been there," there in passing, as passing. Embalmed in its change, maybe, as Bazin would have it, but also dissected in its advance and reassembled for inspection. Like memory as it lays down its own tracks, such a sense of filmic (as against digital) inscription puts us in mind of—or, say, philosophizes—all things no sooner manifest than virtual. In that respect at least, and still in Deleuze's terms, lies narrative film's deepest value as an art of the image: an image always integral and vestigial at once.

Cinema reconstellates the mnemonics of mentality's own reeling film in the aesthetically usable form of a screen projection. Here is an optical field vanishing even as eventuated, manifest by serial effacement, evident only in difference, inscribed by evanescence, secured in the imaginary through intermittence and slippage. When such a correlation is glimpsed or intuited on the narrative screen—this double tracking of photogram and psychic memory trace—it need no longer be entrammeled in the fantastic as a fictive genre in order for it to carry the full charge of virtual reality. Need not, but may still be. For wherever digital process may yet take us in revising such a picture of perception itself, the on-screen drama of narrative is likely to follow. This will in all probability go on involving not only the reinvention but the continual *replotting* of the screen's optical register as the proximate cause for visual events within the cinematic spectacle—even if not exactly for lived action. Among such perceptual events, though only as one genre type among many, we will, in all likelihood, continue to see fantasies of time framed and reframed, made both mutable and portable: plots driven, that is, by the deep logic, as well as the production logistics, of computer memory. Yet this is a logic that often reduces to cartoon the real force of time on the cinematic screen.

Parallels across the Atlantic between experiments in temporal fantasy, in their continuing relation as well to the magicalized chronicles of Asian digital epics, continue apace. But you, the reader of this book, are under no more compulsion to buy such an argument, or the implications these chapters have spun out from it, than is the viewer, even if it's right, to buy into any serious reflection on its consequences with the latest Hollywood ticket price. Once you have taken your seat in the theater, the fun is meant to be mostly free of concern. Many of these films only work if your critical resistance goes on vacation. And in any case, methodological procedures don't stand or fall on particular interpretive claims. It doesn't matter, at least finally, whether one suspects in European films a new transnational imaginary that appears in perverse erotic symptoms and uncanny liaisons, the force fields of ethical affinity and its suppressions. Or whether one finds in recent Hollywood films a flight from geopolitics into a totalized (rather than merely global) fantasy of electronically modeled virtuality. Either way, or on any other credible account, it is still often the case that a hermeneutics of suspicion, so eminently called for at least by the latest Hollywood narratives, stands careful watch only if the medium itself is seen as having been taken on board by the equivocations of plot. Under such internal constraints within a recurrent structure of postponed disclosure, the medial texture of film itself, in a given narrative, may seem to be vexed by more than it wants to say. About the least searching and ambitious of these genre products, I began by suggesting that they can perform what they know even without knowing what they are doing. This partly narrative, partly graphic performance *is* their textuality.

Though attendance and deciphering have involved from the start a broad structural narratology of the trick beginning, we have also been specifically concerned with Hollywood films marked, indeed shaped wholesale—such is this book's particular argument—by the viewer's complicity in a genre-coded nondisbelief (the more actively repressive version of "suspended disbelief"). This is a classic "disavowal" operating at the plot level primarily—or at least to begin with: He can't be dead already. Surely she isn't simply a digital figment of the computer's imagination. He isn't, let's hope and assume, just one more case of an abused child. In view of all this, it might seem right to say of narratography that it locates, in such recent cases, the hermetics (rather than hermeneutics) of *non*suspicion within a more sweeping culture of the virtual. For whether focused by narrative around memory or simulacrum, premonition or ghostly return, it is the category of the virtual that seems finally to embrace all those meanings unlost on analysis precisely in the way they remain unactualized by plot.

If so, and elsewhere too, it is not just an interest in media history but an impulse to cultural critique (as, for instance, in Willemen's emphasis on indexed industrial labor as spectacular surplus value) that will keep us reading as we watch. In focusing attention on such legible spectacle in postfilmic cinema, much ground has been gained if we can agree at this point, retaining analysis's good name, that one useful further name for such watching may well be narratography. In either Lyotardian or Deleuzian terms, this would be a practice of the ocular emergent, where image passes into the *figuring* rather than the picturing of—and this by enlisting the material substrate of its own signifying process. This is where Deleuzian time-images in particular, inchoate images *taking thought*, hold interest even when they betray themselves to cliché.[23] And this is why, in turn, this last chapter's threefold title can be understood dialectically. Especially in the mongrelized environment of recent digital cinema, narratography does more than come to the rescue in local standoffs between ocular archaeology and hermeneutics. It arrives as their mediating third term.

In the decade under investigation in this book, whether the screen illusion in a given moment of the fantastic (often a mnemonic mirage) comes across as uncanny or supernatural, its effects filmic or digital, nonetheless the graphic signals of its apparition repeatedly exceed the scene. At which point the work of optical allusion anchors such imaging within either the rule of visibility that permits it or the cultural field that emits it as symptom—or, most pertinently, both. And something more too. Diagnosis can probe the graphic medium to yet another depth. Not just a symptom or even a sign of the times, what we may encounter there in certain effects, at a further stratum of reciprocity, is both the time of a new digital signage and, implicated with it in screen narrative, a new signification of time itself. It is at this depth, I repeat, with the image system of cinema thinking itself through narrative crux—and as paradoxical as it still sounds—that the message delivers the medium. It does so by delivering it up to reflection on the legible surface of story. Narratography is one way to characterize a readiness to accept such delivery.

APPENDIX: *Precinematics; or, Reading the Narratogram*

This afterpiece can go the way of most: to be read not necessarily in order but maybe first, only, or never. It is not a last chapter in disguise. Nor does it extend my argument forward in time. It is included only as ballast at the other end. It has in fact nothing to do with cinema's second century in a postfilmic and postrealist mode. It mostly concerns, instead, a precinematic manner of prose discourse, though one whose writing effects, I hope, will come clear—under narratographic analysis—as demonstrably *proto*filmic. As an exercise in cross-medial archaeology, if you will, it is primarily about what the greatest theorist-practitioner of the photogram's place in cinema, Sergei Eisenstein, missed in the Victorian writer, Charles Dickens, whose influence on montage he famously celebrated. The closest thing to this appendix, then, in the run of chapters to which it is attached, remains the consideration of Henry James in the first of them, his prose evincing a prehistory of suture in the grammatically knotted and entirely implicit sightlines of dialogue cues. Earlier and on several fronts at once, Dickens wrote a yet more filmic prose than James. We learn something irreplaceable about cinema from contemplating the plastic and differential basis, not just the broad melodramatic legacy, of this precursive technique.

DICKENS, EISENSTEIN, AND THE FILM YESTERDAY

Dickens was born for the movies. That's the truism. The further truth that film was born from Dickens is the burden of the most famous genealogical essay in the literature of cinema, "Dickens, Griffith, and the Film Today" (1944), by the renowned Soviet director Sergei Eisenstein. The accomplishment of that essay, and certain motivated blind spots in its attention, is the topic here.[1] Though never thinking to film a Dickens novel, Eisenstein understood the cinematic strategies—if not the deeper verbal logic—of the novelist's construction as no one before him. His observations offer an endlessly fertile point of departure for what I would call a filmic grasp of Dickensian prose.

The trouble comes mostly with filmed Dickens—or, say, cinema's Dickens. What movies repeatedly ignore in his writing, as they milk locations for his "atmosphere" and dial up his dialogue, is exactly the shared basis of the two media, film and prose fiction: their common reliance on the very dynamo of narration. This is the structural engineering of storytelling itself, operating in Dickens's prose from the level of syllable and word to sentence and paragraph. In stylistic matters, adaptation is usually the graveyard of appreciation. Occasional screen exceptions, to which we will come round in the end, only cement that verdict. For what Dickens secretly willed to film is what no copyright could ever have protected: a whole new mode of kinetic sequencing in which discrepant juxtaposition is submitted to continual resynthesis. Such oscillating effects take shape in ways that Eisenstein, given his allegiance not only to a dialectical or conflictual model of thesis/antithesis but to its frequent figurative resolutions, might have appreciated more fully than he did.

These appended pages, therefore, attempt a comparative media analysis rather than anything resembling the triumphalist narrative of advance and transcendence proposed in Eisenstein's three-tiered title, "Dickens, Griffith, and the Film Today," where the last refers mostly to the recent glory days of Soviet avant-garde practice. D. W. Griffith's transfer of Dickensian technique into a different medium is only two-thirds of the story Eisenstein tells; the rest concerns the outdistancing of Hollywood craft by Soviet artistry. For all of Eisenstein's sincere admiration and detailed celebration of Dickens, it is crucial to note how the Victorian storyteller's narrative technique becomes, toward the climax of the essay, mostly a marker for conceptual limitations in Griffith.

To see beyond this deployment of Dickens requires a return to what I will be calling the filmic rather than the cinematic elements of his work, a dimension of his writing that goes mostly unnoticed by Eisenstein. Filmic Dickens locates the rapid layered succession of his verbal as well as imagistic effects, whereas cinematic Dickens concerns larger, more readily staged (and filmed) blocks of description and plotting. The ideal Dickens movie would have been shot by Griffith at the height of his powers; that's Eisenstein's slightly patronizing gist. My point is that the ideal Dickens movie might have been shot by Eisenstein himself, keyed to the increments and upheavals not only of shot sequence but of disjunctive frame advance: precisely that collision of images so central to Eisenstein's program.

Some rudiments, then, in review. Montage: the piecing together of discrete images to make a film sequence. Call it editing. Writing: the piecing together of discrete signifiers to make a verbal sequence. Call it syntax. But each depends on a preceding level of serial juxtaposition as well. On the screen, automatically recorded and then projected photographic frames displace each other to induce the look of movement. On the page, alphabetic characters compact into words and accumulate toward grammar. Concerning the cellular subunits of the filmic image, however, Eisenstein was interested more in tension than in sheer succession, explosive contrasts that recombine under pressure into new meaning. Image slams against discordant image to produce a metaphoric third term. It is the prose equivalent of this counterpoint and resolution that he consistently overlooks in what we might call Dickens's word advance rather than frame advance. Eisenstein's attention to Dickens rests at the macro rather than micro level, again cinematic rather than filmic. For Eisenstein, that is, Dickens anticipates film narrative mostly in his obsessive descriptive details (especially in close-ups), their intercutting, and the broad formats of parallel plotting.

This deliberately leaves to one side exactly what Eisenstein prides himself on in his own filmmaking practice: namely, "conceptual montage," where editing is designed not for sheer sequencing but for higher-order integration, where disjunction leads to "a new qualitative fusion out of juxtaposition" (238). This intellectual synthesis is what Eisenstein found missing in Griffith, as implicitly in Dickens before him. A clear sense of Eisenstein's limited homage to Dickens—as in fact a rallying cry, by contrast, for his own avant-garde invention—only slowly emerges over the course of an unfolding three-stage argument. First, as Eisenstein demonstrates at length, Griffith may well have taken Dickens as a model in matters of both emphasis and suspense, from shot to sequence level, in everything from melodramatic close-ups to parallel storylines. Second, the practice of Soviet montage, learning from Griffith all that it did in the first flush of enthusiasm for Hollywood technique, participates in a similar lineage. Third and finally, as detailed in the closing pages of Eisenstein's essay (often forgotten by literary critics), these popular origins of cinema had nevertheless to be left behind in the articulation of a revolutionary aesthetic. Here we have one of the oldest stories in criticism: the historical destiny of a formal potential. In this version, Dickensian fiction claims attention mostly, for Eisenstein, as the long-outgrown seed of a modernist initiative fully realized only in another medium.

The reason to revisit this position has little to do with celebrating either the prescience or the influence of Dickens, let alone with smoking out the patronizing undertone of Eisenstein's famous appreciation. It is simply to see what Eisenstein's bias led him to miss. One of the unfollowed leads in his essay is easy to spot. Instead of alluding to the Chinese ideogram as model for cinema's composite signifier, as he had done in earlier position papers, Eisenstein refers this time to the work of Russian linguistics in accounting for the rudimentary compactions of the "word-sentence" (237). This is a primitive utterance housing a latent syntax within its isolated unit of diction. Like the "montage cell" of film (236)—organically mutable into higher orders of form—the word-sentence is the "embryo of syntax" (237). But by this point in the essay, Eisenstein is three pages beyond his last mention of Dickens (234), who was never for him a linguistic experimenter anyway (to be partly explained by the problem of Russian translation, no doubt)—while still eighteen pages from the end. The linguistic anchor for his theory of montage collision as an advance on parallel editing (the simple intercutting between separate narrative strands) would seem to have left Dickens, let alone Griffith, far behind.

This is where we need to put on the brakes. For what seems important about the "word-sentence" as a model for the explosive character of Soviet montage—its exemplary fission between morphology and grammar, word and clause—also typifies the tensed density of Dickensian phrasing, its overcrowding and potential splitting of syllables and linked sounds. I will begin with the most circumscribed examples I can find, at the level of word borders and their phonetic friction. Turning on the sibilant (or *s* sound), the slippages in question range from eroticized sentimentality, at one pole, to outright farce at the other. Late in *Little Dorrit*, we are let in on the hero's abiding image of Amy Dorrit as a "youthful figure with tender feet going almost bare on the damp ground, with spare hands ever working."[2] Aural juxtaposition captures the virtually fetishistic association of vulnerable body parts in the slide from "almost bare" to the phonetic ligature—and lexical resynthesis—of "spare." The word-sentence indeed—or at least the two-word phrase that speaks volumes. Or take the comic variant of such phonetic flicker in *Dombey and Son*, when the bumbling Mr. Toots plants an uninvited kiss on the cheek of Susan Nipper. Rebuffed, "the bold Toots tumbled staggering out into the street,"[3] with Toots's "stumble" heard in the toppling forward of the *s* from his own improbable last name. Imitative pratfall syntax: sound in echo of sense. That might be the conventional description. Another account comes to mind. In this alphabetic "montage by collision" (Eisenstein's preferred format), and its resulting grammatical implosion, the dialectic of difference bursts forth into a new third term. Toots + tumbled staggering = Toots stumbling. Words harbor between them a surreptitious syntax of the (phono)graphic skid.

In a novel like *Dombey and Son*, a harmless erotic lunge of this sort is set in contrast with the lifeless energy of such implacable forces as business and railroads. It is a continuous pitched battle of Eros versus death. *Dombey and Son* is the first great fictional treatment of the railway revolution. It is no accident that it was the industrial locomotive which was later to become a central symbol—and dynamic prototype—for Soviet cinema: the disorienting machine of modernity rescued for the streamlined aesthetics of modernism.[4] In this light, it is revealing to note the emphasis placed by Eisenstein on the preindustrial stagecoach-eye view in *Oliver Twist*'s racing panorama of commercialized London street life (217). What he might have called a *tracking shot* (laterally moving camera) achieves only the nonsynthesized seriality of urban clutter. By the coming of the railroad a decade later, a similar tracking procedure has grown all the more driven and ironic.

In *Dombey and Son*, both diction and syntax capture the speed of the new railroad with such an uncanny aptness that they become filmic almost by default: descriptive flashes that seem at first atomistic, then serial, then fused in a transformative rush, artificially synthesized by the relativities of speed and position. The following is a description from the grieving Mr. Dombey's point of view, in mourning for his dead son. The despoiled cityscape is so completely focalized through his perspective that it becomes almost an hallucinatory "subjective shot," whereby the train sweeps him past life like "a type of the triumphant monster, Death" (20). The whole paragraph is triggered by an adverbial exclamation ("Away!") that seems to vibrate halfway between description and imperative—another "word-sentence." Grammar then grinds its way through several more prepositions in present-tense series: raw momentum without sensed destination. This began three paragraphs back with the locomotive, in a single sentence, "burrowing among . . . flashing out into . . . mining in through . . . booming on in . . . bursting out again" (20). And here is the subsequent surge of force and disorientation:

> *Away*, with a shriek, and a roar, and a rattle, plunging *down into* the earth again, and working *on in* such a storm of energy and perseverance, that amidst the darkness and whirlwind the motion seems *reversed, and to tend* furiously backward, until a ray of light upon the wet wall shows its surface flying past like a fier*ce s*tream . . . sometimes pausing for a minute *where a crowd of faces are, that in a minute more are not.* (20; emphasis added)

All the registers of attention available to narratography are elicited here at once, from ephemeral sibilance to jolted grammatical parallelism. The syntactic torque from past participle to infinitive phrase in "seems reversed, and to tend" administers a wrench to grammar—to narrative editing—comparable to the optical jolt described. Further, the steam-driven hissing collision of "fier*ce s*tream" insinuates the engulfing "dream" that offers the deep logic for this waking nightmare of funereal projection on the perceiver's part, the morbid Mr. Dombey under the sudden onrush of the inevitable. All is vulnerably relativized and fleeting. This is most obvious when the collective noun "crowd," taking "are" as verb, pivots on the comma and—instead of offering further predication ("are" what?)—hustles a glimpsed cross-section of the waiting public away to negation ("that in a minute more are not") in the suspended present tense of this bleak epiphany.

A more figurative variant of the funereal and subjective tracking shot occurs in the implausibly animated stasis of a later turn of phrase in *Little Dorrit*. A deserted street is described as "long, regular, narrow, dull, and gloomy; like a brick and mortar funeral" (bk. 1, chap. 27). The only motion is that limping, uneven burden of one-, two-, and three-syllable adjectives. The result is that the static and depopulated space is converted by simile to a kind of paralyzed motion. The paradoxical effect falls somewhere between a tracking shot of blank fixity and a spectral freeze-frame of immobilized procession. In the eye of its forlorn beholder, the world is caught dead in its tracks under the analytic camera of Dickens's silent cinema.

And even Dickensian dialogue, though unmentioned by Eisenstein, can undergo a serial dislocation that results in a mismatch of utterance and narrative inference. Such discrepant succession is thereby related indirectly to that most radical aspect of Soviet sound technique, the "vertical montage" (256), mentioned in passing at the end of his essay. Here lies Eisenstein's resistance to the seamless realism of standard sound cinema, in particular to the "mechanical parallelism" (256) of image and voice. He champions instead the pasting over of sound upon image in a discomfiting and conceptual manner, contrapuntal rather than naturally meshed. Such a skewed laminate of soundtrack upon image series generates a kind of aural palimpsest that would be entirely divorced from the slavish representation of continuous speech.

In a strictly verbal medium like Dickensian fiction, a related conceptual gap may open up between the flow (or stammerings) of dialogue and the disruptive identification of the speaker by narrative tag.[5] Here is Dombey ponderously inflating himself after the birth of his son and heir in the opening chapter of *Dombey and Son*. "'The House will once again, Mrs. Dombey,' said Mr. Dombey. . . ." Will what? Be a home? Hardly. We immediately suspect Dombey to be referring to the patrilinear Firm, not the family hearth—as is all too clear when he finishes his swelling thought: " . . . be not only in name but in fact Dombey and Son." Only then does he round off this direct address with a revealing stutter of barely checked capitalist possessiveness: "Mrs. Dombey, my—my dear." Across the halting uneasiness of Dombey's speech, that is, as further dislocated by the syntax of exposition, the words of a continuous oral enunciation are caught stripping their gears. In Eisenstein's terms, the effect of this slippage is an overlapping or "vertical" montage in which speech and its narrative presentation are ironically syncopated rather than smoothly synchronized.

So we continue bearing down on the prose of Dickens, as filmic rather than filmable, to ask whether and where, well beyond the sweep of Eisenstein's spotlight, such prose might achieve something like a dialectical synthesis in its own right, sprung from ironic disjunctures of all sorts. In pursuit of this question, we get unexpected mileage out of a concocted example of montage failure dropped on the run by Eisenstein. In its striving for the detonation of seriality into nuclear intensity, Eisenstein readily admits that the attempted "flashing unity of image" may never purposefully fuse. Half-hearted efforts may degenerate instead into "a miserable trope . . . left on the level of an unrealized fusion, on the level of a mechanical pasting together of the type of 'Came the rain and two students'" (253). Pages past his last mention of Dickens, Eisenstein would seem to have no inkling of just what accidental two-headed nail his sarcastic hammer blow has hit on the head. For "came the rain and two students" is close kin to one of the signature effects of Dickensian comic writing—and directly related to his most serious symbolic effects. Usually, in noninverted grammar, this "trope" (or "figure," here of speech rather than image) is a device called *syllepsis*, sometimes *zeugma*. It typically involves one verb governing two objects or modifiers in different senses, rather than two subjects laying claim to a lone verb (as with the "coming" of downpour and students). A classic example in Dickens's first novel, with debts to the eighteenth-century comic writers he grew up on: "Mr. Pickwick fell into a wheelbarrow, and fast asleep."[6] Compare *Little Dorrit*'s "and so to bed, and to sleep" (bk. 1, chap. 28), where the prepositional sense slides over into an infinitive verbal force.

Such sylleptic effects proliferate in *Little Dorrit*. The frequent difference that gets (s)played out in such phrasing—between a figurative and a literal sense—can be carried over to a metaphysical difference between immaterial and material meanings: "He would have taken my life with as little scruple as he took my money" (bk. 2, chap. 20). How can one really call such phrasing, as the schoolbooks would have it, "faulty" parallelism? The whole logic of grammar seems under interrogation at such moments. Or consider this self-referential example, again from *Little Dorrit*, about the normative constraints of script and their occasional defiance. Writing about writing, Dickens even reads the aggressive flourishes of calligraphy through the double focus of a sylleptic shift, so that the decorated capitals of certain manuscripts "go out of their minds and bodies into ecstasies of pen and ink" (bk. 2, chap. 15). It is a matter of psychic and linear transgression in a single wrenched idiom, dovetailed with the shifting of sense from "go out" to "go into": a case, all told, of writing overstepping its own syntactic rather than graphic bounds. Comparable to the narratographic reading of larger plot trajectories, these are the miniature equivalents of that "garden-path" format discussed above in recent Hollywood twists, where a line of development must correct itself en route. At other times, Dickens takes as much delight in elaborating the divergent senses of such forked grammar by open repetition as he does in collapsing them into one split syntax, so that *Little Dorrit* also offers summary descriptions of debtors "being left behind, and being left poor" (bk. 1, chap. 36) and of a climactic banquet where, in a dreary doubling of space and time, "the table was long and the dinner was long" (bk. 2, chap. 19).

Eisenstein may have had more of a sense of this verbal license in Dickens than he lets on. As it happens, his first example of the human "close-up" in his essay links the "icy cold moral face" of Griffith's villains with the detailed physiognomy of Mr. Dombey in the early chapters, "revealed through cold and prudery" (199). At least in the English translation of Eisenstein's essay, that phrase "cold and prudery" has something of a sylleptic feel in its own right: the coldness of both house and heart at the novel's opening, chilling all desire to the bone. A later "close-up" on Dickens's part recruits the sylleptic double take even more directly, when Mr. Dombey, as pillar of capitalist society, is seen "stiff with starch and arrogance" (chap. 8): uptight linen and rigid class pride seized in the same phrasal breath. What is such an effect but a "montage trope" (240) at the level of the unreeling syntactic strip? Quite apart from his immediate discussion of the close-up, Eisenstein acknowledges that "montage thinking" (253), in becoming the essential language of Soviet cinema, had to pass through—and well beyond—a kind of primitive phase of naked metaphor, more pictogrammatic than cinematographic. His analogue from the history of language is the centaur as primordial glyph: a composite image of human and animal properties synthesized to indicate those "fast in the race" (246). We might say that sylleptic juxtaposition in Dickens's satire is a kind of sophisticated centaur speech in the syntax of description: a melding of incompatibles, combusting into a third or higher-level term. Starch and arrogance: stiff (f)rigidity in a single figurative close-up.

Though arguably filmic, syllepsis doesn't necessarily feel cinematic even when it is more visual to begin with. Take the first lawyer we meet in the famous opening panorama of *Bleak House*. As Dickens's choppy prepositional satire activates the lightest form of comic syllepsis (the three objects of the preposition *with* all evoking different aspects of possession), the portrait is then capped by a legalistic oxymoron: "a large advocate with great whiskers"—rather than a great advocate with large whiskers—"a little voice, and an interminable brief."[7] In the idiomatic paradox of an endless "brief," legal adjudication in the law courts of Chancery stands accused as a charade and a joke. The deeper "montage thinking" in this famous opening passage, however, arrives a paragraph earlier, where the mist and slipperiness of a drizzly London day offer an objective correlative for the obscurity and muddle of the law courts. And more: a reciprocal troping, so that each figures the other in a triggered explosion of new meaning. "Never can there come fog too thick, never can there come mud and mire too deep, to assort with the groping and foundering condition" of legal process. The tacit metaphor is visualized as a kind of stark parallel montage, eroded into total overlap: "Come the fog and the befuddle-ment together." That's the "trope," the *turn*, from material to "conceptual" plane, atmospheric sludge to bureaucratic muck: in a more modern idiom, from smog to smoke screens.

In this dialectic of indictment, the move toward ironic synthesis—or, in other words, toward exposed commonality—erupts in the form of prose superimposition. Fog upon fog: murkiness thickening to metaphor. Yet such figural mockery carries its own inoculation. Confusion at law is as inevitable as bad weather to a Londoner, hence insurmountable. Beyond political remedy. Culture has thereby been naturalized, however bleakly. This is the very work of ideology, un-raveled only momentarily here by being made brazen and overexplicit. Critique stops far short of a call to arms. This is the famous conservatism of Dickens's radical instincts in a nutshell. The senseless Chancery litigation is not only seen for what it is. It is also seen for the way it makes us sense it as an intractable force of nature. The system can only be ironized, not revolutionized.

So let's look, for contrast, at a genuine revolutionary image, perhaps the most famous in all Soviet cinema: the sequentially lying, crouching, and sitting statues of three marble lions rapidly intercut—and so cartooned into motion—in Eisenstein's own masterpiece, *Potemkin*. Rather than a metaphor for uprising—a direct translation of image into visual pun—Eisenstein intended the image to cut deeper into the materiality behind the metaphor. For the three shots are "merged into *one* roaring lion and, moreover, in another *film-dimension*," where they operate as "an embodiment of a metaphor: '*The very stones roar!*'" (253; Eisenstein's emphasis). Such is the most fully achieved form of the synthesized image as trope: a complex double translation of sculpture into animal life, silent cinema into defiant outcry.

Those stones and that roar are Dickensian too, in more ways than one. I am thinking of gloomily animated cobblestones, the ones that pave the streets on a forlorn Sunday in *Little Dorrit*, which seem to wear a "penitential garb" (bk. 1, chap. 3) of soot. Rapid montage multiplies the sameness of these streets into a metaphor of tedium. "Nothing to see but streets, streets, streets. Nothing to breathe but streets, streets, streets." Cut, cut, cut. Synthesis achieves only the fusion of the morbidly unvaried: the multiplied dead weight of Sunday desolation. Chapters later, it is into the "roaring streets" (bk. 2, chap. 34) of this same mercantile and dehumanized urban world that Arthur and Amy make their marital way at the end of the novel, but not until she retrieves him from imprisonment across a strange recapitulative temporality in which the novel's long day of suffering is clocked symbolically to its finish: "And the day ended, and the night ended, and the morning came, and Little Dorrit"—vessel of the novel's redemptive vision, with her twin dwelling in spirit and flesh—at last arrives. She enters as the living incarnation, no less, of sylleptic—and here synthesizing—resolution. For not unlike those students and their rain in

Eisenstein, she "came into the prison with the sunshine" (bk. 2, chap. 34). Literally, the heroine makes her final entrance at day's first light. Figuratively, she embodies in her own person the brightness she brings with her—and will soon take back into the world at large.

After the couple sign their marriage register, prose achieves perhaps the greatest "loop" effect in precinematic imaging. In three iterations, the verb "went down" precipitates a series of grammatical fragments that materialize the couple's described future scenes like superimposed vistas of selfless service: "Went down into a modest life. . . . Went down to give a mother's care. . . . Went down to give a tender nurse and friend" (bk. 2, chap. 34). These doings are less chronological than simultaneous, operating upon each other in mutual reinforcement like a cinematic overlay that releases the principals from the once-upon-a-time of the *fabula* to the happily-ever-after of their recompense.

The whole loosened and levitated syntax of this final overlapping effect is for the heroine, in one sense, the rectifying double vision (of now and to come) for a life previously trapped in a kind of hallucinatory limbo of retrospect. After sudden unearned wealth has released her family from the Marshalsea prison, Little Dorrit on the mandatory Grand Tour kept refiguring the decayed relics of the former Roman Empire as the foregone domesticity of her life in debtors' prison: "The ruins of the vast old Amphitheatre, of the old Temples," and so forth, through three more parallel phrases, "besides being *what they were, to her, were* ruins of the old Marshalsea—ruins of her own old life—ruins of the faces and forms that of old peopled it—ruins of its loves, hopes, cares, and joys" (bk. 2, chap. 25; emphasis added). Each scene is caught slipping past—and surrendered again—in the flitting serial syntax, which pits two time frames, epochal and familial, against each other: the olden and the "of old." Or not so much pits as superimposes them in the tracked itinerary of review. And does so across the self-dissolving phonetic evanescence of "were, to her, were ruins." Ultimately: "Two ruined spheres of action were before the solitary girl," the subjective flashback and the fabled historical panorama, so that "she saw them both together." By such means does parallel editing come under psychic constriction as a sustained double exposure. Here, in sum, is the mirage laminate of the heroine's ocular and spiritual displacement—as redeemed finally by the alleviated moment of future overlay at novel's end.

Compared with that closing "loop," that palimpsest of omniscient narration, Dickens can just as drastically reverse his narrational point of view to a lone subjective vantage. A single flamboyant example of such a focalized POV shot from this same novel can help specify one of the most oblique of Eisenstein's unexemplified technical allusions—to the actual deployment in prose of "special lenses" (213), in this case a refracting telephoto effect—perhaps even a tacit zoom. The exemplary scene in *Little Dorrit* is set entirely by premonition. The international banker and crook Mr. Merdle has tortured his daughter-in-law with boredom during a brief visit, his own dialogue caught up in a sylleptic doubleness when he apologizes for "equally detaining you and myself" (bk. 2, chap. 24). Here is the fracturing internal montage of identity itself (as both subject and object of consciousness), just before he borrows the penknife with which to execute his suicide in the break between chapters. Exit Merdle.

But exit Merdle—more to the immediate point—in the eyes of the latest uncomfortable victim of his repressed panic, who watches him literally pass away in the street from a removed perch on her moneyed balcony and cage. At which point we can indeed see the sort of thing Eisenstein may have meant by including in the "optical quality" of Dickens's "frame composition" the "alternation of emphasis by special lenses" (213). With Fanny's proleptic and unconsciously mortal sense that "this was the longest day that ever did come to an end at last," we watch Merdle recede in a bird-eye's shot filtered and distorted through her rippling tears: "Waters of vexation filled her eyes; . . . making the famous Mr. Merdle, in going down the street, appear to leap, and waltz, and gyrate, as if he were po*ssessed* of *several devils*" (bk. 2, chap. 24). Even blurred vision finds its aural correlative in the thick of hissing phonetic ricochets. Prophetic cinematography, by any other name. Or think of it as prose's own "special effect," abetted by every linguistic twitch available to it. The mercantile cynosure of all eyes is subjected here to the private disclosures of the gaze. In the buckling perspective of the image, whereby she half sees in him what he feels, objectivity and subjectivity succumb to each other across the nervous wreckage of a transferred point of view.

So now let's take what we have seen back to the cinema—not in the spirit of judged and ranked "translations" but of comparative narratography. A "subjective shot" like Fanny's, anchored by a POV, is the kind of technical inflection of Dickensian prose that gets readily transposed to screen, offering directors an effective, if limited, way of dynamizing their narratives. Two years after Eisenstein's essay, David Lean achieves an ambitiously distilled moment of subjectivity in the first-person bildungsroman of *Great Expectations* (1946). Succumbing to fever after the death of Magwitch, Pip is disoriented in the London street by the heaving sea of glinting satin top hats. They manifest en masse a dizzying sign of that faceless gentlemanly order into which Pip has never successfully inserted himself. Hot flashes appear before his eyes like the rotating shimmers of a flickering hypnotic wheel, eventually blurring his own intercut image as well. All is swallowed up in the fever pace—and pointless round—of social mobility. In the shot/reverse shot pattern, these blades of light continue to distort the hero's own POV until, even when he is safe in his bedroom and falling toward his pillow, the image still pulses with these glintless bursts

A.1–A.2

of light—like the migraine auras of his social station's false allure (fig. A.1). At which point the screen fades to black, and the track to silence, for a full ten seconds. As we dissolve back into the image of Joe watching over Pip at bedside after an elliptically treated bout of nursing, the film, alas, veers from the novel in giving us the moment in a more or less full resumption of omniscient narrative—so that Joe is seen (as if through the blurred vision of the story rather than its hero) slightly from the side (fig. A.2) rather than as a strict POV of hard-won recognition.

In the more cinematographically accomplished *Oliver Twist*, filmed by Lean in 1948, a similar subjective fade-out develops across several scenes as an actual motif of marginal consciousness. POV shooting thus links the inward life of certain focalizing characters and sets them into ultimate contrast to the spiritual evacuations of the scapegoated villain Sikes. On her birthing bed turned deathbed, Oliver's unwed mother drifts into consciousness for the last time to hold down a subjective shot that can't quite keep in focus the intercut flickering candles of her charity room as her own life is rapidly snuffed out. Oliver later goes unconscious in a POV shot after a blow in the street (fig. A.3), and yet again after being dizzied by confrontation with his false accusers in police court (fig. A.4), the camera toppling with him to the floor this time before cutting to the next scene. In both these latter cases, the lapse in consciousness is an elision of narrative time as well. This temporal effect is, of course, the most common use of the fade-out in the syntax of cinema. In its overt form as flashback transition, it is just the sort of thing Eisenstein hopes to be "excused" for finding in Dickens when unable to resist identifying a "dissolve" in *A Tale of Two Cities*. (To which he adds, in a speculative spirit that has certainly motivated my own discussion: "How many such 'cinematic' surprises must be hiding in Dickens's pages!" [213])

A.3–A.4

A.5

In a striking dissolve from Lean's *Oliver Twist*, two separate temporal trajectories are involved, along with a spatial arc. From a close-up of his mother's recovered locket in a lowlife setting of the detective counterplot, we quickly fade from Oliver's dead maternal image to its genetic double (the pretty look-alike son)—as the boy comes sweeping into the foreground on a garden swing in his new pampered setting at Brownlow's (fig. A.5). Her static image seems sprung to breathing life in an allegory of the generations: a time-image, one might say, giving thought to a Deleuzian difference within repetition. Juxtaposition becomes fusion: image plus spitting image synthesized in the biological transcendence of duration itself. It is as if the mother—whose own death scene was inscribed by a cinematic fading of focus—is consoled in absentia by the complementary magic of transitional montage.

But turn now to a last sequence from Lean's film: a fleeting and ingeniously Dickensian effect that not only avoids the subjective POV shot (as also in the disembodied transitional trope of the cameo/swing) but actively overrules it in the moment of death. I am thinking of the accidental suicide—become in the film the public execution—of the murderer Sikes, who slips into a noose wrought by his own hand. "Staggering as if struck by lightning," writes Dickens, "he lost his balance and tumbled over the parapet. The noose was at his neck. It ran up with his

A.6–A.7, A.8–A.9

weight, tight as a bow-string, and swift as the arrow it speeds."[8] After a deep-focus establishing shot of Oliver at the roof's peak, Sikes in the middle distance, and the ravening mob beneath (fig. A.6), we cut to a close-up of Sikes's panicked face. At which moment it is as if the metaphor of "lightning" has been overexplicitly spelled out as a musket shot, sending Sikes's body backwards in recoil out of the screen frame (fig. A.7). At once a disorienting shift of focus steals definition from the roof's tiles behind (fig. A.8). All that remains within the forefront of the frame is the lethal rope rapidly snapping tight across the angled expanse of tiles (in the bottom right of the frame [fig. A.9]). It does so not just as a function of the camera's disoriented depth of field in the abrupt removal of its human object, but also as a shuddering trope, all but subliminal, for the last spasms of the instantaneously removed villain.

    "Tight as a bow-string, and swift as the arrow it speeds": a kind of normalized forking

A.10

grammar taking the form of exaggerated parallelism. In prose, the elegantly ironic concisions of syllepsis would have seemed out of place for this melodramatic come-uppance. As with Dickens's aggressively compressed simile of cause and effect, in Lean's film Sikes's lifeline gives out in visual simile before our eyes. All the while we hear the rope whipping over the roof until the shot cuts—on the sudden jerk of the body's dead weight—to the chimney above (fig. A.10). This is where the movie has stationed Oliver—as if to ground its version of Dickens's final grue-some personification: "The old chimney quivered with the shock, but stood it bravely" (chap. 50). Holding firm against horror as well as mechanical tension, the inanimate formation of bricks is braver than the coward Sikes. The jump cut is itself the near cousin of a metaphoric (Eisensteinian) syllepsis: came the jolt and unflinching justice together.

In my attempt at a precinematic narratography of filmic prose, there is at least this much to say in sum. By such cognitive overlays and new resolutions, we come upon the two-made-one of juxtaposition as fusion, linkage as reconfigured sense, ironic disparity as triggered synthesis. Regardless of whether a given cinematic adaptation may be said to get it right, such lexical and syntactic oscillation remains the governing filmic energy of Dickensian technique: its intrinsic flicker effect and its resultant conceptual montage. This is the dictional, grammatical, and fig-ural dialectic of his hypercharged style. This is, in short, the way we screen his sentences in the synthesizing mind's eye. Such is one dimension of narratography's abiding place, then, more tensed and unsettling than other approaches to Dickensian writing, in a general narratology of his fiction.

ONTOGRAPHY

And a brief word more about subterranean lexical syntheses—including their links to the betweenness of cinematic mediation: in either the inherent slippage of strip/track/screen or,

recurring to the graphic pacing of our first narrative example, the running succession of photo/
slide/film. Having rewound the medial clock in our look back at Dickens, we can now round out
this discussion—and without leaving behind entirely the topic of parafilmic writing—with two
versions of a canonical scene of sacrifice in film history, two incarnations of Joan of Arc at the
stake: once via Deleuze and Rossellini; once, earlier in both cases, via Carl Dreyer and Stanley
Cavell. For each philosopher, Cavell as well as Deleuze, virtuality can grant us heuristic, even
therapeutic, even perhaps cathartic engagement with *a* world on the way back to *the* world.
Virtuality at this level (and this is my point, not theirs) should be neither annoyed nor spoiled
by the recognition of its strictly graphic manufacture. To use the metaphors of immortality
in a Bazinian thinking so close to Cavell's, the fact that time is mechanically sampled on the
strip doesn't at all prevent its change from seeming to be "mummified" by the filmic medium,
preserved for on-screen reanimation. Since in the literary medium wording doesn't ruin the
narrative it realizes, why should photograms, when recognized, derail the world space of film
that in fact they track? That the discrete flickering imprints rushed past on the strip are all that
keeps such a world moving before us: this fact should be no more disabling, no more likely
to dissipate the illusion, than the fact that Shakespeare's Othello is a name made up in part by
the hell within, or Desdemona by the overridden presence of a demon (points dear to Cavell's
reading, elsewhere, of that verbal and performance text).

For me, then, it is not even an irony—but merely a sliding evocative homage to the intermedial nature of temporal increments across lexical and photomechanical representation
alike—when Cavell's own prose waxes filmic. Where his subtitle to *The World Viewed* offers
Reflections on the Ontology of Film, his prose is able to offer reflections *of* it in a kind of distilled
ontography. It does so, in one grand case at least, by all but consciously signaling its own flicker
effect of syncopated progression. One paragraph away from the end of *The World Viewed*, Cavell
reminds us how the camera leaves behind a close-up of Joan at the stake, in Dreyer's silent
masterpiece, for the sighting of birds that "wheel over her with the sun in their wings."[9] *In*
their wings, not *on* them: a world of light itself registered not just optically but somatically.
Yet most of all, what is on view is film's general indexical record of a world surviving human
death—as well as the marked symbol of a personal resurrection. In the immolation of one life,
the immanence of a larger life of which the sainted heroine has until now been a part. And
more—which mostly goes unsaid, intuited only between the lines, between the words. In
death, there is always continuance. An immortality machine, film is the true medium of this
secularized perpetuity. Those birds go on holding the attention of prose as well as camera in
this tacit four-word ontology of all projected screen presence: "They, there, are free" (159). The

instantaneously eroded grammatical space between the nominative and the locative, between pronominal subject and its free and separate adverbial placement, arranges that one word should get phonetically detached from the other as the very microdrama of release in a monosyllabic theater of words. In the further swift gust of the verb across the cadenced swoop of "ey/ere/ are," we audit on the underside of writing a pervasive "air," the subliminal breath of airiness itself, all but spelled out as the medium of uplift.

Cinema for Cavell, with his title's play on *Weltanschauung*, is nothing less than the World Viewed, specified throughout in its medial function as "automated world projections." With Dreyer's flapping birds, flickering in transit almost as an oscillatory as well as ontological synec-doche for the intermittent generation of the world's continuance, the scene lifts into the sphere of the metafilmic. A flutter of organic propulsion is both taken down piecemeal ("automated") and given back whole (a projected "world"). Such ongoingness survives the self in extremis, the sacrificial victim, the scapegoat.[10] The associated filmic ritual is one of reconviction. If the world persists beyond the heroine, then her willingness to die for its good, and the higher truths that sanction it, has not been in vain. Skepticism is defeated by a faith to which—beyond all doctrine—film everywhere testifies, in Cavell's view, and in screwball comedies as much as in silent tragedies.

For Deleuze, this same cinematic effect is indeed called "belief." It participates in and con-tinues a crucial "turning point in philosophy, from Pascal to Nietzsche," which amounts to the rescue of epistemology from its own dead ends in skepticism.[11] In this line of thought, Deleuze might be rehearsing (though without ever mentioning) book after book by Cavell, right down to the time of his writing (*The Time-Image*, 1985; *Pursuits of Happiness*, 1984). "With Dreyer, then Rossellini, cinema takes the same turn," writes Deleuze: the "turn" from epistemologi-cal skepticism to "belief" (172). So that it is only "from the height of eternity" in *Joan of Arc at the Stake* that the heroine can, looking down on the "tatters" of this life, somehow once again "believe in this world" (172). For Cavell's related example from Dreyer rather than Rossellini, however, the point of view is as much ours as that of the transcendent dead: a view grounded and familiar even if upward. That rootedness is what makes it so quintessentially cinematic for Cavell: the world viewed from a point we can recognize (though not presently inhabit). But the other philosopher's claim, about the later Joan of Arc film, is scarcely far off from this emphasis. "Whether we are Christians or atheists, in our universal schizophrenia," italicizes Deleuze, "*we need reasons to believe in this world*" (172). Such belief for Deleuze—and here lies a Foucauldian stress that seems in some measure to distinguish his terms from Cavell's—marks film's way of reaching for "the body before discourses" (172). And not just the body outside the theater. My own stress across two books on film has been on that heightened sense of reception where one particular "before" of narrative "discourse"—its medial precondition on the strip—can be somatically registered even when not directly recognized as such.

It is in this same spirit (carnal, subvocal) that we hear in Cavell's "They, there, are free" a final launch from "ey/ere/are" into the chiastic switch—and reach—of "ree." At such a moment the ordinary-language philosopher shows the modernist allegiance of his own craft by writing what amounts to a photogrammatic prose. More to the point, he does so in service to a mediation that seems—and never more so than when recaptured in this capping flash of highly wrought vocabular inspiration—at once immanent, spontaneous, and free. Like the film image itself in this described shot: virtually transparent, and vice versa.

We began with a philosopher's definition of film: Lyotard's sense of cinema as a sequential writing with movement. We have come full circle to the filmic movement within writing. That Dickens and Cavell, a Victorian novelist and a modern philosopher, can each write filmically is a testament to the ingrained energies of lexical and phonetic succession itself—and, in response to it, of the body's attentive durations. Quite apart from the new desktop conventions of hypertext, what it might mean to write digitally—when the binary differentials are no longer those of phonetics rethought by phonology (Ferdinand de Saussure) but are mathematical functions imperceptibly distributed rather than serially elided—might, if determined, make for another and quite different appendix to a book on the postfilmic moment. Simply—and without prejudice—to rule out the very possibility of such a digital script would, of course, be another way to close this study as well. For we have, after all, been concerned with the timed effects of screen imaging in the shift from photomechanics to digitization. More with each passing year, the figurations to which narratography is alive—though still graphic, and still coded, to be sure—are no longer evocative of writing (Lyotard) in the least, but rather of mutation. Wherever morphing replaces morphology, a new paradigm is upon us.

INTRODUCTION

1. Garrett Stewart, *Between Film and Screen: Modernism's Photo Synthesis* (Chicago: University of Chicago Press, 1999).

2. Jay David Bolter and Richard Grusin, *Remediation: Understanding New Media* (Cambridge, MA: MIT Press, 1999).

3. This photo series by Sugimoto, called *Theaters*, is persuasively offered, along with a fine reprinted instance, as opening exemplum by David Green in his introductory essay, "Marking Time: Photography, Film and Temporalities of the Image," in *Stillness and Time: Photography and the Moving Image*, ed. David Green and Joanna Lowry (Brighton, UK: Photoforum/Photoworks, 2006), 9–21. In summarizing the way Sugimoto's photographs digest the film track whole, as if in one ocular swallow, Green draws deliberately on the metaphoric register of much film and photography theory when he writes: "Life is given to the photograph through the death of the film" (10).

4. A succinct polemic formulation, occurring to me too late for *Between Film and Screen*, was used in summary on the back cover: "Photography is the lost cause of cinema, gone in projection and too soon forgotten."

5. With reference especially to previous work by Lev Manovich in *The Language of New Media* (Cambridge, MA: MIT Press, 2002), this revolutionary aspect of the digital field, linking it to the "sequential scanning" of radar rather than to the enchainment of cinematography, is reviewed by Mark B. N. Hansen, *New Philosophy for New Media* (Cambridge, MA: MIT Press, 2004), 8–9.

6. This, for instance, is the way digital special effects are treated by Thomas Elsaesser and Warren Buckland in "Realism in the Photographic and Digital Image (*Jurassic Park* and *The Lost World*)," from their book *Studying Contemporary American Film: A Guide to Analysis* (New York: Oxford University Press, 2002), 195–219. In a similar vein, and overlapping by title with the same decade covered by my book, no publication could any more clearly mark off cinema scholarship as cultural studies from its counterpart as media studies than the former emphasis in *The End of Cinema as We Know It: American Film in the Nineties*, ed. Jon Lewis (New York: New York University Press. 2001), whose nearly three dozen contributors offer diagnoses—and mostly negative prognoses—more concerned with the Hollywood industry than with filmic or computer technology, more institutional than constitutional. With the exception of prominent digital effects in a recurrent touchstone like *The Matrix*, the relations among sociological, political, corporate, and thematic issues are

pursued in these essays independently of the shifting material basis of recording, generation, and projection, let alone the technical and stylistic inflections they facilitate in a given screen narrative.

7. In one of those coveted summary moments of colloquial clarification in Gilles Deleuze, *Cinema 2: The Time-Image*, trans. Hugh Tomlinson and Robert Galeta (Minneapolis: University of Minnesota Press, 1989), we find that in postwar modernism "the viewer's problem becomes 'What is there to see in the image' (and not now 'What are we going to see in the next image')" (272).

8. Sean Cubitt, *The Cinema Effect* (Cambridge, MA: MIT Press, 2004).

9. Gunning's wide-ranging consideration of this topic began with "The Cinema of Attractions: Early Film, Its Spectator and the Avant-Garde," in *Early Cinema: Space, Frame, Narrative*, ed. Thomas Elsaesser with Adam Barker (London: BFI, 1989), 56–62. A more voluminous and multiauthored literature on suture began with Colin MacCabe's translation of Jacques-Alain Miller, "Suture (Elements of the Logic of the Signifier)," *Screen* 18 (Winter 1977–78): 24–34.

10. Discussed in my *Between Film and Screen*, 260, via Tom Gunning's emphasis, concerning the earliest cinematographic screenings, on the single water-cooled frame held in abeyance before the disclosed trick of motion was unleashed.

11. Mary Ann Doane, *The Emergence of Cinema Time: Modernity, Contingency, the Archive* (Cambridge, MA: Harvard University Press, 2002).

12. Mentioned in passing by Doane (ibid., 133) and discussed at length in my *Between Film and Screen*, 111–15, where comparison is made between the actual phrase in Benjamin's 1931 "A Short History of Photography," in *Classic Essays on Photography*, ed. Alan Trachtenberg (New Haven, CT: Leete's Island Books, 1980), 203, and a return to the possibilities of "unconscious optics" in his more famous essay of 1936, "The Work of Art in the Age of Mechanical Reproduction," in *Illuminations*, trans. Harry Zohn, ed. Hannah Arendt (New York: Schocken, 1969), 237.

13. On this see Stewart, *Between Film and Screen*, 87, with further implications detailed in the appendix here.

14. This conference paper appeared as "Crediting the Liminal: Text, Paratext, Metatext," in *Limina/ Film's Thresholds*, ed. Veronica Innocenti and Valentina Re (Gorizia, Italy: Forum, 2003), 51–68.

CHAPTER ONE

1. Preoccupied in considerable part with cinema's electronic aftermath, the coming chapters, while constituting a sequel in themselves, are also a companion volume—to an as yet unpublished book on literary methodology. After having practiced, and tacitly advocated, a fairly consistent mode of intensive analysis for several years, it seemed time to say so more directly: by giving it, and so finding for it, a name. Literary narratography, then, is what I have been elsewhere illustrating with extended examples from Victorian fiction. Its first published installment is to be "Dickens and the Narratography of Closure," forthcoming in *Critical Inquiry*. The scale of its approach to the microarticulation of prose narrative is comparable to that taken in these chapters toward both—and the difference between—a filmic and an electronic cinema.

2. The case, it should be noted, can easily be reversed. Having minted the term (I thought) for my own microstylistics of narrative, I discovered the coinage already in use in a quite difference sense. Building on his earlier work in *Music as Cultural Practice, 1800–1900* (Berkeley and Los Angeles: University of California Press, 1990), 186–89, Lawrence Kramer returns to the concept in *Classical Music and Postmodern Knowledge* (Berkeley and Los Angeles: University of California Press, 1995), where he distinguishes narrativity from the actual writing practices that generate it (including in the clefs, staves, tempi, and notational codes of musical composition), which he calls "narratography." Though I have not found the name attached in either sense to literary analysis, it is important to stress my terminological difference from Kramer's musicology (musicography?). His nominalized sense of the concept is obviously closer to a self-sufficient substantive like *photography*

or *holography* than to a bidirectional term like *geography*. My own deployment of the term means to bring out instead not the *instance of* so much as the *study of*: the mode of approach more than the local provocation. Concerning verbal or visual narrative, what is methodologically abstracted by the term *narratography*, for our use here, is not therefore the inscription per se but the principles—and practice—of decryption.

3. The classic texts in the effort to theorize film as a linguistic system are Christian Metz, *Film Language: A Semiotics of the Cinema*, trans. Michael Taylor (New York: Oxford University Press, 1974) and Metz, *Language and Cinema*, trans. Donna Jean Umiker-Sebeok (The Hague: Mouton, 1974). More recently, I have suggested in *Between Film and Screen* (see the introduction, n. 1) that there is a deep structural affinity between photogram disappearing into action on the track (cinema's filmic "photo synthesis") and phoneme subsumed (though often not entirely) to word formation in literary language. But acquiescence in this proposal isn't at all necessary in order to credit the ocular force of the frame increment as part of a *textual indication* (as well as a racing discontinuous index) of space and motion.

4. David Bordwell, *Narration in the Fiction Film* (Madison: University of Wisconsin Press, 1985), 53; his emphasis.

5. Tzvetan Todorov, *The Poetics of Prose*, trans. Richard Howard (Ithaca, NY: Cornell University Press, 1977).

6. Tzvetan Todorov, *The Fantastic: A Structural Approach to a Literary Genre*, trans. Richard Howard (Ithaca, NY: Cornell University Press, 1975).

7. Michael Riffaterre, *Fictional Truth* (Baltimore: Johns Hopkins University Press, 1990).

8. William Henry Fox Talbot titled his 1844–46 portfolio of London-published photographs *The Pencil of Nature*. Appropriately appearing in the journal titled *Ecran Français* just over a century later, in 1948, Alexandre Astruc's "The Birth of a New Avant-Garde: La Camera Stylo" (rpt. in *The New Wave*, ed. Peter Graham [New York: Doubleday, 1969]), lent its seminal term for cinematic inscription to active discussion in French theory for a couple decades to come. In contrast with Fox Talbot's mystification of nature's book as transcribed by light, Astruc's polemical interest was in the elevation of film imaging to a discursive system equivalent to auteurist production in literary writing.

9. Originally published in Paris by de Seuil in 1983, 1984, and 1985, respectively, under the title *Temps et Récit*, the books appeared in English as *Time and Narrative*, vols. 1–3 (Chicago: University of Chicago Press, 1984, 1985, 1988), the first two trans. by Kathleen McLaughlin and David Pellauer, the third by Pellauer and Kathleen Blamey.

10. Jean-François Lyotard, "L'acinéma," in *Des Dispositifs Pulsionnels* (1973; Paris: Galilée, 1994), 57–69; trans. as "Acinema" in *The Lyotard Reader*, ed. Andrew Benjamin (Oxford: Blackwell, 1989), 169–80. Though Lyotard's thinking helps launch film scholar David Rodowick's study of "the figural" as a breakdown in the aesthetic distinction between the linguistic arts of time and the plastic arts of space (as this rescinded distinction paves the way for a philosophy of new computerized media), even his book doesn't call up this early essay on the "writing" of the image. See "Lyotard's Leap in the Void: Aesthetics before the New Media," in *D. N. Rodowick: Reading the Figural; or, Philosophy after the New Media* (Durham, NC: Duke University Press, 2001), 4–44.

11. The verbal distinction in French shifts the differential somewhat (without exactly changing the sense tried for by the English translators): "Le cinématographe est l'inscription du mouvement. On y écrit en mouvements" (57). An inscription of, a writing in. Only the English translation stresses the writing *with*. Yet the main point is retained.

12. Jakobson's definition of the "poetic function" as that textual mechanism which "projects the principle of equivalence from the axis of selection into the axis of combination" appears in "Linguistics and Poetics," in *Essays on the Language of Literature*, ed. Seymour Chatman and Samuel R. Levin (Boston: Houghton Mifflin, 1967), 303.

13. As discussed below, see Peter Brooks, *Reading for the Plot: Design and Intention in Narrative* (New York: Knopf, 1984).

14. All quotations from *The Golden Bowl* are from the Penguin Classics edition (Harmondsworth, UK, 1985), 580.

15. This scene is discussed and illustrated, along with several others for which there are no frame captures here, in Stewart, "Citizen Adam: the Latest James Ivory and the Last Henry James," *The Henry James Review* 23, no. 1 (Winter 2002): 1–24.

16. On the matter of photographed artworks and the new scholarship they permitted, see André Malraux, *The Voices of Silence*, trans. Stuart Gilbert (Garden City, NJ: Doubleday, 1953), 13–130.

17. See "Film's Victorian Retrofit," the sixth chapter of Stewart, *Between Film and Screen*, 225–63, for a full discussion of this tendency. The more recent Jane Austen adaptations find, of course, no historical context for filmic self-reference, even in precinematic optical toys and other modes of Victorian visual spectacle, and for the most part avoid such interpolations. But history is no constraint on metacinematic inside jokes for the filming of the "unfilmable" *Tristram Shandy*, with its subtitle *A Cock and Bull Story* (2006), a movie about the cobbling together of its own low-budget production. Directed by Michael Winterbottom, veteran of the somberly reflexive *Jude* (see Stewart, *Between Film and Screen*, 251–52), the movie version of the Laurence Sterne novel is justified by the novel itself in bravura anachronisms of every sort, including a digital freeze at one point followed by an electronic composite (instead of filmic superimposition) of the narrator-hero in "face-over" (rather than voice-over) upon the scene we're watching. Among such frame-breaking devices, all fully licensed by the spirit of Sterne's novel, is the midstream filming of an interview with the star for the eventual DVD, where he explains that the novel was "postmodern before there was anything to be post about." True to form in this regard, the film goes so far as to give us a black screen during the script conference devoted to whether an audience would be interested in such an update of Sterne's totally blacked-out mortuary page. Yet elsewhere—in the footage we see of the historical comedy as it is actually shot, rather than drafted and debated—reflexivity stops short of the kind of celluloid "arrhythmies" one finds in the frameline disjunctures of *Jude*, whose effect was to bracket cinema in a kind of Victorian electromechanical prognosis. Rather, whenever the *Tristram* narrator admits to "getting ahead of himself," the whole framed scene, rather than having its movements vertically rewound in a spooled blur, slips visibly to the right as if in a tray of prephotographic slide transparencies, moving us back to the preceding black-framed (but still magically live-action) tableau.

18. For Tom Gunning on the "cinema of attractions," see the introduction, n. 9.

19. The debatable relations among photography, film, temporality, and embalming, famously explored by André Bazin, are discussed in Stewart, *Between Film and Screen*, 341n11.

20. Defined along with the *opsign* in the glossary to *Cinema 1: The Movement-Image*, the *sonsign* is constituted by audiovisual material that "breaks the sensory-motor links, overwhelms relations and no longer lets itself be expressed in terms of movement, but opens directly on to time."

21. Veering from anything in their sources, this tendency is clearly manifested down through adaptations in the last decade of such literary staples as *Bram Stoker's "Dracula," The Age of Innocence, The Portrait of a Lady, Jude the Obscure, The Secret Agent*, and the children's classic *The Secret Garden*. Beyond literary adaptations, the range of films I have in mind (as discussed in *Between Film and Screen*) includes everything from B gothics like *The Asphyx* (1973), action romances like *Time after Time* (1979), and psychological melodramas like *Somewhere in Time* (1980), to such supernaturalist fantasies as *A Faery's Tale* (1999).

22. Edith Wharton, *The House of Mirth* (New York: Bantam, 1984), chap. 14, p. 337.

23. I am quoting from the director's commentary on the DVD release (Sony Pictures Classics, 2000).

24. Paolo Cherchi Usai, *The Death of Cinema: History, Cultural Memory and the Digital Dark Age* (London: BFI, 2001).

CHAPTER TWO

1. Manovich, *The Language of New Media*, 307 (see the introduction, n. 5).

2. As discussed in the fifth chapter, "The Photographic Regress of Science Fiction Film," of my *Between Film and Screen* (see the introduction, n. 1).

3. In the chapter "On Soviet Magic Realism" in Fredric Jameson's *The Geopolitical Aesthetic: Cinema and Space in the World System* (Bloomington: Indiana University Press; London: BFI Publishing, 1992), 87–111—concentrating as it does on the 1970s work of Sokurov and Tarkovsky, and stressing the "instructively different" (87) relation of Soviet sci-fi to its Western counterparts—we find that the fantastic is much more closely entwined with the "instrumental marvelous" (Todorov's terms for gadget-heavy plots) in this mode of filmmaking than in the separate spheres of their narrative experiment since then in European versus American production. Jameson's earlier essay from 1986, "On Magic Realism in Film," in *Signatures of the Visible* (New York: Routledge, 1990), operating in a rather different vein, singles out Polish, Venezuelan, and Colombian historical films whose use of foregrounded color in association with both violence and sexuality serves to rupture the norms of realist filmmaking even as it sets their preternatural storytelling forms apart from the nostalgic "sheen" of postmodern retro cinema. In contrast both with sci-fi allegory and with such a magicalized historicism, then, my interest lies more with the way history can be said to have caught up with magic in the contemporaneous plots of the European fantastic, where the uncanny infiltrates the everyday psychodrama of contemporary social, erotic, and cross-national encounters.

4. Benedict Anderson, *Imagined Communities: Reflections on the Origin and Spread of Nationalism* (London: Verso, 1983).

5. Franco Moretti, *Modern Epic: The World System from Goethe to Garcia Marquez*, trans. Quintin Hoare (London: Verso, 1995), 6. That this is a position from which Moretti, for his recent work in the mapping of publication and distribution patterns for fiction and film, has mostly retreated doesn't detract from the fine balance his earlier writing managed to strike between hermeneutics and social contextualizing.

6. Todorov, *The Fantastic* (see chap. 1, n. 6).

7. To say this is to render more specific a tantalizing intuition, almost a digression, in the film work of Fredric Jameson. Indeed, where his speculations on the magic realist element in historical melodrama come into close conjunction with a structural hypothesis of the present study is in his call for further theorizing, in effect, on the liminal moment of narrative inauguration: "I feel that we must sharpen our consciousness of the shock of entry into narrative" (*Signatures of the Visible*, 131). In a remarkable and extended passage on cinematic beginnings, a passage unusually impressionistic for Jameson, he expatiates on "the body's tentative immersion in an unfamiliar element, with all the subliminal anxieties of such submersion" (131). He then sums up his call for "a historical phenomenology of such entry-points, an inventory of curtains that part in bravura, or of the various flaps and apertures through which we are asked to introduce our heads" (131). What is estranging about such inaugural and ushering images, for my examples as well as for his, including such disorienting devices as "flash-forwards" (131), is only exaggerated in the ambiguous suspensions of what I am calling "trick beginnings" (on their way to frequent twist finishes).

8. Todorov, *The Poetics of Prose* (see chap. 1, n. 4).

9. "First, the text must oblige the reader to . . . hesitate between a natural and a supernatural explanation of the events described" along a spectrum indicated as follows: "uncanny/fantastic-uncanny/fantastic-marvelous/marvelous" (Todorov, *The Fantastic*, 44). In sum: "The fantastic in its pure state is represented here by the medial line . . . a frontier between two adjacent realms" (44).

10. Christian Metz, "*Trucage* and the Film," *Critical Inquiry* 3, no. 4 (Summer 1977), where Metz puts the historical mutation this way: "What is experienced as a simple figure of speech today was quite frequently, for the first spectators of the cinematograph, a magic 'trick,' a small miracle both futile and astonishing" (665).

This leads to Metz's sense that "montage itself, at the base of all cinema, is already a perpetual *trucage*, without being reduced to the *false* in usual cases" (672).

11. Henri Bergson, *Matter and Memory*, trans. N. M. Paul and W. S. Palmer (New York: Zone, 1991), where Bergson adduces "what happens in cases of sudden suffocation, in men drowned or hanged. Such a man, when brought to life again, states that he saw, in a very short time, all the forgotten events of his life passing before him with great rapidity" (155). For the larger view of this compacted, detemporalized moment—as its inferences cut across much of Bergson's work—see the appendix to Georges Poulet, *L'Espace Proustien* (Paris: Gallimard, 1963): "Bergson—Juxtaposition and the Theme of the Panoramic Vision of the Dying" (137–77). On Bergson's consequent sense of cinema as more like the deathly reconstruction than the lifelikeness of human movement, there is this, in sum, from Gilles Deleuze, *Cinema 1: The Movement-Image*, trans H. Tomlinson and B. Habberjam (Minneapolis: University of Minnesota Press, 1986): In contrast to phenomenology, "Bergson condemns the cinema as an ambiguous ally in a completely different way. For if the cinema misconceives movement, it does so in the same way as natural perception and for the same reasons: 'We take snapshots, as it were, of passing reality. . . . Perception, intellection, language so proceed in general'" (57). Deleuze sets out famously to disagree, believing that movement is immanently given rather than artificially constructed in cinema.

12. Deleuze, *Cinema 2*, 68 (see the introduction, n. 7).

13. In what follows, I am consolidating some of the guiding terms from Riffaterre's last three books: *The Semiotics of Poetry* (Bloomington: Indiana University Press, 1978); *Text Production*, trans. T. Lyons (New York: Columbia University Press, 1983); and most recently, *Fictional Truth* (Baltimore: Johns Hopkins University Press, 1990)—where "matrix" and "model" recede into the background for an emphasis on the "intertext" and its recursive "subtext" in longer narrative works.

14. I refer here to the loaded last sentence of *Fictional Truth*: "Like the unconscious described by Freud, the unconscious of fiction and, therefore, its truth, stands outside the realm of time and is impervious to its ravages" (111).

15. At least such was the claim made by Philippe Dubois in response to my screening of a clip at the Udine conference—and in connection with a paper of his own that was to appear in the published volume, "Au seuil du visible: La question du figural," in Innocenti and Re, eds., *Limina/Film's Thresholds* (see the introduction, n. 14), 137–50. In the saturation of fabric by flow that opens Nolan's *Insomnia*, Dubois believed that we entered the same territory of figuration where, in his example from Sjöström's *Le Vent* (1927), the spreading of water through sand is a "subnarrative" or "prenarrative" example of the pure image, before it becomes an image *of*. I am (and was) happy to agree, indeed eager to do so, since the liminal image at a film's opening often seems to precede the explanatory grip of the narrative mechanism altogether. My only proviso is that once this figuration gets harnessed by plot, as it so frequently and so quickly does, then it may well become available as a "model," however abstract, of the ensuing diegesis and its subtext. Until then, its hovering figuration points only elsewhere, perhaps toward an unrealized matrix, rather than forward toward story.

16. André Bazin, "The Ontology of the Photographic Image," in *What Is Cinema?* trans. Hugh Gray (Berkeley and Los Angeles: University of California Press, 1967), 15.

17. Arthur Symonds, "The World as Ballet," in *Studies in the Seven Arts*, vol. 9 of *The Collected Works of Arthur Symonds* (London: Martin Secker, 1924), 246.

18. For Bergson's critique of this view of successional temporality in the cinematic model, and Deleuze's response to it, see *Between Film and Screen*, 85–87.

19. Slavoj Žižek, *The Fright of Real Tears: Krzysztof Kieślowski between Theory and Post-Theory* (London: BFI, 2001), where Žižek's leading question, in the chapter "Back to the Suture," is as follows: "What, then, happens when the exchange of subjective and objective shots fails to produce the suturing effect? Here enters

the function of *interface*" (39). In Kieślowski, and uniquely so, "these magic moments of interface are not staged by means of standard Gothic elements (apparitions in the fog, magic mirrors), but as part of an ordinary, everyday reality" (39). This specialized concept of blocked suture may seem to have borrowed the terminology of "interface" from the work of Francesco Casetti, *Inside the Gaze: The Fiction Film and Its Spectator*, trans. Nell Andrew with Charles O'Brien (Bloomington: Indiana University Press, 1998), where the term locates a "counterfield" in the very "conditions of visibility" through which, by the reverses of enunciation, "once seized, the *you* can become *I*" (131). As signalized by the shot/countershot pattern, and more broadly applied by Casetti to the whole ambit of spectation and interpretation at the switch point of screen and viewer, the term, in Žižek's narrower deployment, marks an interplay that has dropped back, improbably, into the diegesis itself: as a variable (and often preternatural) reciprocity between subject and image in the same plane of view.

20. Or to put it again in Dubois' terms (n. 15): from a noisy on-screen blackness, the camera pulls back to disclose the outline of a revolving auto tire and some roadway sweeping past. Pure figuration, maybe—but only as long as it can stave off the narrative vector of fatality.

21. Žižek, *The Fright of Real Tears*, 39.

22. Todorov, *The Fantastic*, 77. When such a literalizing of psychic dissociation turns on the mechanism of photographic alienation in *The Double Life of Véronique*, it recalls Friedrich Kittler's medial corrective to Todorov's claim that psychoanalysis is the dead end of the nineteenth-century fantastic. In the essay "Romanticism—Psychoanalysis—Film" from his *Literature, Media, Information Systems*, ed. John Johnston (Amsterdam: Overseas Publishers Association, 1997), Kittler insists that the genre of the fantastic "is only imploded by the two-pronged attack of science and industry, of psychoanalysis and film" (95). This is most obvious in the fact that the motif of the double, so pervasive in fantastic literature, is no longer a strictly supernatural or magical phenomenon "ever since media were also able to substitute for central nervous systems" (97). In this light, Kittler's main screen evidence is the rethematizing of the visual double in German silent cinema and his chief "scientist," the psychologist (turned film theorist) Hugo Münsterberg. The subsequent place of the "interface effect" in such a development, we might add, is often to return film to its constitutive division—for once within a single frame—between breathing subject and its two-dimensional simulacrum.

23. Metz, "*Trucage* and the Film," 672.

24. Todorov, *The Poetics of Prose*, 42.

25. See David Green and Joanna Lowry, "From Presence to the Performative: Rethinking Photographic Indexicality," in *Where Is the Photograph?*, ed. David Green (Kent, UK: Photoworks/Photoforum), 47–60, an essay that importantly returns us to an aspect of indexicality in Peircean semiotics that has been obscured, as the authors see it, by Roland Barthes's elegiac emphasis on the "that-has-been" of the objectified past body. Peirce wanted from the indexical function a sense not just of a trace left but of its point and manner of inscription, what Green and Lowry stress as the performative index, simultaneous of course with its referential counterpart. Together, these functions delimit photography's motor as well as mortal trace, or what we might distinguish as the inscribed presence *to* as well as *of* the captured object, both moment of record as well as recorded moment.

26. See Tom Gunning, "Tracing the Individual Body: Photography, Detectives, and Early Cinema," in *Cinema and the Invention of Modern Life*, ed. Leo Charney and Vanessa R. Schwartz (Berkeley and Los Angeles: University of California Press, 1995), 37.

27. For Poulet's view of Proust contra Bergson, as the spatialization of time, see the *L'Espace Proustien*.

28. See n. 15 above.

29. I refer again to Philippe Dubois' notion (following Lyotard's distinction of *figure* from *discourse*) of a visual presence slipping outside representation (see above, n. 15), an "image event" apart from narrative action, a plasticity in the seen that is not yet (or no longer) rigidified by textual logic. One form this takes is what Dubois has called elsewhere, in connection with abstract superimpositions in Chris Marker's *La Jetée*,

the "cinematogram": a minimal cinematic (rather than photogrammatic) unit of cinema, or in other words the transitional fluctuation that underwrites the discourse of plot with the open energy of figure. See Philippe Dubois, "*La Jetée* de Chris Marker ou le cinematogramme de la conscience," *Théorème* 6, *Recherches Sur Chris Marker* (Paris: Presses Sorbonne Nouvelle, 2002), 9–45. I am supposing that many of the images reprinted and discussed in my *Between Film and Screen*—images meant to illustrate the surfacing of the filmic against the grain of the cinematic—might also be understood in productive relation to this category of Dubois'. The photogram in its minimal and prenarrative process of frame advance, once synthesized as a "cinematogram," may then be commandeered as what I am calling "narratogram," the smallest unit susceptible to narratographic analysis.

30. See Manfred Jahn, "Speak, Friend, and Enter: Garden Paths, Artificial Intelligence, and Cognitive Narratology," in *Narratologies: New Perspectives on Narrative Analysis*, ed. David Herman, 167–94 (Athens, OH: Ohio State University Press, 1999), a category of which the death-moment false lead of flight and conjugal reunion in *An Occurrence at Owl Creek Bridge* would be a classic film, as well as fictional, example. Such illusory narrative lines are to be distinguished from, as it were, the garden variety of forked-path trajectories, less deceptive and self-corrective, discussed by David Bordwell in "Film Futures," *SubStance* 97, vol. 31, no. 1 (2002), 88–104, where he lifts the term for such coaxial narratives from Borges' story "The Garden of Forking Paths." Among his examples, apropos of this chapter, is *Run Lola Run*.

CHAPTER THREE

1. Todorov, *The Fantastic*, 49 (see chap. 1, n. 6).

2. Marshall McLuhan, "The Medium Is the Message," in *Understanding Media: The Extensions of Man* (New York: Mentor, 1964), with the second paragraph's bald pun: "The instance of the electric light may prove illuminating in this connection."

3. In casting this distinction between *filmic* and *cinematic*, I recur to *Between Film and Screen* (see the introduction, n. 1), where the filmic dimension is understood to tap the programmatic run of single cellular imprints on the photochemical strip, while the cinematic effect is concerned with the edited world space of the story.

4. See Jameson, "Totality as Conspiracy," in *The Geopolitical Aesthetic*, 9–86 (see chap. 2, n. 3).

5. Laura Mulvey, *Death 24x a Second: Stillness and the Moving Image* (London: Reaktion, 2006). See especially her last three chapters: "Delaying Cinema," "The Possessive Spectator," and "The Pensive Spectator."

6. Manovich, *The Language of New Media*, 29 (see the introduction, n. 5).

7. When "the history of the moving image thus makes a full circle," the following historical dialectic grows clear: "*Born from animation, cinema pushed animation to its periphery, only in the end to become one particular case of animation*" (ibid., 302; Manovich's italics). Despite Manovich's sense that the photographic constituent is thus an unfelt element in screen projection, one sees that the original purveyors of the cinematographic apparatus had it right in boasting of the new medium as "animated pictures." The material base in the photograph was not thereby forgotten. Its automation was simply recognized within a broader category of illusionism.

8. A hypertrophic split-screen experiment like Mike Figgis's *Timecode* (2000), where the narrative field is drawn and quartered into four separate image planes running continuous footage of separate but simultaneous event streams, might be seen as the rare case of time spatialized: a clear if strained instance of Manovich's "spatial montage" (as a deviation from the claimed exclusive temporality of ordinary cinematic montage). Yet this is so only in the sense that Figgis's exacerbated format exposes—by imposing on the viewer's own selectivity (its tagline "What do you want to watch?")—the kind of contrapuntal attention usually channeled for the spectator by the "libidinal economy" of narrative intercutting.

9. Thomas Elsaesser, "Digital Cinema: Delivery, Event, Time," in *Cinema Futures: Cain, Abel or Cable?—The Screen Arts in the Digital Age*, ed. Elsaesser and Kay Hoffman, 217 (Amsterdam: Amsterdam University Press, 1998).

10. It was only just before this manuscript went to press that my interim assistant, Glenn Brewer, curious to give this downbeat film a second try after proofreading my chapter, hastened to alert me to this early thematic tethering point—and precisely in connection with the triple pun on the hero's name: Sy (rather than Seymour) Parrish, as he is always called, who tries to make of the circuit of images a kind of secular parish of the domestic faithful, complete with mediated communion rites in the effort always to "see more."

CHAPTER FOUR

1. Doane, *The Emergence of Cinematic Time* (see the introduction, n. 11). The uncited quotations three paragraphs back from Doane's summary remarks on a constructed temporality meant to "protect the subject" from the medium's own "fear of . . . total representation" come from the opening chapter, "Temporality, Storage, Legibility," 48. Her book is perhaps closest of all to the thinking of the present one, as well as to my previous study of film, when she compares what I stress as the on/off pulsional transit between film and screen to Freud's description of the "flickering up and fading away" of consciousness in its tolerance for stored impressions: "The description of this process is strikingly similar to that of intermittent motion in the cinema (Freud refers to the 'periodic non-excitability of the perceptual system')" (43).

2. For Bazin's metaphors, see again chap 1, n. 19.

3. For Doane's initial use of Kittler, see *The Emergence of Cinematic Time*, 45.

4. In chapter 6, I relate this new fillip of digital compositing to the history of the traditional zoom effect, cinematographic rather than electronic, in a recent essay by Paul Willemen, where it is shown to be an invasion of diegesis by rhetoric.

5. Elsaesser, "Digital Cinema: Delivery, Event, Time," in *Cinema Futures*, ed. Elsaesser and Hoffman, 217 (see chap. 3, n. 9), where (with echoes of Manovich on the overthrow of identification) "the identificatory potential of film-narrative time (made up of suspense, surprise, the uneven distribution of knowledge and how to maintain it from one plot situation to the next) is replaced by the temporality of the count-down" (219). By way of critique rather than description, T. J. Clark's *In The Sight of Death: An Experiment in Art Writing* (New Haven, CT: Yale University Press, 2006) laments in passing an entirely hackneyed "discourse" of electronic images bombarding attention in "doses more and more self-administered by interactive subjects, each convinced that the screen" offers a "realm of freedom" (285). It is by closeted analogy with such "freedom" that many of the fantastic plots discussed here, with their frequently escapist endings, seem to have been fabricated, participating thereby in the utopic passivity Clark scorns as a "democracy of the virtual" (185).

6. This pivotal photograph from *Blade Runner* is reprinted and discussed in *Between Film and Screen*, 10 (see the introduction, n. 1), where I note the way it comes forward to fill the screen as a tricked mechanical trace of a biomechanical being—a trace briefly and fantastically reanimated as film image, complete with passing shadows, but no more real for all that.

7. A similar overview of more openly narrated rather than encoded media rivalries—and fantasies—informs Paul Young's *The Cinema Dreams Its Rivals: Media Fantasy Films from Radio to the Internet* (Minneapolis: University of Minnesota Press, 2005), where a discussion of television—as responded to by the cinematic imaginary—is followed by a canvassed reaction to the early stage of global electronics and the Internet by Hollywood films from 1982's *Tron* through 1992's *Lawnmower Man*. Young sees the first phase of Hollywood's cyberphobic narratives coming to a close in the mid-1990s. Where emergent digital technology had once been staged to pose a global threat (as in *WarGames* [1983]), its themes give way to a seemingly more tolerant and porous response since approximately 1995 (a year marked for both our arguments, if in different ways, by *Johnny Mnemonic*). If you can't beat 'em, join 'em. The result is that the anxieties accompanying a "mass fantasy about an intersubjective digital future" have gone "almost entirely underground" (226), masked by plots seemingly otherwise concerned than with what Young lifts (with a twist) from André Bazin to call

"the myth of total media" (200; rather than Bazin's "total film"). Tracking the sequential touchstones of telegraphy, radio, and TV, Young's dialectical history—amounting to a media archaeology of contested audiovisual modalities—is therefore mostly over just where this book begins. My hypothesis about a double strand of American cinema over the last decade—including fantasies, less or more gothic, of lives both drastically revisable and consciously outlived, plots of temportation and posthumous agency alike—amounts to claiming that what Young calls "underground" can be specified further as the subtextual displacement at work in the operation of a digital intertext, where the aberrant action of a schizophrenic hero, for instance, reads as an encoding of the normal amusement—and artificial empowerment—of a video gamer.

8. It is as if this film is well aware of the cultural intertext it is so narrowly skirting. At one point *Frequency* goes out of its way to raise, even while diverting central attention from, the specter of the high-tech revolution. It does so by separating the economical from the technological aspects of that electronic boom. A pal of the hero's, kicking himself over not buying into the dot-com industry years back, seems to have had a second chance afforded through the time-loop interventions of his friend. With nothing said explicitly about this alternate future, when we see the taillight of his oversized Mercedes smashed by a stray sandlot baseball in the last utopian scene, we also notice its YAHOO license plate. The new technology becomes the commercial reward system of its own audiovisual repression in the rest of plot.

Repression is repeatedly the key to a digital intertext. In a parallel historical trajectory to that of Paul Young's account (n. 7 above), though without Young's steady emphasis on cinema's reflexively medial response to such alternative technologies of apparition and transmission, Jeffrey Sconce, in *Haunted Media: Electronic Presence from Telegraphy to Television* (Durham, NC: Duke University Press, 2000), stresses a "paranormal continuity" (202) in the discourse surrounding electronic media, from telegraphy through wireless communication to the spread of radio—and then on from the advent of television to the dawn of cyberspace and virtual reality—all accompanied by figures as well as fictions of the uncanny or the supernatural (though Todorov's dialectical genre of "the fantastic" is not adduced). Neither Young nor Sconce mentions the film *Frequency*, but its relevance to their argument is clear. Stress is placed by Sconce on the recurrently ghostly or hypnotic aura associated with instantaneous and disembodied mediation, wireless short-wave radio included. In my most recent film example of Hollywood fantasy, *The Lake House* (discussed in chapter 6), we see this legacy of spectralized media updated by being upended, so that—via the frequent digital intertext of latter-day Hollywood gothic—the narrative premise of supernatural communication is "haunted" in reverse by the threat of its deflating literalization as cell-phone messaging.

9. This is the terminology introduced by Marie-Laure Ryan, "Cyberage Narratology: Computers, Metaphor and Narrative," in *Narratologies: New Perspectives on Narrative Analysis*, ed. David Herman (Athens, OH: Ohio State University Press, 1999), 113–41.

10. Marie-Laure Ryan, *Narrative as Virtual Reality: Immersion and Interactivity in Literature and Electronic Media* (Baltimore: Johns Hopkins University Press, 2001).

11. See N. Katherine Hales and Nicholas Gessler, "The Slipstream of Mixed Realities: Unstable Ontologies and Semiotic Markers in *The Thirteenth Floor, Dark City*, and *Mulholland Drive*," *PMLA* (May 2004): 482–99.

12. I am referring here to the director's commentary on the film's DVD release (Focus Features, 2004).

13. Many of these career-defining topics are anticipated by Deleuze in a 1968 book only posthumously translated into English by Paul Patton, under the title *Difference and Repetition* (New York: Columbia University Press, 1994).

14. Elsaesser and Buckland, *Studying Contemporary American Film* (see the introduction, n. 6), devote a chapter to the generational ironies of desire in *Back to the Future*.

15. Brooks's model has been previously discussed. Bordwell's appears in "Film Futures," *SubStance* 31, no. 1 (2002): 88–104.

16. Branigan's response appears in the same issue of *SubStance* (see n. 15 above) under the title "Nearly True: Forking Plots, Forking Interpretations," 105–14, where he introduces, as an example of the radically "disnarrated" (an extreme form here of the hermeneutics of suspicion), his thesis—developed more fully in a conference paper (see chap. 5, n. 5)—that the child in *The Sixth Sense*, so alert to the ghostly appeals of other molested or poisoned children, has been abused by his own mother.

17. Ibid., 110, citing Gerald Prince, "The Disnarrated," *Style* 22, no. 1 (Spring 1988): 1–8.

18. Notable in Branigan's response to this paper by Bordwell is his effort to extend narrative shape to hermeneutic form, the forking plot to the forking interpretation. His gesture thus resembles what we have seen Todorov doing with the anomalies of the fantastic as benchmark of literary ambiguity in general. Branigan suggests that the peculiarities of a bivalve plot might be erected into a paradigm of cognitive ambivalence and analytic decision making *tout court*.

19. See Manfred Jahn's consideration of the "garden-path" miscues in chap. 2, n. 30.

CHAPTER FIVE

1. I collect now, for easy reference, the borrowings from Deleuze made in advance of explicitly citing him. Following Antonioni on "the division of western cinema into European humanism and American science fiction," see Deleuze, *Cinema 2: The Time-Image*, 17 (see the introduction, n. 7); on the new media images, "video" and "numerical," 265; on the modernist character as "viewer," 3; on the indiscernible borders between imaginary and real, 7; on time as malady, where there is "no other sickness but the chronic," 24; on the pure time-image come back as "phantom" to haunt postwar cinema after the decline of action and movement, with the result that virtual and real are no longer opposed, 24; on the Bergsonian containment of the subject by time rather than vice versa, 82; and on film's power to restore our belief in the world through its virtual images, 172. Subsequent references to this volume are given in the text.

2. In Bernard Stiegler, "Our Ailing Cultural Institutions," one of the few translated pieces from the third volume of his *Le Technique et le temps, Le temps du cinéma* (Paris: Galilée, 2001), trans. Stefan Herbrechter in the online journal *Culture Machine*, vol. 5: www.culturemachine.tees.ac.uk.

3. Without reference to cinema history (and the commercial monikers of its early devices), Mladen Dolar, in *A Voice and Nothing More* (Cambridge, MA: MIT Press, 2006), follows Aristotle and Agamben by distinguishing two forms of somatic reality, two forms of life: *zoe* as naked animal existence versus *bios* as communal presence, the species life of raw being versus the existence of a person in a polis. Only *bios* has the kind of autobiographical shape that leads to memory or guilt, repression and its returns. Only *bios*, in Dolar's terms, has voice, that abstraction of inwardness Dolar finds in such phrases as "the voice of conscience." In this exact vocabulary, the organic implant of a Zoe device is an etymological misnomer, an intrusion of the biograph, one might say, at the merely zoological level.

4. On the "aura" of the art object foregone by mechanical reproduction, see Walter Benjamin, "The Work of Art in the Age of Mechanical Reproduction," in *Illuminations*, 217–52 (see the introduction, n. 13); and on the mortal vantage of narration, in the same volume, "The Storyteller: Reflections on the Work of Nikolai Leskov," 83–110.

5. Edward Branigan develops his argument about a narrative suppression of the abusive-mother plot, as mentioned in passing in his response to David Bordwell (see chap. 4, n. 16), in an unpublished paper entitled "Which Sixth Sense?" delivered at the ELTE Institute for Theory of Art and Media, Budapest, in September of 2002.

6. For most of a decade, and at least since my 1999 chapter on the stop-action frame as trope of mortality, I've been expecting eventually to see, in narrative cinema, the equivalent for digital rhetoric of the freeze-frame death moment in watered-down modernist practice, that is, a decomposition of the pixilated field to

match the default of photogrammatic succession in the freeze-frame. Though with one foot heavily in the sci-fi diegesis, Hakman's closing POV shot may be the first.

7. This is where my earlier film book was forced to part company with Deleuze. Combining the dialectical model of Eisenstein with the psychoanalytic model of apparatus theory, I had set out to assess the way in which the medium is textualized through and through, all the way back and down to the strip—which I called the photogrammatic "undertext" of the screened action. Viewing was, therefore, always a kind of reading. This is what I had hoped for in stressing the flickering grain of the filmic against the continuous glimmer of the cinematic. Deleuze, as we know, is entirely indifferent to this undeniable (if invisible) fact of film, pressed into exaggeration on this score, no doubt, by his debate with Bergson. In my demurral from Deleuze on the matter of immanence putting entirely from mind the intermittence of the strip (*Between Film and Screen*, 86–89 [see the introduction, n. 1]), I note David Rodowick's similar resistance to the logic of this intended correction of Bergson in *Gilles Deleuze's Time Machine* (Durham, NC: Duke University Press, 1997), 22.

8. A similar generational irony occurs in Haneke's earlier film, *Benny's Video* (1992), when a teenage son turns his own obsession with video recording against his parents, and his camera on them, by secretly recording an incriminating conversation later disclosed to the police in the film's surprise ending.

9. In regard to French postcolonial immigration in particular, Haneke's film relentlessly dissects the failure of national "double occupancy" that Thomas Elsaesser (writing just before its appearance) poses as a partly utopian alternative to "Fortress Europe" and that appears in the plots of so many hyphenated national film productions. In contrast with the "ImpersoNations" of resolutely national cinema in an earlier chapter, see Elsaesser's 2005 essay (and fourth chapter), "Double Occupancy and Small Adjustments: Space, Place, and Politics in the New European Cinema since the 1990s," in *European Cinema: Face to Face with Hollywood* (Amsterdam: Amsterdam University Press, 2005), 108–30. What he understands as the inevitably homeless and deterritorialized European consciousness—the always diasporic and orphaned sharing of a contestable fatherland—is exactly what the bourgeois Frenchman in Haneke's film cannot bear in the stepbrotherhood (the Algerian boy's adoption) that he has prevented as a child. He thus refuses the "mutual interferences" (126) between subject and Other that Elsaesser finds a film like *Amélie* designed to allegorize in the mode of the fantastic (127). Georges' hosting of a literary talk show, his role as a gatekeeper of high French culture, thus shores up those nationalist investments that Elsaesser sees television and film otherwise working to "dis-articulate" (114–15).

10. A conception of medial reorientation that pervades Mulvey's new study, *Death 24x a Second* (see chap. 3, n. 5).

11. A loaded, microcosmic turn of dialogue enfolds a single late moment in the largest ironies of plot. Georges' last line in the film, while on the phone to his wife and meaning for her to make sure the son doesn't disturb him in his sleep, is this: "Tell him not to be too hard on his old man." The line operates as *pars pro toto* for the generational legacy of communal recrimination that the ambiguous last scene will then perform in pantomime—where the two sons, of Algeria and France, meet and exchange words on the schoolhouse steps. Here again is the wish fulfillment, drugged this time, with which Georges tries vainly to shut out the real.

12. Given all the surveillance cameras from which the hero's drugged final sleep is meant as respite, when his reliving of the past (the banishment of the unwanted colonial Other) is focalized through a static camera at the far back of the fateful family barn, the effect cannot remain purely "stylistic." Instead, this fixated POV seems to be a highly charged allegorical version of what Edward Branigan calls "projecting a camera"—or, in other words (to remotivate his general cognitive theory for narratographic edge), a politically implicated version of screen viewing at large, which by its optical nature always tethers the experience of a moving picture to its focal imaging. See Branigan, *Projecting a Camera: Language-Games in Film Theory* (New York: Routledge, 2006). In Haneke's film, in short, the nightmare filter of Georges' last scene overdetermines the self-consciousness of camera emplacement to a unique degree.

13. Also outside the realm of the uncanny altogether, though evolving into the ironies of real-time mediation, another angle of contrast highlights the difference between Haneke's film, with its medial epistemology of self-inspection, and other modes of trick simulation in American cinema. The Hollywood prototype goes back well before video. In *The Sting* (George Roy Hill, 1973), a false *mise-en-scène* of a speakeasy is constructed, within the plot, for a final con—not only of the rival gambler but of the film audience itself, climaxing in a staged murder of one hero by the other for which none of us has been prepared. Upgrading this irony of deceptive plot space to video mediation, there is the remake of *Oceans Eleven* (2003). The movie audience thinks it is watching a surveillance recording of a casino heist in progress. In fact, what the digital monitors show, we later find, is the playback of a simulated break-in to a fake vault constructed off site by the thieves—a recording smuggled in to override the closed-circuit surveillance transmission. These are the same thieves who can then escape with the money in real time by pretending to be the SWAT team called in to intercept the robbery.

14. Brooks, *Reading for the Plot*, 100 (see chap. 1, n. 13). This bidirectional reading of plot's seeming linearity explains why Brooks speaks of "the *anticipation of retrospection* as our chief tool in making sense of narrative" (23; emphasis in the original).

CHAPTER SIX

1. See Wolfgang Iser, *The Fictive and the Imaginary: Charting Literary Anthropology* (Baltimore, MD: Johns Hopkins University Press, 1993).

2. Deleuze, *Cinema 2*, 262 (see the introduction, n. 7).

3. Bergson, *Matter and Memory*, 249 (see chap. 2, n. 11).

4. In *New Philosophy for New Media* (see the introduction, n. 5), Mark B. N. Hansen's critique of Manovich is concentrated on pp. 32–42.

5. Benjamin speaks only of photography, not cinema, in this regard. But motion pictures can have their own "optical unconscious" too, somatically sensed. This is the point on which I closed *Between Film and Screen* (see the introduction, n. 1), stressing the body's position at exactly that juncture, that *between*. In replacing the plenitude of the view with the flux of the planar image in the filmic "undertext," such moments offer, or better solicit, a "return of the body to the scene of reception" (349). This is therefore a "body immersed in the working upon me of the film work, unalienated for a change from the stuff of the given" (349). However subtle or arresting its ramifications, nothing digital art does to, for, or with the body, except when it *simulates* (rather than merely stimulates) either its form or its perceptions, goes farther than filmic cinema—interstitial and synthetic as it is, or in other words intervalic—in destabilizing the holistic View.

6. Hansen, *New Philosophy for New Media*, 259.

7. Addressing Cavell's multifaceted claims in *Between Film and Screen* (see esp. 128–30), I had wanted to suggest, against his general Bazinian predisposition, that the undertext of the strip, overcome by a bracketed conviction in the scene, might also be part of the skeptical therapy of the medium, philosophically conceived; part of what the embodied brain is willing to override in order to keep in mind the world picture.

8. For a discussion of Cavell's interest in the freeze-frame, see ibid., 126, 128–30.

9. Rick Altman, *Film/Genre* (London: BFI, 1999).

10. At the descriptive rather than narrative level, a similar point is made by Michelle Pierson in *Special Effects: Still in Search of Wonder* (New York: Columbia University Press, 2002) concerning the contrast in *Star Wars* between computerized digital creatures, supposedly organic, and the marked unnaturalness of included optical technology: "Is it any wonder that in this context it is the digital simulation of a special photographic effect, the flickering blue image generated by a holographic communicator, that still looks special, its simulated degradation of an analog image a rare curiosity?" (154).

11. The valuable set of readings by Elsaesser and Buckland, devoted to the analysis of Hollywood cinema in part under shifting medial conditions, a book called *Studying Contemporary American Film* (see the introduction, n. 6), certainly makes good on its own actual subtitle not only by offering, chapter by chapter, a *Guide to Analysis*, but by defending such interpretive work in the conclusion. For these co-authors, neither hermeneutics nor a more objective or technical analysis is necessarily divorced from theory, as it tends to be for David Bordwell and Noel Carroll (whose positions are encountered in the course of discussion). Still, what I am calling narratography brings the tangent spheres of theory and image into closer interpretive overlap. Thus my use of frame captures, where none are necessary for Elsaesser's and Buckland's level of response (as "analysis").

12. Thomas Elsaesser, "The New Film History as Media Archaeology," *CiNéMAS* 12, nos. 2–3 (2004): 78.

13. Rejecting what he sees as the too-easy historical reversal by which on-screen computer effects return us to an earlier cinema of the spectacle, reestablishing an aesthetics of display rather than situated telling, Elsaesser, in "The New Film History as Media Archaeology," minimizes such technological genealogies in favor of a new history of audience orientation—which he then includes in his broadened notion of diegesis. It matters absolutely for him, in short, whether a story is watched (transacted) in a theater, on television, or on an interactive computer screen. In this sense, his work complements that of Rick Altman (n. 9 above) in what we might call the archaeology of "presence" negotiated by the audience orientations of genre in its evolution from home movies.

14. The closest Elsaesser comes at least to reorienting the question of fantasy's recent upsurge, if not addressing it, is this, near the end: "I wanted the reference to diegesis to overcome several kinds of dichotomies: the one between documentary and fantasy as well as realism and illusionism, but also the one between the 'cinema of attractions' and 'the cinema of narrative integration'" ("The New Film History as Media Archaeology," 108). He ultimately wants to adjust or supplement Tom Gunning's "aesthetics of astonishment" with a "*hermeneutics* of astonishment" (113), but it is not clear, at least from this one essay, what this adds to our understanding of the turn to fantasy as spectacle, let alone as genre.

15. Paul Willemen, "The Zoom in Popular Cinema: A Question of Performance," *New Cinemas: Journal of Contemporary Film* 1, no. 1 (2003): 6–13.

16. And beyond such speculative histories, there are entirely fantastic ones. Evoking the tacit optics of the zoom, an artist in another medium altogether seems to have glimpsed the importance of its innovation as one break point between still photography and the true charge of motion pictures. For a kind of archaeological erotics of the zoom appears in a wry conceptual piece by Sol LeWitt called *Muybridge 1*, from 1964, in which the famous lateral-motion studies of nineteenth-century chronophotography have been rotated ninety degrees for an even more direct—if imaginary—anticipation of cinema. What we see, through a row of peepholes, is the frontal view of a nude woman in successive close-ups, her body not so much analyzed in its motion as fetishized in its incremental proximity to the camera. See the piece illustrated in *Open Systems: Rethinking Art c. 1970*, ed. Donna De Salvo (London: Tate Modern, 2005).

17. Here Elsaesser's call for a "hermeneutics of astonishment" seems to be answered *avant la lettre*, if only by a year—even as Willemen's essay drives a wedge precisely into the always renegotiated borderline (in Elsaesser's rethinking: the shifting diegetic horizon) between public and private reception. On Willemen's account in "Inflating the Narrator: Digital Hype and Allegorical Indexicality," *Convergence* 10, no. 3 (2004), our submission to digital spectacle as an overt celebration of capitalist outlay (as well as of technological input) "reactivates the question of the public sphere and its reconfiguration consequent to the triumph of capitalism in the second half of the twentieth century" (13). Yet in the spirit of the essay on the zoom effect (n. 15 above), and in ways Elsaesser has thought in general to write off, Willemen takes this as evidence that we have returned to a "better targeted and more frenetically calibrated cinema of attractions" (14) based on thrill of the image rather

than a concentrated attention to it. Again, though, given the prominence of the zoom, and therefore in a way that lends support at the same time to Elsaesser's revised diegetic parameters, Willemen takes this to be, in a spin on the hucksterism or "hype" indicated by his title, a "hyper-inflation of the narrator" (16).

18. It is worth noting a comparable allegorical instinct in Elsaesser's thinking when he is operating in a more hermeneutic vein. With a temporal fantasy like *Back to the Future* and its sequels, with their franchised nest of time frames, "such concepts as 'deferred action' and 'repetition compulsion'" operate like displaced symptoms of what Willemen would call the corporate unconscious of Hollywood production. "Rather than referring to a possible logic of the subject, these achronological temporalities now stage the logic of the film industry, with its remakes, sequels, prequels, and revivals." See Elsaesser and Buckland, *Studying Contemporary American Film*, 292.

19. In this emphasis Willemen ("Inflating the Narrator," 17–20) follows Philip Rosen's *Change Mummified: Cinema, Historicity, Theory* (Durham, NC: Duke University Press, 2001).

20. Willemen, "Inflating the Narrator," 19.

21. As at the end of *Syriana* from the same year, so at the opening of *The Jacket*: providential narrative—the suggestion may run—is never far from the comparable (high-tech) unnaturalness of aerial surveillance as well as the electronically aided violence of a gun-sight annihilation. In the new global unconscious of televisual mediation, death is always only an eyeline match away. Or, say, that the logic of a searchable database (and satellite download) like that of GoogleEarth (in this sense the everyday interactive variant of screen spectacle's cosmic zoom) is reversed in such a case from magic cartography to the electrographic mechanism of stealth targeting.

22. Bernard Stiegler, *Technics and Time, 1: The Fault of Epimetheus*, trans. Richard Beardsworth and George Collins (Stanford, CA: Stanford University Press, 1998).

23. In the free-form grammar of cinema's instigating "will to art" (Deleuze's rallying cry), filmic procedure loosens the nominalized status of philosophic "thought" into a new participial dispensation. The term becomes in this manner the very epithet for a filmic episteme, the continuous inflection of the seen toward meaning. For screen art deals not just in the image viewed or the image felt but in the image *thought*. Or, allowing for a further emergence into nominalization, the thought-image: Deleuze's *noosign*.

APPENDIX

1. Sergei Eisenstein's 1949 "Dickens, Griffith, and the Film Today" is reprinted in Eisenstein, *Film Form: Essays in Film Theory*, ed. and trans. Jay Leyda (New York: Meridian, 1957), 195–256; my fuller account of Dickens's "filmic" prose appears in *Dickens on Screen*, ed. John Glavin (Cambridge: Cambridge University Press, 2003), 122–44.

2. Charles Dickens, *Little Dorrit* (Harmondsworth, UK: Penguin, 1985), bk. 1, chap. 27.

3. Charles Dickens, *Dombey and Son* (Harmondsworth, UK: Penguin, 1985), chap. 22.

4. See on this point Lynne Kirby, *Parallel Tracks: The Railroad and Silent Cinema* (Durham, NC: Duke University Press, 1997), 8.

5. See Mark Lambert, *Dickens and the Suspended Quotation* (New Haven, CT: Yale University Press, 1981), where the "aggressive and disruptive" effects (133) of dialogue interrupted by such narrative markers as "said Dombey" (though without the present example) are given a full hearing.

6. Charles Dickens, *The Pickwick Papers* (Harmondsworth, UK: Penguin, 2000), chap. 19.

7. Charles Dickens, *Bleak House* (Harmondsworth, UK: Penguin, 1985), chap. 1.

8. Charles Dickens, *Oliver Twist* (Harmondsworth, UK: Penguin, 1982), chap. 50.

9. Stanley Cavell, *The World Viewed: Reflections on the Ontology of Film* (New York: Viking, 1971; enlarged. ed., Cambridge, MA: Harvard University Press, 1976), 159; this and subsequent references are to the

enlarged edition. On the force of this closing stylistic (or "ontographic") turn in *The World Viewed* against the backdrop of a certain resistance to his work on the part of literary and film scholars alike, see my essay, "The Avoidance of Stanley Cavell," in *Contending with Stanley Cavell*, ed. Russell B. Goodman (Oxford: Oxford University Press, 2005), 140–56.

10. A similar effect is achieved, riding the crest of an overt recurrence, by the last shot of *The Constant Gardener* (Fernando Meirelles, 2005), when the sacrificial murder of the diplomat-hero is altogether elided from the closing moment. Gunmen approach, and as he turns away from them back to the isolated Kenyan lake, we cut from his certain death to the liminal shot of the opening credit sequence, repeated more than once en route: birds traversing the desolate beauty of the watery expanse. As the doomed interloper seems to have anticipated, they, there, outlive him in the freedom of their native habitat.

11. Deleuze, *Cinema 2*, 172 (see the introduction, n. 7). Without exploring the connection with Cavell, this matter of "belief" in Deleuze is taken up by Ronald Bogue in *Deleuze on Cinema* (New York: Routledge, 2003), 179–81.

arrhythmy (Lyotard): the "acinematic" irruption of filmic discrepancies into the flow of realist representation (as in **freeze-frames** or accelerated motion), with narrative discourse thereby seen converting a **medial** aberration to a signifying figure (as in a "figure of [filmic] speech").

CGI: computer-generated image(s).

cimnemonics: temporal succession and its mnemonic chain, understood as cognate with the serial manifestation and vanishing—and potential replay—of cinematic projection; a portmanteau coined to evoke the inextricable relation of cinema to memory as modes of the **virtual**.

cosmic zoom: a technique of digital rhetoric capable of drastic shifts in scale—as when plummeting from a satellite-range scan to a facial close-up, or lifting back out again—that also appears lately, in muted variants, for such radical shifts (in temporal rather than spatial orientation) as a precipitous tunneling from present into biographical past.

crystal-image (Deleuze): otherwise (etymologically) **hyalosign**, an image in which the given and the virtual are so closely faceted together, and so transparent to each other, that the imaginary and the actual appear as two coordinate planes of the real.

diegesis: story space in which the events of a plot are imagined as executed by characters; the world of the **fabula**; as opposed to extradiegetic features like a musical score.

digital freeze: contemporary equivalent, through binary coding, of the classic **freeze-frame**, where the **photogram** is repeated long enough to project stasis.

digital intertext: designating, within certain fantastic plots, a tacit cross-reference from paranormal phenomena to their technological "naturalization" by computer electronics—as, for instance, when the fantasy of first-person time travel may seem to be a cover story for a youth culture of interactive video addiction.

digitime: used here to capture the blurred distinction—in the algorithmic plane of the digital image—between duration and change.

exposure time: required for the registration, frame by frame, of the photomechanical **index** on film; contrasted with composure (or compositing) time in **CGI**, where the duration of imaging is entirely the result of keyboard input.

fabula (Bordwell): the story events constructed (ex post facto) by the viewer when cued both by a plot-generating structure (**syuzhet**) and by localized stylistic device.

the fantastic (Todorov): as genre rather than mode, a narrative system sustained by a spectrum of indecision on the part of reader or viewer (usually but not always of the hero as well) concerning the explanation for anomalous events, whether uncanny (psychological) or marvelous (supernatural), as for example a neurotic obsession with—versus a ghostly possession by—specters from beyond the grave.

fantasy: a narrative mode inviting suspension of disbelief in magical events, now increasingly niche marketed in Hollywood as a high-tech special-effects genre, as represented by such films as the *Lord of the Rings* trilogy.

figure/discourse (Lyotard): for use here, the deconstructed border, and disclosed overlap, not just between an always partly verbalized image and an always partly visualized linguistic referent but also between showing and telling, depicting and inscribing, within a given visual medium, including anything from a graphic rebus to an entire film narrative; see Deleuze's **opsign** for comparison.

filmic cinema: at base photomechanical, and to this extent indexical (however much it may be fictionalized or tricked), as contrasted *not* with digital camera work per se—merely another way of indexing (or recording)—but with the possibility of the image's digital manufacture in the first place, where serial record is altogether superseded by binary generation.

frame time: the imperceptible seriality of all on-screen sequence, with each single imprint subsumed on the run to its disappearing act in projected motion.

framed time: the sense here, by way of a postfilmic **narratography**, that time, captured within a single pictorial field, can then—via certain aberrant plot devices—be morphed like a freestanding image rather than undergone like a real duration.

frameline: the filmstrip in motion—the image track—on the way to its projection as continuous screen action.

freeze-frame: the geometric bracket or border of a shot (its "frame," in this sense) as well as its contained image—when both are held "still" by the rapid duplication of a single **photogram** on the strip. See **digital freeze**.

hermeneutics: the practice of interpretation, whether in deference to or in "suspicion" of the received assumptions behind a given textual meaning (whose overall structure, rather than import, may otherwise be analyzed by **narratology**).

hyalosign (Deleuze): see **crystal-image**.

index: among the three sign functions identified by C. S. Peirce, with the icon serving to picture by resemblance; the symbol serving to denominate only by arbitrary marks like alphabetic signifiers; and the index more immediately indicating real contact with the thing signified, as when a footprint takes the trace of a past foot or smoke is contiguous with a present fire to which it alerts us—or as when, in **filmic cinema**, the run of **photograms** re(as)sembles the split seconds of a profilmic event that was once present in front of the camera.

instrumental marvelous (Todorov): related to the "scientific marvelous," though with emphasis more on gadgetry than on abstract laws; a category of the "supernatural" (rather than the merely "uncanny") that overlaps with what we usually think of as science fiction, especially in recent films of digital dystopias and the ontological gothic.

interface effect (Žižek): a moment breaking with the shot/countershot pattern of traditional narrative **suture** so that self and Other—or a self and its own representational Other, like a photograph or a mirror double—are anomalously framed within one and the same shot.

lap dissolve: short for "overlapping dissolve," the classic technique of scene change that "fades" one scene into the next by gradual superimposition.

lectosign (Deleuze): facet of the screen image whose attenuated ratio of visible to interpretable material favors the reading of a shot rather than the reception of a view.

liminal shot: a threshold moment, often an inaugural image later echoed on narrative exit, in which some unforeseen plot development gets tacitly modeled at the level of a purely visual figure awaiting its full discursive entailments as story. See **matrix**.

matrix (Riffaterre): in the predigital sense of the noun, a text's unseen or unspoken source, retroactively inferred from the way it generates a **model** as the first appearance of an ongoing **subtext** designed to unfold the linked **hermeneutic** cues of a reading or viewing.

media archaeology: resisting the linear evolution (genealogy) of a single medium from technical precursors and experimental dead ends; a concern instead for the cultural and cognitive grounds of its possibility and its reception, as for instance the consideration of institutional cinema in light not just of photography and nickelodeons but—in a wider sense of the **protofilmic**—of Victorian panoramas, stereoscopes, railroad vistas, and urban street illumination.

medial: identifying (by back-formation from *intermedial*) the technological specificities of a given representational form—as with the lexemes that make up a literary text; the photomechanical process as "median" between recorded scene and its projection in **filmic cinema**; the binary algorithm as mediating between (or otherwise rendering unnecessary by **CGI**) an ocular event and its digitally scanned and rearrayed screen image.

model (Riffaterre): initial and paradigmatic form of an ongoing **subtext**; see **matrix**.

movement-image (Deleuze): rudimentary facet of on-screen action or event (generated at the "plane of immanence") from which duration and the **time-image** arise by secondary visual inference.

narratogram: the smallest unit of graphic effect in a given medium, filmic or lexical, that is capable of taking a narrative charge.

narratography: the tracking of narrative in operation (rather than the sorting and classifying of its functions) at the **medial** as well as structural level, where visual "style" has its own microplot; a mode of response conceived in these pages as the sharpest point of intersection between **narratology** and a medium-specific **hermeneutics**, driving its wedge thereby into **media archaeology** as well.

narratology: the comparative study of narrative as such, in a tradition going back through structuralism to Russian formalism, and characterized by an attempt to determine the underlying functions and structures of narrative process across different media.

noosign (Deleuze): an image that must be thought rather than simply or strictly seen, whose **virtual** status is more conceptual than optical.

opsign (Deleuze): breaking with the sensorimotor schema and delinked from the normal sense of a view, an image where the seen is no longer a scene of movement, no longer extended into action (just as the **sonsign** no longer emanates from action as its recorded aural **index**).

paratext (Genette): material outside the narrative text itself but held in close association with it, as credit sequences, advertising taglines, and, most recently, DVD supplements.

photogram: in French theoretical usage, the unitary imprint, or *photogramme*, of the moving photomechanical strip, the serial frame before it disappears by projection into the screen frame, and thus the constitutive subunit of cinema that, by a false English etymology, puts the single "cell" back in celluloid; to be distinguished from the nineteenth-century use of the term for the prephotographic tracing of images onto treated paper by light alone from contiguous objects, without lens or other mechanical procedures.

POV: a camera's point of view (or, in narrative terms, "focalization") as it guides the viewer's ocular identification with a given narrative agent, through whose eyes or mind's eye—objective or subjective—the recognition of a given event is channeled.

protofilmic: preliminary to film or cinema in historical and technological terms, as opposed to the profilmic (meaning "before" the camera in the other sense, as recorded object).

sonsign (Deleuze): an aural effect independent of the **movement-image** (of objects, bodies, or lips) that produces a kind of semiotic counterpoint to "natural" or "realist" sound.

stacking and popping (Ryan): derived from computer models, especially Windows formats, the turn in cognitive **narratology** by which the traditional notion of an inset or embedded or framed narrative is reconceived instead to overlay—rather than be contained by—its delimiting structure (now electronic platform), with all immediate effects commanding attention only until "popped" or canceled, at which point the grounding level returns from beneath our immersion in the upper stratum of the "stack."

subtext (Riffaterre): recurrent manifestation of an organizing principle that is kept tacit even in its own sequential ramification across the text; see **matrix**.

suture: the folding into narrative of spectatorial identification by way, especially, of editing devices that displace the interanimating exchange of gazes within a plot space onto the viewer's own line of sight, thus "stitching" spectators into the filmic articulation in order to seal up the **diegesis** as a self-coherent (yet at the same time paradoxically accessible) space.

syuzhet: see **fabula**.

temportation: aside from time travel more generally, this book's term for the impossible moment when temporality and portability, time and spatial motion, are seen to collapse into each other via some ambivalent magic of transformation, whether the flight from a traumatic memory trace or its revisitation and erasure.

time-image (Deleuze): following upon the exhausted classic genres of action cinema in the alienating spaces of postwar modernism, a dimension of screen viewing that registers less a scene or event, less a picture at all, than a more abstract image—and hence the image, or *figure*, of an event's discursive conversion to mental trace within the flow of time: the very birth, for instance, of memory as a **lectosign**; see **virtual**.

timespace: the implosion of spatial and temporal categories (including the whole Deleuzian dialectic of **movement-image** and **time-image**) into a **fantastic** zone—either uncanny or supernatural—in which, for instance, the plane of the past can be morphed by retraversal (or *trans*-formed) rather than just remembered; see **temportation**.

trucage (Metz): optically conceived, the cinematic "trick," once filmic, now digital too, by which the image plane is either imperceptibly or quite visibly enhanced by technical artifice; historically conceived, a touchstone in the evolution of **medial** codes by which the initial magic of cinema's attractions was gradually assimilated to unmarked devices of transition, as with the technique of "ghostly" overlay becoming the everyday grammar of the **lap dissolve**.

virtual (Deleuze): a facet or avatar of consciousness where time on film, for instance, appears under the sign of its memory or its anticipation rather than its presence; as opposed *not* to the real but merely to the actual (the latter as monopolized by the **movement-image**).

VR: virtual reality in the electronic sense of simulated perceptual environment, as distinct (but not always) from the Deleuzian **virtual**.

INDEX

Abre Los Ojos (Open Your Eyes) (film), 94–96. *See also* Vanilla Sky
acinema, 216. *See also* Lyotard
Adaptation (film), 83, 106, 107, 134; counterplot of, 112; *trucage* in, 112
affection-image, 121
Agamben, Giorgio, 277n3
The Age of Innocence (film), 270n21
A.I.: Artificial Intelligence (film), 90, 107, 170, 214, 226; coda of, 99, 100; interface sequence in, 99; liminal shot in, 98
Alien Resurrection (film), 172
allegory: and fantastic, as enemy of, 198
Almodóvar, Pedro, 83, 176, 181, 182, 183, 192, 194, 212; hypervisual imagination of, 178; style of, 173
Altman, Rick, 220, 228, 280n13; and domestic imaging, 233; and remote consumption, 221, 222
Amélie (film), 278n9
Amenábar, Alejandro, 94, 96
American cinema, 16, 166; vs. European cinema, 8, 18, 57, 60, 67, 12, 124, 125, 129, 133, 153, 165, 166, 167, 170, 172, 192, 195, 205; movement-image, undermining of in, 82; plot devices of, 168; and supernatural, 100; temporality, annihilation of in, 61; time-warp plots in, 8. *See also* Hollywood cinema
anamorphosis, 181
analog holography, 56
anime, 130
Anderson, Benedict, 57
animated pictures, 6
animation effects, 114
Antonioni, Michelangelo, 170, 277n1
Aristotle, 277n3
arrhythmy, 28, 31, 32, 112, 176, 181, 217; as defined, 283; and optical illusion, 148; and time-images, 219

Asian films: special effects, as driven by, 124
The Asphyx (film), 270n21
Astruc, Alexandre, 269n8; and caméra-stylo, 26
Austen, Jane: adaptations of, 270n17

Back to the Future (film), 18, 137, 150, 169, 276n14, 281n18
Bad Education (La Mala Educación) (film), 83, 174–79, 191, 195, 199; child abuse in, as motor, of plot, 192; death drive in, 175, 192; digital in, as elegiac virtual, 173; digital ingenuity of, 179, 180, 181; digital morphing in, 180, 181; digital *trucage* in, 183; and 8 ½, as predecessor, 174; elegiac cast of, 184, 192; lap dissolve in, 178; last shot of, 182; narrative discourse of, 183; opening scene of, 175; as screened illusion, 174; slow motion in, 181, 182; time in, 171; *trucage* in, 178, 179; typescript in, 182; uncanny in, 177, 178, 180, 181, 192; virtual memories in, 193
Bakhtin, Mikhail, 202
Ballet mécanique (film), 67
Barthes, Roland, 62, 65, 273n25
Bazin, André, 43, 65, 66, 89, 245; photography, metaphors for, 128; photography, as mummy complex, 240; total film, myth of, 275–76n7
A Beautiful Mind (film), 91
Being John Malkovich (film), 156, 163; digital effect in, as portal, 157; double twist in, 160; ending of, 157, 160, 162; as photomechanical parable, 158; photographic montage in, 158, 159; plot of, as fantastic, 157, 158; plot twist in, 159; POV in, 157, 160, 162; predigital imagery in, 162; trick beginning in, 157
Being and Time (Heidegger), 243
Benjamin, Walter, 187, 216, 279n5; and optical unconscious, 13
Bennett, Alan, 1
Benny's Video (film), 278n8

Bergman, Ingmar, 83, 168

Bergson, Henri, 13, 14, 113, 144, 191, 214, 244, 278n7; and center of indetermination, 215; and Deleuze, 61, 215; dying, panoramic vision of, 61; and movement, 272n11; and paramnesia, 168, 169

Between Film and Screen: Modernism's Photo Synthesis (Stewart), 4, 9, 267n4, 268n10, 269n3, 273–74n29, 274n3, 275n6, 279n5, 279n7

binary imaging, 15

Biograph, 186, 222

Blade Runner (film), 90, 136, 275n6

Bleak House (Dickens): comic syllepsis in, 256; montage thinking in, 256; and prose superimposition, 257; reciprocal troping in, 256

body: as medial term, 217; and text, performance of, 218

Bogue, Ronald, 282n11

Bolter, Jay David, and Richard Grusin, 4, 9, 94

Bonnie and Clyde (film): Liebestod ambush in, 81

Bordwell, David, 23, 25, 112, 231, 274n30, 277n18, 277n5, 280n11; alternate-universe plots, forking paths structure of, 156, 161; style, as adjunct, 25

Borges, Jorge Luis, 161, 274n30

Brainstorm (film), 212

Brakhage, Stan, 241, 243

Bram Stoker's Dracula (film), 270n21

Branigan, Edward, 161, 232, 277n5, 277n18; disnarrated, example of, 277n16; and projecting a camera, 278n12

Brazil (film): vs. *Johnny Mnemonic*, 131; plot of, 131

Brewer, Glenn, 275n10

Brokeback Mountain (film): final POV shot in, 237

Brooks, Peter, 24, 30, 204, 205, 279n14; and death drive, 156

Buckland, Warren, 232, 267–68n6, 280n11

Bullock, Sandra, 223

Burch, Noel, 233

Butler, Samuel, 105–6

The Butterfly Effect (film), 156–57, 186, 202, 215, 221, 223, 229, 237, 239, 240; child abuse, as traumatic norm in, 188; coda of, 151, 204; Deleuzian moment in, 152; digital editing in, 203; DVD version of, 153; ending of, 153, 155; film as medium in, 149; last shot of, 154; original ending of, 151, 153; past in, as remaximized, 151; as pushed, 151; self-infanticide in, 155; subplot of, 149–50; time in, as popped and pushed, 151; timespace, morphing of in, 204

The Cabinet of Dr. Caligari (film), 214

Caché (Hidden) (film), 165, 171, 196–99, 201, 210, 221, 223, 229; as allegorical, 197, 198; as coproduction, 195; as

humanist temportation, 195; model shot of, 196; plot twist in, 196; remediation in, 199; switch ending of, 195; time, spatialization of in, 204; time-image in, 197, 199; as uncanny, 198, 199

Cage, Nicolas, 106

calendarity, 171

cardinality, 171

Carroll, Lewis, 234

Carroll, Noel, 280n11

cartography, 22

Casetti, Francesco, 272–73n19

Cavell, Stanley, 66, 210, 217, 229, 265, 266, 279n7; and death-drive plots, 220; filmic prose of, 264; and freeze-frame finishes, 220

Cavett, Dick, 137

celluloid order, 219, 231; and frame time, 32; representational and narrational systems, interplay of, 32

CGI (computer-generated image) effects, 51, 114, 283, 285

Chatman, Seymour, 23

chronophotography, 13, 280n16; and trace methods, 133

chronotope, 171; in postrealist form, 202

cimnemonics, 3, 181, 193; as defined, 283

cinema, 10, 17, 128, 245; beginnings of, 222; and computer simulation, 129; and cut, 13; dance, affinity with, 67; digital, encroachment in, 1, 55, 141, 155; and fantasy, 57; as filmic medium, 129, 217, 223; as filmic operation, end of, 154; and frame, 67; haptic in, 216; institutional fears about, 162, 163; as institutional medium, evolution of, 221; and memory, 245; as memory machine, 173; movement in, as given, 194; movies, as nickname, 6; and narrative sequence, 98; photographic base, erosion of, 3; photograph, as index of real, 56; photograph, as touchstone in, 56; and photography, 5, 267n4; photography, remediation of, 4; presence, vs. nonpresence in, 228; presence, restoring of, 222, 223; *realité*, roots in, 222; and reality TV, 104; and single framed image, 129; spiritual automaton of, 214; and still image, 146; and temporality, 170; as threatened, 163; and time, 13, 14, 125, 126, 129, 162, 167, 169, 194, 245; time, writing of, 194; time effects, and virtual, 125; time-warp genre, special effects of, 135; transmission, medium of, 47; and video pause button, 111; and virtual, 167; and virtual replay, 202; and visuality, 17; as writing, with movement, 219. *See also* early cinema; film; narrative cinema

Cinema 1: The Movement-Image (Deleuze), 270n20

Cinema 2: The Time-Image (Deleuze), 14, 277n1

The Cinema Dreams Its Rivals: Media Fantasy Films from Radio to the Internet (Young), 275–76n7

The Cinema Effect (Cubitt), 10

cinematogram, 273–74n29

cinematograph, 15

cinematography, 15, 22, 26, 241; image, as superimposable, 6; movement, as inscription of, 29

cinephilia: and epistemophilia, 229

Citizen Kane (film), 51; Rosebud theme in, 132

City of Lost Children (film), 171, 215; as uncanny, 172

Clair, René, 80, 128

Clark, T. J., 275n5

Classical Music and Postmodern Knowledge (Kramer), 268–69n2

computer games: and Hollywood cinema, 136

computerized screen effects: industrial capitalism, link to, 235

computer memory, 245

conceptual montage, 263

consciousness, 209; of lived duration, polarity of, 166; and memory, 190

conspiracy thrillers, 145; ontological gothic, as transformation of, 104

The Constant Gardener (film): last shot in, 282n10

cosmic zoom, 203; as defined, 283; and digital rack focus, 237; in sci-fi, 134; and swish tracking, 237. *See also* zoom

The Crazy Ray (Paris Qui Dort) (film), 128

critical theory: textural turn in, 230

Crouching Tiger, Hidden Dragon (film): history and fantasy, marriage of in, 124

Crowe, Cameron, 94

Cruise, Tom, 95

Cruz, Penélope, 95

crystal-image: as defined, 283

Cubitt, Sean, 10, 11, 12

Cusack, John, 108

cut, 10, 13, 177; and photogram, 11; and vector, 11–12

cybernetics, 142

Dark City (film), 56, 89, 90, 133, 156, 212

Darwin, Charles, 112

Dasein, 244

Davies, Terence, 48, 49

The Death of Cinema (Usai), 51

Death 24x a Second (Mulvey), 111

decryption, 268–69n2

Deleuze, Gilles, 2, 3, 9, 14, 16, 18, 30, 31, 58, 66, 67, 109, 111, 115, 134, 135, 140, 161, 167, 171, 174, 177, 178, 183, 191, 199, 203, 204, 205, 216, 229, 237, 245, 264, 276n13, 277n1, 278n7, 281n23, 282n11; and attentive memory, 168; belief, in world, 265; belief, as freely willed, 217; and Bergson, 61; and center of indetermination, 215; cinema, as spiritual automaton, 170; and cinema studies, 194; cinematic virtual, 210; and digital, 208; and flashbacks, 169; and forking points, 168; and *hyalosign*, 139; and *lectosign*, 194, 219; memory, as virtual, 169; and mirror-image, 169; modernist actor, as spectator, 218; on movement, 272n11; movement-image, and time, as measure of movement, 15; and New Wave, 169; and *opsigns*, 219; and plane of immanence, 23; and postwar cinema, 125, 172; and *sonsigns*, 219; strip, and time-image, 219; telepath, importance of, 168; temporality, and frameline, 231; and time-image, 8, 15, 125, 127, 169; time-image, fate of, 210; time-image, as invented figure, 209; virtual and actual, fold between, 169; and virtual image, 189, 190; and virtuality, 193, 194, 217

Deleuze on Cinema (Bogue), 282n11

Demme, Jonathan, 91, 92

Deux Frères (film), 199

Dickens, Charles, 259, 261, 264, 266; cinematic adaptations of, 250; comic writing of, 255; dialogue of, 254; and Eisenstein, 249, 251, 255, 256, 260; and film, 249; filmic prose of, 249, 250, 254, 263; and kinetic sequencing, 250; montage, influence on, 249; phrasing, density of, 252

"Dickens, Griffith, and the Film Today," 249, 250

diegesis, 227, 229, 232, 236, 237, 244, 286; as defined, 283; narrative action, and narrative interaction, 233; and narratography, 236; and space of narration, 233; and syntax, 60

diegetization: and digitization, 233

Difference and Repetition (Deleuze), 276n13

digital compositing: and zoom effect, 275n4

digital editing, 240; eroded reality, as metaphor for, 147; interactive video, as electronic fantasy, 136; special effects of, 134; time, consciousness of, 203. *See also* cosmic zoom; morphing

digital freeze: as defined, 283

digital image, 3, 6, 14, 15, 19, 53, 56, 113, 114, 232; filmic cinema, death of, 53; and pixilation, 216

digital intertext, 275–76n7; as defined, 283; repression, as key to, 276n8

digital zoom, 238

digitime, 8, 181, 200; as defined, 283; narratives of, 236

digitization, 10, 52, 88, 154, 208, 241; and diegetization, 233; and time, 181; and *trucage*, visible vs. imperceptible, 60

Discours, Figure (Lyotard), 28

Des Dispositifs Pulsionnels (Lyotard), 27

Doane, Mary Ann, 13, 14, 127, 128, 154, 160, 166, 275n1

Dolar, Mladen: on somatic reality, forms of, 277n3

Dombey and Son (Dickens), 254; railway revolution, fictional treatment of, 252, 253

Donnie Darko (film), 136, 137, 141, 203, 215, 220, 240; ending of, 134–35; iris shot in, 135; time-loop narrative of, 134

La Double Vie de Véronique (*The Double Life of Véronique*) (film), 57, 73, 107, 168, 181, 273n22; doppelgänger motif of, 74; interface effect in, 70–71; Other in, 72, 74; as uncanny, 72

Dreyer, Carl, 170, 264, 265

Dubois, Philippe: and cinematogram, 273–74n29; and figuration, 81, 272n15, 273–74n29, 273n20

duration: as inscription, 133; technics of, 243

durée, 9, 96, 232; and timespace-image, 171

DVD: film fantasy subgenre, as supplements to, 226, 227

Dylan, Bob, 96

early cinema: as photofilmic machination, 67; as spectacle, 11. *See also* cinema

Ecran Français (journal), 269n8

Edison, Thomas, 94

8 ½ (film), 83, 168; and *Bad Education*, 174

Eisenstein, Sergei, 66, 257, 258, 278n7; on cinema, 14; and conceptual montage, 251; and Dickens, 249, 250, 251, 252, 255, 256, 259, 260; and Griffith, 251, 252, 256; mechanical parallelism, resistance to, 254; and montage collision, 252; montage thinking, and metaphor, 256; *Oliver Twist*, stagecoach-view in, 253; standard sound cinema, resistance to, 254; and word-sentence, 252

electrical cinema: vs. electronic cinema, 6

electrograph: and psychological automata, 15

electronic fantasy: commercialization of, 236

electronic imaging, 2; global traffic in, 57; and plotted temporality, 56

Elsaesser, Thomas, 227, 232, 267–68n6, 278n9, 280n11, 281n18; cinema of attractions vs. cinema of narrative integration, 280n14; cinema history, as media archaeology, call for, 234; diegesis, emphasis on, 236, 280n13, 280n14; hermeneutics of astonishment, call for, 280n14, 280–81n17; narration vs. navigation, 115, 136, 233; narrative space, concept of, 233

The Emergence of Cinematic Time (Doane), 13, 127, 275n1

The End of Cinema as We Know It: American Film in the Nineties (Lewis), 267–68n6

Enyedi, Ildiko, 79

epistemology, 217; and skepticism, 265

Erewhon (Butler), 105–6

Eternal Sunshine of the Spotless Mind (film), 126, 145, 147, 148, 150, 157, 172, 202; as Deleuzian, 149; trick beginning, and trick ending, collision with, 146

European cinema, 16, 17, 209–10; agency in, 166; vs. American cinema, 8, 18, 60, 67, 124, 125, 129, 133, 153, 165, 166, 167, 170, 172, 192, 195, 205; and coproductions, 57; historical memory and cultural identification, anxiety over, 57; as humanism, 205; temporality, annihilation of in, 61; time loop in, as Deleuzian fold, 149; and transnational imaginary, 246; transnational uncanny of, 57, 211. *See also* European uncanny

European uncanny, 63, 72, 100, 183; and Hollywood ontological gothics, 124, 165, 195; and Hollywood virtual, 60; and neurosis, 102; paranormal filiations of, 68; and transnational guilt, 210. *See also* European cinema

exceptionalist narratography, 161

exposure time: as defined, 283; digital compositing, displaced by, 208

fabula, 23, 24, 89, 258; as defined, 283; and *syuzhet*, 25, 62, 220, 228, 239

A Faery's Tale (film), 270n21

The Fallen Idol (film), 11, 12

Fanny and Alexander (film), 168

the fantastic, 16, 24, 25, 32, 67, 85, 91, 208, 210, 247; and allegory, 198; appeal of, 212, 213; in contemporary film, resurgence of, 59; as defined, 283–84; and detective fiction, replacement of with, 89; double in, 273n22; ethics of epistemology, 210; as genre, definition of, 59, 60; as genre, structure of, 145; and instrumental marvelous, 271n3; literary, as purest state of, 73; liminal shot, emphasis on, 100; magic realism, influenced by, 57; and narrative cinema, 73; and neurotic disturbance, 101; and New Europe, 57; photography in, 61; plot in, 61, 103; plot time, articulation over, 58; plots, morphing in, 7; and psychotic break, 101; as right wing, 212; self, and Other, difference between, 102; and spatial form, 204; and subtext, 63; supernatural belief in, 212; temporality, narratology of, 58; time, reframing of, 100; trick beginnings in, 59; and uncanny, 101; uncanny vs. supernatural, tension between in, 25, 60; and undecidability, 60. *See also* fantasy

The Fantastic (Todorov), 101

fantasy: mode, as defined, 284

Fellini, Federico, 83, 168

Fictional Truth (Riffaterre), 24, 272n13, 272n14

Fight Club (film): *Jekyll and Hyde*, as version of, 90; liminal scene in, 91

Figgis, Mike, 274n8

figuration, 28, 272n15, 273n20

figure/discourse: as defined, 284

film: as cinematic rudiment, 129; digital incursion into, 124, 155; duration, as visual object, 146; as immortality machine, 264; as indexical capture, 216; indexical record, and death, 264; as mass cultural object, remote consumption of, 221; medial texture of, 246; narration, reliance on, 250; photographic basis of, 15, 53. *See also* cinema; filmic cinema

film editing, 114; and digital, 135

film history: lost presence, as quest for, 222; visual attraction, and plot action, dichotomy between, 237

filmic cinema, 4, 5, 6, 19, 210, 214, 231, 285; cellular basis of, and narrative possibility, 114; as defined, 284; digital, and death of, 53; between strip and projected image, 9; medial operations of, 9, 53; as movement inscribed, 31; and photographic index, 113–14; sampling, vs. simulation of time, 52; and temporal change, 3

filmic imaging, 11; digital imaging, shift to, 233

filmic narratography, 161; image, textualization of, 27. *See also* narratography

filmic reception: optical unconscious, 216

filmic storage, 153–54

The Final Cut (film), 156, 173, 194, 199, 212, 220, 222, 227, 228; as allegorical, 189, 190; child abuse in, 188; and digitime, 193; DVD of, 184–85; death drive in, 192; plot of, as expository, 185; plot twists in, 186–89, 191–92; POV shot in, 195; "rememory" in, 192; sci-fi and social satire, as blend of, 185; temportation plot, as cultural inversion of, 187; time in, 171; trick ending of, 192; *trucage* in, 190–91; video-log image in, 190

Fincher, David, 90

flash frames, 243

flicker effect, 242, 263, 264; vs. digital violence, 243; early cinema, effect on, 113

The Forgotten (film), 133, 169; plot twist of, 168

form: as content, 63, 236

Fox Talbot, William Henry, 269n8; and pencil of nature, 26

framed time, 2, 5, 27, 33, 67, 163; as defined, 284; film history, as replayed, 201; and narrational order, 32; as narrative inflection, 2; as psychic topography, 2

frame enlargements, 5

frameline, 11, 23, 55; as defined, 284; and movement, 231; recognition of, 31

frame narrative: as misnomer, 141; as popped, 141; and vertical stacking, 141

frame time, 119; and celluloid order, 32; as defined, 284; framed time, as yielding to, 7, 236

Frankenheimer, John, 91

freeze-frame, 182, 183, 283; death moment, and digital rhetoric, 277–78n6; as defined, 284; versus pause button, 111

Frequency (film), 169, 225, 234, 276n8; time-loop plot of, 137, 138, 139, 140; trick beginning of, 137

Freud, Sigmund, 62, 100, 120, 275n1; cinema, resistance to, 14; consciousness, sense of, 154; and unconscious, 204, 272n14

The Fright of Real Tears: Krzysztof Kieślowski between Theory and Post-Theory (Žižek), 272–73n19

Fujimoto, Tak, 188

"The Garden of Forking Paths" (short story), 274n30

Genette, Gérard, 23, 153

The Geopolitical Aesthetic: Cinema and Space in the World System (Jameson), 271n3

German silent film: visual double in, 273n22

Gertrud (film), 170

Gibson, William, 129, 130

Gilles Deleuze's Time Machine (Rodowick), 278n7

Gilliam, Terry, 131

Godard, Jean-Luc, 61, 167, 169

The Golden Bowl (film), 4, 17, 34, 45, 48, 49, 51, 55, 64, 208; coda of, 222; film-within-a-film in, 40, 46, 50, 52, 53, 207; filmic execution of, 52; framing-within-the-frame in, 44; modernist devices in, 47; and narratography, 52, 53; photomechanical genealogy in, 39, 40; screenplay of, as different from novel, 47; slide show scene in, 41, 42; and two-shot, 35

The Golden Bowl (James): diagrammed, as suture, 36–37

Gondry, Michel, 149

GoogleEarth, 281n21

Gordon, Douglas, 217

grammatology, 236, 244

Great Expectations (film): shot/reverse shot in, 259–60

Green, David, 267n3, 273n25

Greene, Graham, 12

Griffith, D. W., 251, 256; and Dickens, 251; Dickensian technique, transfer of, 250; innovations of, 201

Gunning, Tom, 221, 268n10; and aesthetics of astonishment, 280n14; and cinema of attractions, 11; and *optogramme*, 77

Haneke, Michael, 3, 165, 171, 195, 202, 205, 221, 278n8, 278n9, 278n12, 279n13; themes of, 210

Hansen, Mark B. N., 217, 218; and affection-image, 216

Happy Accidents (film), 234–35

haptic optic, 215–16

Harry Potter films, 90

Haunted Media: Electronic Presence from Telegraphy to Television (Sconce), 276n8

Heart of Darkness (Conrad), 141

Heaven (film), 80–82

Heidegger, Martin, 243

He Loves Me, He Loves Me Not (film), 242; electrocardiograph in, 133

heritage film, 41, 42, 48

hermeneutics, 219, 231, 232, 247, 285; as defined, 284

Hero (film), 124–25

The History Boys (Bennett), 1

Hitchcock, Alfred, 167; and scissored jump cuts, 217

Hollywood cinema, 65; and alternate realities, 103; characterization in, as depleted, 103; child molestation, as predictable plot turn in, 121; chronocides in, 212; and classic genre narrative, 102–3; and computer games, 136; digital competitor, repression of, 136; digital sci-fi and posthumous gothics, 125; *fabula*, as illusory, 103; and fantastic, 57, 85; forbidden sexuality in, and death drive, 193; geopolitics, flight from in, 246; home movies, closural reversion to, as domestic emplacement, 221; and mediation, 229; and mnemonic gothic plots, 136; and MTV format, 103; multilayered temporality of, 59; nonbelief, viewer's complicity in, 246; and nonelectronic virtuality plots, 232; ontological subterfuge in, 64, 68, 92, 105, 209, 220; plot level, disavowal of, 246; plots, as allegories of the unreal, 59; plots, paranoia in, 210; political disengagement of, 213; as postsubjective virtuality, 211; presence, crisis of in, 229; and religious Right, 212; self, externalization of in, 102; temporal transmigration in, 127; temportation, subgenre of, 205; temportation, trope of, 127; thrillers, digital technology in, 181; time-travel narratives, and digital thematic, 136; trick endings in, 63, 102, 103; unreality in, as pervasive, 145; video games, and role-playing, 102, 136; virtual, aesthetic of in, 133, 211, 246; virtual self in, 89. *See also* American cinema

holography, 268–69n2

House of Mirth (film), 48–49

humanism: vs. science fiction, 193; vs. technofuturism, 171

hyalosign, 139, 283

hypertext, 266

iconography, 21

Identity (film), 106, 107, 116, 192; anti-hero of, as parody, 115; child abuse, as traumatic norm in, 188; freeze-frames in, 109, 110, 111; interface effect in, 108; memory in, 126; narrative reversal in, 109; prologue of, 109; revelation scene in, 108; time in, 126; trick ending in, 112

illusionism, 274n7

image, 34, 245; digital morphing of, 154; figuring of vs. picturing of, 246; as metaphor, 29; as signifier, 154; and suture, 154; temporal transit of, 2; as thought, 281n23

image theory, 65

imaginary: and temporality, 173

imagination: as fictive, 209

index, 229, 241, 244, 283, 285; as defined, 284; icon, in contrast with, 236, 242

indexical interface: vs. virtual interface, 229

indexicality, 233, 236, 273n25

In the Sight of Death: An Experiment in Art Writing (Clark), 275n5

inscription, 26, 220, 266; and cinematography, 29; and memory, 245

Inside the Gaze: The Fiction Film and Its Spectator (Casetti), 272–73n19

Insomnia (film; Nolan), 272n15; credit sequence in, 65; twist in, 65

Insomnia (film; Skjoldbjaerg), 65

installation art, 217

instrumental marvelous: as defined, 284; and fantastic, 271n3

interface effect, 70, 81, 139, 181, 211, 273n22; and anamorphosis, 70–71; as defined, 284; and European fantasy, 124; narrative idiom, invasion of, 156; suture, breakdown of, 69; and suture theory, 160; and time, 162; and uncanny, 69, 70; video-games, as reduced to, 116

intermedial, 9

interpretation, 230. *See also* hermeneutics

intertext, 154, 214. *See also* digital intertext

interval: interface, replaced by, 205

I, Robot (film), 170

Irreversible (film), 202, 243; and narratology, 244; and postnarrative foreclosure, 244; reverse chronology of, 242–43; strobe finish of, 244; time-images of, as *syuzhet*, 242

Iser, Wolfgang: fictive vs. imaginary, 209

Ivory, James, 23, 34, 35, 38, 39, 40, 41, 46, 48, 49, 51, 52, 55, 64

The Jacket (film), 90, 92, 105, 165, 169, 185, 212, 232, 242, 281n21; and American avant-garde, 241; digital sequences of, 239, 240; final credits of, 238, 240, 241; as homage, to *Mothlight*, 241; optical allusion in, 240; plot of, 238, 239, 240; and postnarrative foreclosure, 244; real in, 241; trick beginning in, 238–39

Jacob's Ladder (film), 90, 92, 162, 169, 212, 238; model shot in, 105; plot of, 104; trick ending of, 105. *See also* An Occurrence at Owl Creek Bridge

Jahn, Manfred, 274n30

Jakobson, Roman: poetic function, definition of, 269n12; and structuralism, 30

James, Henry, 18, 23, 33, 34, 38, 52, 222, 249; and closure, 35–36; and *The Golden Bowl*, 35

Jameson, Fredric, 102; on cinematic beginnings, 271n3, 271n7; on magic realism, 271n3; 271n7; paranoid logic, as narrative category, 145

Jekyll and Hyde: and *Fight Club*, 90

Jeunet, Jean-Pierre, 172

Joan of Arc at the Stake (film), 265

Joe (film), 27–29; death scenes in, 219; and irrational editing, 31; and narratography, 30; and timing, 30; violence in, 220

Johnny Mnemonic (film), 90, 115, 129–32, 156, 185, 188, 221, 223, 228, 275–76n7; vs. *Brazil*, 131

Jonze, Spike, 83, 106, 156, 159

Joseph Andrews (Fielding), 30

Jude (film), 270n17, 270n21

Jude the Obscure (Hardy), 270n21

Jules and Jim (film), 95

jump cut: and syllepsis, 263

Kauffman, Charlie, 106, 157, 159

Kauffman, Donald, 112

Kelly, Richard, 134

Kidman, Nicole, 96

Kieślowski, Krzysztof, 17, 56, 57, 69, 70, 72, 74, 75, 79, 80, 82, 125, 194; interface effect in, 211, 272–73n19

King, Stephen, 83

Kittler, Friedrich, 128; on the fantastic, 273n22

Kramer, Lawrence, 268–69n2

Kubrick, Stanley, 100, 167, 243

Lacan, Jacques, 181

The Lake House (film), 234, 276n8; cell phones in, 225; parallel-universe plot of, 223, 224; photography in, 235; and subtext, 225; supernaturalism in, as normalized, 225; as wish-fulfillment fantasy, 224

The Language of New Media (Manovich), 12, 267n5

langue, 230

lap dissolve, 178, 286; as defined, 284

Last Year at Marienbad (film), 167

Lawnmower Man (film), 275–76n7

Lean, David, 259, 260, 263

lectogram, 194–95

lectosign, 170, 184, 187, 197, 199, 215, 219, 225, 286; as defined, 194, 284; and figures, 194; and movement, 194; and narratography, 140

Leger, Fernand, 67

Leibniz, Gottfried, 149

LeWitt, Sol, 280n16

Libidinal Economy (Lyotard), 27

Lichtenstein, Roy, 113

liminal shot: as defined, 284–85; and film narrative, 63–64; as ironic threshold, 195

literary narratography, 268n1

Literature, Media, Information Systems (Kittler), 273n22

Little Dorrit (Dickens), 252, 257, 259; overlapping effect in, 258; parallel editing, as double exposure, 258; POV in, 258; syllepsis in, 255; tracking shot in, 254

Longo, Robert, 129, 132

Lord of the Rings trilogy (film), 284

The Loved One (Waugh), 185

Lovers of the Arctic Circle (film), 75–76, 77; photography, as mourning in, 76, 135

Lowry, Joanna, 273n25

Lumière, Louis, 10, 221, 222, 223

Lyotard, Jean-François, 29, 30, 31, 81, 109, 110, 176, 177, 266, 269n10; and acinema, 216; and arrhythmies, 217; cinema, as writing with movement, 194, 266; and discourse subsuming figure, 194; and figure/discourse dichotomy, 218, 273–74n29; imprint, invisibility of, 28; narrative drive, and image track, 27; and motor writing, 182; narrative discourse, and antirealist figure, 28; representation, and narration, distinction between, 220; stasis vs. variable pace, 182; on violence, somatic affect of, 219

The Magic Chase (film), 79

magic realism, 8, 75

Malkovich, John, 157, 158, 159, 160, 162

The Manchurian Candidate (film), 105; digital illusion in, 91; plot reversal in, 92

Mangold, James, 106, 126

Mankiewicz, Joseph: and flashbacks, 169

Manovich, Lev, 12, 13, 113, 114, 191, 216, 218, 227, 267n5, 274n7, 275n5; cinematic identification and computer activity, duality between, 132; and kinetic reframing, 115; kino-eye to kino-brush, move from, 55; and memory, 115; on new media, 115; and spatial montage, 274n8

Man with a Movie Camera, 15

Il Mare (film), 223, 225

Marey, Étienne-Jules, 13, 14, 127, 128; and electrocardiogram, 133

Matter and Memory (Bergson), 61, 214, 272n11

matrix, 17, 65, 73; as defined, 285; and subtext, 64

The Matrix (film), 82, 88, 132, 136, 168, 212, 215, 217, 267–68n6; reverse-zoom in, 134

The Matrix (film) trilogy, 56, 90, 136, 170

Maybury, John, 165, 238, 240, 241, 242

McLuhan, Marshall, 94, 274n2

Medem, Julio, 56, 82, 135, 169, 194; fantastic cinema of, 75; Proustian images, deployment of, 77

media archaeology, 34, 208, 218, 221, 233, 234; as defined, 285; and diegetic space, 232

media history, 22

medial, 9, 21, 30, 84, 208, 230, 236; as defined, 285

media studies, 18, 208, 218, 219

mediation, 12, 41, 140, 146, 156, 219, 230, 266; shift in, 53; as temporal, 220; *unheimlich* of, 202

Memento (film), 68, 167, 202, 242; and model shot, 64; reverse plot chronology in, 64, 65

memory, 2; and cinema of mind, 193; and consciousness, 190; as filing system, 114; and film, 115; and inscription, 245; as virtual, 169

Ma Mère (film), 199

metafilmic, 265; perspective of, as optical illusion, 51

metalepsis, 110

metempsychosis, 162

metonymy: and metaphor, 30

Metz, Christian, 60, 84, 118; and imaginary signifier, 194; on montage, as *trucage*, 271–72n10; and trick effects, 73; and *trucage*, 106;

Minority Report (film), 90, 162, 168, 170, 215, 225–27; hologram in, 230; *mise-en-scène* in, 230; and time-image, 228; *trucage*, as digital illusion, 230

mirror-image, 169, 182

mnemonic gothic, 115

model: as defined, 285; and imprint, 66; and liminal shot, 64; and semiosis, 205; and subtext, 204–5

Modern Epic: The World System from Goethe to Garcia Marquez (Moretti), 271n5

modernism, 253; and movement-image, 194

montage, 61; as apparition, 146; as articulated temporal sequence, 114; as defined, 114, 250–51; and narrative time, 114; temporal violations of, 177; as *trucage*, 60, 73, 84, 118

montage cell, 252

montage trope, 256

Monteil, Sara, 179

Moreau, Jeanne, 95

Moretti, Franco, 271n5; and centaur criticism, 58

morphing: as metaphor, 192; as metaphysical transformation, 192; morphology, replacing of, 266; time, figure of, 203–4. *See also* digital editing

Mothlight (film), 241

movement, 251; in cinema, as given, 194; and framed image, 205; and time, 172, 205

movement-image, 8, 31, 34, 52, 77, 286; actor, as viewer, 194; as defined, 285; and postwar cinema, 126–27; and temportation, 205; and time-image, 219, 242; time-image, replaced by, 167, 204; and timespace-image, 136

moving images: archaeology of, 223; family snapshot, as extension of, 221

Mulvey, Laura, 111; and delaying cinema, 197, 227

mummification: and photography, 43

Münsterberg, Hugo, 273n22

Music as Cultural Practice, 1800–1900 (Kramer), 268–69n2

Mutoscope, 186

Muybridge 1 (LeWitt), 280n16

Muybridge, Eadweard, 13, 127

My Twentieth Century (film), 79

Naim, Omar, 173

narration: vs. representation, 30, 31, 218, 220

narrative: and interactive zone, 236; and metatext, 227; and narratography, 26; and paratext, 227; and plot, 27; and structure, 24, 25, 59; and subtext, 62; temporal causation, as artifice of, 59; and time, 27, 155; and virtuality, 67, 142, 144–45

narrative cinema: as art, of image, 245; and cybernetic memory, 220; as fantastic, 73; and hermeneutics, 219. *See also* cinema

narrative structure, 62; and digital templates, relation between, 141; and human temporality, 129; as protective function, 127; and textuality, 217

narrativity, 22; as phantom, 161

narratogram, 273–74n29; as defined, 285

narratographic perception, 3–4

narratography, 7, 9, 16, 17, 18, 23, 26, 27, 29, 32, 34, 38,

65, 85, 109, 116, 121, 138, 149, 155, 162, 163, 204, 208, 219, 222–23, 230, 231, 236, 246, 247, 253, 280n11, 284, 286; as archaeology, 53; articulation, between tempo and structure, 27; of biographic loop, 141; and catharsis, 37; as cognitive operation, 141; as defined, 285; and diegesis, 236; and fiction, 263; figurations, and mutation, 266; filmic base, nostalgia for, 120; and flashback structure, 184; as inscriptive process, 22, 52; as intermedial, 32; and interpretation, 39; interval, as graphic interference, 177–78; and *lectosign*, 140; and match cut, 144; and medial tension, 30; as mediating third term, 247; and mediation, 41; as medium specific, 156; narrative, transtextual function of, 26; and narrative process, 26; vs. narrativity, in music, 268–69n2; vs. narratology, 23, 24, 26, 27, 30, 32, 35, 37, 144, 156, 184, 244; and optical allusion, 124; and photography, 26, 30, 242; and plot, 35; and psychopoetics, 184; and reading, 33; and remediation, 221; scene and meaning, tension between, 31; and storytelling, 26; and Thanatos, 184; and time, 194; and time-image, 112; of virtual, 165

narratological gothic, 214

narratology, 29, 84, 149; as cognitive approach, 141, 142; as defined, 285; and narrative, 26; vs. narratography, 23, 24, 26, 27, 30, 32, 37, 144, 156, 184, 244

Narratologies: New Perspectives on Narrative Analysis (Herman), 274n30

Neuromancer (Gibson), 129

New Europe: as libidinal imaginary, 211; and national memory, 211; and uncanny, 195

new media, 126, 128, 133, 141

New Wave, 169

Nietzsche, Friedrich, 265

Noé, Gaspar, 242, 243

Nolan, Christopher, 64, 65, 167, 272n15; style of, 66; and uncanny, 68

noosign, 281n23; as defined, 285

An Occurrence at Owl Creek Bridge, 104, 162; death-moment false lead in, 274n30. *See also* Jacob's Ladder

Ocean's Eleven (film), 279n13

ocular archaeology, 247

Oliver Twist (Dickens), 261; tracking shot in, Eisenstein on, 253

Oliver Twist (film): climactic image in, 261, 262, 263; POV in, 260

One Hour Photo (film), 117–19, 124, 149, 162, 187, 191, 192, 195, 223, 235; child abuse, as traumatic norm in,

188; digital intertext of, 120; end of, 121, 145; hero's nickname, pun on, 275n10; last shot, as falsified index, 116, 147; photography, fantastic of, 116; plot turn of, 121; self-identity in, 121; trick beginning in, 116

ontological gothic, 213; and family photograph, 223

opsign, 138, 219, 270n20, 284; as defined, 285

optical allusion, 19, 33, 68, 119, 209, 247; and arrhythmy, 148; diegesis, and index, 244; and narratography, 124; and remediated sneak ending, 228; viewing, act of, 233

optogramme, 77

The Others (film), 56, 89, 168, 212; artificial verticality, *trucage* of, 96; finale of, 97–98; title sequence of, 117; trick ending of, 97

Ozon, François, 82

paramnesia, 168, 169

paratext, 153; as defined, 285

Pascal, Blaise, 265

Peeping Tom (film), 117, 192, 195

Peirce, Charles Sanders, 273n25

The Pencil of Nature (Talbot), 269n8

period cinema, 41, 46

Persona (film), 83, 84

photogram, 4, 6, 10, 14, 16, 38, 53, 55, 56, 121, 182, 216, 218, 240, 241, 244, 264, 283, 284; and cut, 11; as defined, 285; as definitive, 66; and image, 146; and memory trace, 245; and narratogram, 273–74n29; and numerical algorithm, 233; as overexposed, 148–49; phoneme, structural affinity with, 269n3; vs. pixel, 123; as signifying increment, 147; on strip, 69; and textual indication, 269n3; and virtual real, 120

photography, 96, 268–69n2; and cinema, as lost cause of, 267n4; cinema, and temporal form, 126; and digital, 90; and fantastic, dimension of in, 146; as fantastic temporality, 73; as fixed, 66; memory, as form of, 74; as motionless backtravel, 235; and mummification, 43; and narrative virtuality, 67; and narratography, 26, 242; and ontological gothic, 90; and photographic index, 90; snapshot vs. analytic serial imaging, 127; as spatiotemporal communications medium, 223; and temporality of past, 9; time, stopping of, 181

Pierson, Michelle, 279n10

Pinocchio (film), 98

Pitt, Brad, 90

pixel, 10, 11, 14, 53; and photogram, 123; as pinpoint, 135

pixilation, 22–23, 191; and digital image, 216

plot, 27, 124, 170, 231; as ambivalent, 204; and atemporal
    semiotic pattern, 24; bullet time, suspensions of, 125;
    character driven, as eroded, 102; computer-assisted
    narratives, trend toward, 124; false photographic
    witness, reliance on, 90; and fantastic, 141; and
    hermeneutics of suspicion, 246; human time in, as
    under siege, 125; liminal onset of, 123; and liminal
    shot, 59; manipulations of, 63; parallelism of,
    204; superimposition, as ghostly memory, 60; and
    temporal progression, 166; and theme, 64; and time,
    170; and virtuality, 205; and visual technique, 46
political thrillers: paranoia plots in, 102
The Poetics of Prose (Todorov), 24, 59
The Portrait of a Lady (film), 270n21
postfilmic cinema, 6, 163, 194, 208, 246; and
    narratography, 236; playback loop in, 203; and
    postrealist narrative, 3; time, reshuffling of, 215
postmodern fantastic, 16
postrealist cinema, 124, 163, 194, 202, 209
postwar cinema, 127–28, 168, 268n7; aesthetics of, 125–26
Potemkin (film), 257
Poulet, Georges, 77
POV, 15, 223; as defined, 285; logic of, 190
Powell, Michael, 117, 192
presence: archaeology of, 280n13; and nonpresence, 228;
    restoring of, 222, 223
prewar cinema: and intervallic movement, 15
Prince, Gerald: and disnarrated, 161
Projecting a Camera: Language-Games in Film Theory
    (Branigan), 278n12
protofilmic, 249; as defined, 285; and motion effects, 216
Proust, Marcel, 77, 107, 169
psychoanalysis, 221; desire, otherness of, 101; and
    neurosis, 101; and psychosis, 101; and self, themes of,
    101; and time, 14; and uncanny, 211
Psycho (film), 217

Reading for the Plot (Brooks), 24, 204, 279n14
real, 89, 103, 210; death of, 67; and uncanny, 211; as
    unequivocated, 58
The Red Squirrel (film), 75, 169
Reed, Carol, 11
Reeves, Keanu, 132, 220, 223
remediation, 6
Remediation: Understanding New Media (Bolter and
    Grusin), 94
representation, 32–33; vs. narration, 30, 31, 218, 220; and
    narratography, 33
Resnais, Alain, 167

Ricoeur, Paul, 27
Riffaterre, Michael, 17, 26, 63, 70, 88, 142, 160, 204,
    231, 272n13; and matrix, 64, 73; and model, 64; and
    semiotics, 64, 85; and subtext, 28, 61, 62, 64; and
    ungrammaticality, 24
Rodowick, David: on Deleuze, 278n7; figural, study of,
    269n10
Rossellini, Roberto, 167, 264, 265
Ruiz, Raul, 77, 82, 107, 169
Run Lola Run (film), 161, 168, 242; time, and photography
    in, 68–69
Ryan, Marie-Laure, 142, 143, 160

Saussure, Ferdinand de, 266
science-fiction cinema, 110, 203; cautionary plots of,
    186; digital implant in, 133; and digitization, 89,
    90; and formulaic matrix, 132; narrative in, 205;
    and photographs, 56, 89; plot turns of, 173; second-
    generation imaging, as ontological benchmark, 56
Sconce, Jeffrey, 276n8
screen imaging: digitization, shift to, 4, 266; as medium,
    manipulation of, 145; and movement, 67
screen narrative: editing, as filmic trick of, 84; as fantastic
    deviance, 146; and modernist art film, 210; optical
    register of, 245; temporal modes, emergence of, 166
The Secret Agent (film), 270n21
The Secret Garden (film), 270n21
Secret Window (film), 83
semiosis, 64, 205
semioptics, 204
The Semiotics of Poetry (Riffaterre), 272n13
Shakespeare, William, 264
Sharits, Paul, 113
The Shining (film), 83, 168
Simon the Magician (film): trick beginning of, 79–80; trick
    ending of, 80
simulacrum, 162
The Sixth Sense (film), 56, 82, 89, 90, 97, 103, 168, 212,
    220, 221, 223, 224, 228; child abuse in, 277n16;
    lightbulb, in slow-motion opening, 92, 93, 94, 105;
    photographic uncanny in, 93; and trick ending, 229;
    trucage in, 93
Sjöström, Victor, 272n15
skepticism: and epistemology, 265; and film, 265;
    narrational paranoia, as form of, 104
Sliding Doors (film), 161, 168
Slow Motion (film), 61, 169
Sokurov, Aleksandr, 271n3
somatic reality: *zoe* vs. *bios*, 277n3

Somewhere in Time (film), 270n21

sonsign, 45, 138, 219, 270n20, 285; as defined, 286

Soviet cinema: locomotive, as central symbol of, 252; and montage thinking, 256

Soviet montage, 251

Special Effects: Still in Search of Wonder (Pierson), 279n10

Spielberg, Steven, 90, 100, 215, 226, 230, 231

stacking and popping: as defined, 286

Star Wars (film), 279n10

Steadicam, 11

Sterne, Laurence, 270n17

Stiegler, Bernard, 171, 243, 244

still photography, 39, 125; zoom mechanism of, 234, 280n16

The Sting (film), 279n13

story, 62–63; vs. discourse, 220; and structure, 53; and technique, 53

storytelling, 250; and deathbed reflection, 187; and narratography, 26

Stranger than Fiction (film), 85

Straub-Huillet, 167

strip, 10, 23, 29, 31; and narratography, 141; and time-image, 219

structure: and content, 23; and narrativity, 22; and story, 112; and style, 112

The Structure of Crystals (film), 169

structuralist semiotics, 17, 235

Studying Contemporary American Film: A Guide to Analysis (Elsaesser and Buckland), 267–68n6, 280n11

style, 25; as import, 53

subtext, 24, 26, 28, 29, 69, 100, 285; as defined, 61, 286; and the fantastic, 63; and liminal shot, 64; and model, 204–5; narrative time, relation to, 62; as timeless, 62

Sugimoto, Hiroshi, 5, 11, 267n3

superimposition, 70

supernatural thrillers: plot surprises in, 103

suture, 10, 11, 33, 249, 284; as blocked, 272–73n19; as defined, 35, 286; Henry James, literary equivalent in, 36–37; and image, 154; and interface, 69, 160; and narrative space, 33

Swimming Pool (film), 82–84, 88, 168

syllepsis, 255, 256; and jump cut, 263. *See also* zeugma

Symonds, Arthur, 67, 245

Syriana (film), 281n21; alternating narrative of, 201; climax of, 200, 201; digital mediation in, 200

syuzhet, 23, 24, 283; and characters, 228; and *fabula*, 25, 62, 220, 228, 239; and style, 32

A Tale of Two Cities (Dickens), 260

Tarkovsky, Andrei, 271n3

temporality, 138–39, 237, 243; annihilation of, 61; and cinema, 170; as edited, 203; and escapism, 124; and imaginary, 173; of image, and story time, 161; as morphed, 205; and narrative forms, 6, 27, 58; and photography, 73; as plasticized, 114; as reconfigured, 167; and retroactive initiation, 87; and screen fantasy, 232; of screen narrative, 166; as sign function, 161; and time frame, 203; transformation of, 66; virtual, drift to, 236

temportation, 59, 127, 136, 141, 143, 146, 155, 162, 185, 203, 218, 223; as defined, 286; as dialectical crisis, for cinema, 171; fantasy of, 214; logic of, 32; and movement-image, 205; narratives of, as intertextual cluster, 220; paradoxes of, 18; repairing, as kind of recovery, 163; and time, 134; and time-image, 171, 205

Le Temps du Cinema (Stiegler), 244

The Terminator (film), 18, 150

Text Production (Riffaterre), 272n13

textual model, 33

textual unconscious, 24

textuality, 26, 34, 230, 246; and narrative, 218, 220; and narratography, 220; and representation, 218, 220; and unconscious, 24

Theaters (Sugimoto), 267n3

The Thirteenth Floor (film), 56, 90, 136, 203, 215, 237, 239; doppelgänger plot, as upgrade of, 144; as popped, 143; surprise ending of, 144; title of, 142; trick beginning of, 142

thought-image, 281n23

Three Colors: Blue (film), 72; and liminal image, 70

Three Colors: Red (film), 79; and liminal image, 72

threshold shot: liminal irony of, 195

time, 244–45; and cinema, 13, 14, 66, 115, 125, 126, 129, 162, 167, 169, 194, 245; and computers, 115; as differential, 144; and digital thriller, 203; and digitization, 181; as effect, 126; as embalmed, 244; and the fantastic, 100; filmic medium, as mummified in, 264; as framed, 171; and framed image, 205; imaging of, on narrative screen, 3, 27, 165; and irrational editing, 31; as lived, 14; as malleable medium, 2, 7; as measure of movement, 15; and memory trace, 3; as memory's construct, 114–15; as mobile, 204; morphing of, 237; and movement, 15, 111, 172, 205; movement, writing with, 114; and narrative, 27; and narratography, 194; new signification of, 247; optic field, as reduction to, 8; and photographic tracing, 43; and photography, 181; as process, 126; and psychoanalysis, 14; as read,

time (continued)
194, 195; reassemblage of, 55; as reformatted, 218; reframing of, 217; as reimagined, 203; reshuffling of, 215; retention, and protention, oscillation of, 244; spatialization of, 204; as stilled, 146; as symptom, 126; as traversed, 171; and uncanny, 181

Time after Time (film), 270n21

Timecode (film), 274n8

time-consciousness, 217

time-image, 2, 8, 15, 30, 32, 33, 52, 78, 82, 96, 115, 125, 138, 139, 174, 232, 285; and arrhythmies, 219; break with, 59; as defined, 286; fate of, 210; and humanism, 184; as invented figure, 209; and irrational cut, 203, legacy of, 169; and movement-image, 204, 219, 242; and narratography, 112; as postmodern, 129; and postwar cinema, 126–27, 161; as space deterritorialized, 149; and strip, 219; as taking thought, 247; and temportation, 205; and time, 170; and the virtual, 198

The Time-Image (Deleuze), 177

Time Regained (*Le Temps Retrouvé*) (film), plot of, 77, 78, 79

timespace, 205; and chronotope, 202; as defined, 286; and fantastic, 286; and temporality, 202

timespace-image, 18; and *durée*, 171; and hybrid mediation, 127; and movement image, 135–36; and occupied time, 171

time-travel narratives: Oedipal twists in, 150

Todorov, Tzvetan, 23–24, 25, 58, 59, 62, 85, 108, 142, 144, 158, 160, 194, 198, 211, 271n3, 277n18; and detective fiction, 89; and fantastic, 73, 101, 276n8; fiction, as fantastic, 73, 101; and Freudian dualism, 102; and instrumental marvelous, 170; psychoanalysis, and fantastic, 273n22; psychoanalysis, turn to, 101; and unreal, 193

To Kill a Mockingbird (film), 95

trick beginnings, 17, 80, 82, 87, 123, 228, 246, 271n7; closural reversal, as link to, 63; and fantasy, 59

trick endings, 17, 63, 82, 87; ticket sales, as aid to, 103

trick plots, 214

Tristram Shandy: A Cock and Bull Story (film), 270n17

Tron (film), 275–76n7

trucage, 77, 80, 85, 98, 138, 139; as defined, 286; as digital, 90; of filmic aging, 51; illusionism of, 60; and medial codes, 286; as montage, 60, 73, 84, 118, 181; as subterfuge, 181; as symbolic grammar, 178

twist endings, 141

2001: A Space Odyssey (film), 15, 167; last shot of, 154; Star Child in, 243

Tykwer, Tom, 56, 68, 80, 82

uncanny, 60, 102; and ego, 166; as elegiac, 170; past, as spectral return of, 74; and plot devices, 168; and psychoanalysis, emergence of, 211; and psychology, 101; and time, 181. *See also* the fantastic

Understanding Media: The Extensions of Man (McLuhan), 94, 274n2

undertext, 278n7, 279n5, 279n7

unreal, 90, 193, 214; acceding to, as political act, 145; digital spectacle, as icon of, 236

Usai, Paolo Cherchi, 51, 52

van Gogh, Vincent, 234

Vanilla Sky (film), 56, 89, 94–96, 107, 145, 148, 162, 169, 217, 223; montage flashback in, 61

vectorization, 10; and cut, 11–12

Le Vent (film), 272n15

vertical montage, 254

Vertov, Dziga, 23, 219

video games: and alternate realities, 162; and Hollywood cinema, 136; as interactive, 162, 163; as role-playing games, 136; travel function of, 136

virtual, 18, 245, 283; as defined, 286; and fictive text, 194; and memory, 169; and plot, 246; technologizing of, 190; and time-image, 198; and venture capitalism, 236

virtuality, 27, 56, 57, 84, 165, 166, 193, 213, 264; and bad science, 172; of death, 232; discontinuities of, 167–68; in Hollywood plots, as instrumental tool, 210; and imagined time, 205; and narrative, 142; nonelectronic modes of, 3; and past, 205; and plot, 205; and thought, 209

virtuality theory, 160

visuality studies, 18, 33, 208, 218

Vitagraph, 186, 222

A Voice and Nothing More (Dolar), 277n3

virtual reality (VR), 142, 214, 245; as defined, 286

WarGames (film), 275–76n7

Waugh, Evelyn, 185

Welles, Orson, 51, 168

Western culture: residual supernaturalism of, 211

Wharton, Edith, 48, 49

Where Is the Photograph? (Green), 273n25

Willemen, Paul, 232, 233, 234, 237, 242, 243, 246, 275n4, 280–81n17; and computerized screen effects, 235; Hollywood production, corporate unconscious of, 281n18; indexicality, concentration on, 236

Williams, Robin, 116, 163, 187, 191, 220

Willis, Bruce, 93, 103
Winterbottom, Michael, 270n17
word-sentence: syntax, as embryo of, 252
The World Viewed: Reflections on the Ontology of Film (Cavell), 217, 264, 281–82n9
Wuthering Heights (Brontë), 141

Young, Paul, 275–76n7, 276n8
You've Got Mail (film), 223

Zanussi, Krzystof, 169
zeugma, 255. *See also* syllepsis
Žižek, Slavoj, 82, 120, 124, 272–73n19; and interface effect, 69, 70, 71, 139, 160, 181, 211
Zoetrope, 186, 222
zoom, 280n16; as lens effect, 234; as visual event, 234. *See also* cosmic zoom